Advances in Large Scale Flood Monitoring and Detection

Advances in Large Scale Flood Monitoring and Detection

Editors

Salvatore Manfreda
Caterina Samela
Alberto Refice
Valerio Tramutoli
Fernando Nardi

MDPI • Basel • Beijing • Wuhan • Barcelona • Belgrade • Manchester • Tokyo • Cluj • Tianjin

Editors
Salvatore Manfreda
University of Naples "Federico II"
Italy

Valerio Tramutoli
University of Basilicata
Italy

Caterina Samela
Consiglio Nazionale delle
Ricerche (CNR-IMAA)
Italy

Fernando Nardi
University for Foreigners
of Perugia
Italy

Alberto Refice
Consiglio Nazionale delle
Ricerche (CNR-IREA)
Italy

Editorial Office
MDPI
St. Alban-Anlage 66
4052 Basel, Switzerland

This is a reprint of articles from the Special Issue published online in the open access journal *Hydrology* (ISSN 2306-5338) (available at: https://www.mdpi.com/journal/hydrology/special_issues/Flood_Monitoring).

For citation purposes, cite each article independently as indicated on the article page online and as indicated below:

LastName, A.A.; LastName, B.B.; LastName, C.C. Article Title. *Journal Name* **Year**, *Article Number*, Page Range.

ISBN 978-3-03943-525-8 (Hbk)
ISBN 978-3-03943-526-5 (PDF)

Contents

About the Editors

Salvatore Manfreda (Prof.) is a Full Professor of Water Management, Hydrology and Hydraulic Constructions at the University of Naples Federico II. He is currently the Chair of the COST Action HARMONIOUS and the Scientific Coordinator of the Research Grant aimed at the Development of the Flood Forecasting System of the Basilicata Region Civil Protection. His research activities focus on distributed modeling, flood risk, stochastic processes in hydrology and UAS-based monitoring.

Caterina Samela (Dr.) is a Researcher at the National Research Council of Italy—Institute of Methodologies for Environmental Analysis (CNR—IMAA). She has a PhD in the fields of Hydrology and Hydraulics and her primary interests are related to hydroinformatics, geomorphic analysis, machine learning, GIS and data-scarce environments. Her research activity focuses on the definition of data-driven methods with applications in multiple fields, including fluvial and pluvial flood risk assessment, hydrological models and river monitoring.

Alberto Refice is a researcher at the Italian Consiglio Nazionale delle Ricerche (CNR)—Istituto per il Rilevamento Elettromagnetico dell'Ambiente (IREA). His main research interests concern advanced processing techniques for remotely sensed data, especially synthetic aperture radar (SAR). He is currently working on remote sensing applications such as environmental hazard monitoring and management, geo-hydrological process monitoring and modeling, and landscape evolution.

Valerio Tramutoli (Prof.) is an Associate Professor at the School of Engineering of University of Basilicata where (since 1997) he teaches courses on Satellite Remote Sensing and General Physics. He has been the national coordinator of the SEISMASS Project (funded by ASI) PI, and responsible for DIFA participation in several international projects funded by NATO and by EC in the framework of the Science for Peace, FP6-IST, FP6-INTAS, FP6+FF7 GMES initiatives. During the period 2010–2011, he was the coordinator of the European project PRE-EARTHQUAKES funded by the EC in the framework of the FP7-GMES-Space Program. His research activity has been focused on the development of new satellite sensors and techniques for natural, environmental and technological hazards forecast, monitoring and mitigation.

Fernando Nardi (Prof.) is an Associate Professor at the University for Foreigners of Perugia where he has also served, since 2016, as the Director of the Water Resources Research and Documentation Centre (WARREDOC). Since 2017, Dr. Nardi has been the chair of the Citizens and Hydrology (CANDHY) working group of the International Association of Hydrological Sciences (IAHS). Since 2019, he has been a Courtesy Affiliate Professor at the Institute of Environment and at the UNESCO Chair on Sustainable Water Security of Florida International University (FIU). His scientific interests focus on hydrologic–hydraulic modeling, hydrologic prediction in ungauged basins, geomorphic floodplain mapping, 2-D flood modeling, geospatial information systems and algorithms, Open/Big Data and Citizen Science.

Preface to "Advances in Large Scale Flood Monitoring and Detection"

Changes related to climate, land use, population growth, and urbanization are deeply affecting river basin hydrology. The evolution of such factors is making more and more complex the understanding of hydrological processes amplifying impact of extremes. In this context, floods are increasing in number, magnitude and impact, because several of the mentioned factors are concurring in a systematic growth of floods risk.

Among these aspects, climate change is certainly one of the most controversial and difficult to quantify. At the same time, the steady growth of impervious surfaces and reduction of forested areas amplifies undoubtedly the magnitude of floods. Moreover, the exponential expansion of urban areas, frequently placed nearby rivers, makes this issue even more critical.

As a result, in the last few decades floods have exhibited a rapid upward trend worldwide. This is inducing the international community to invest significant effort for understanding the changed dynamics and projecting them in future flood frequency, in order to guarantee proper planning, management and real-time forecasts.

As the environment evolves, strategies and methods of analysis related to flooding events and their impacts also have to keep up with the changing scenarios. In this framework, the challenge of this book is to describe the state-of-the-art on flood studies using innovative methods and identify the frontier of this research branch. With this aim, we stimulated a discussion on this topic collecting a number of manuscripts recently published on the journal Hydrology which focused on the benefit obtained by the use of new algorithms, new measurement systems and EO data for flood assessment, monitoring, and management.

Salvatore Manfreda, Caterina Samela, Alberto Refice, Valerio Tramutoli, Fernando Nardi
Editors

Editorial

Advances in Large-Scale Flood Monitoring and Detection

Salvatore Manfreda [1],*, Caterina Samela [1], Alberto Refice [2], Valerio Tramutoli [3] and Fernando Nardi [4]

[1]   Department of European Culture and the Mediterranean (DICEM), University of Basilicata, 75100 Matera, Italy; caterina.samela@unibas.it

[2]   Istituto per il Rilevamento Elettromagnetico dell'Ambiente, Consiglio Nazionale delle Ricerche (CNR-IREA), 70126 Bari, Italy; alberto.refice@cnr.it

[3]   School of Engineering, University of Basilicata, 85100 Potenza, Italy; valerio.tramutoli@unibas.it

[4]   Water Resource Research and Documentation Centre (WARREDOC), University for Foreigners of Perugia, 06123 Perugia, Italy; fernando.nardi@unistrapg.it

*   Correspondence: salvatore.manfreda@unibas.it; Tel.: +39-0971-205139

Received: 28 August 2018; Accepted: 30 August 2018; Published: 3 September 2018

Abstract: The last decades have seen a massive advance in technologies for Earth observation (EO) and environmental monitoring, which provided scientists and engineers with valuable spatial information for studying hydrologic processes. At the same time, the power of computers and newly developed algorithms have grown sharply. Such advances have extended the range of possibilities for hydrologists, who are trying to exploit these potentials the most, updating and re-inventing the way hydrologic and hydraulic analyses are carried out. A variety of research fields have progressed significantly, ranging from the evaluation of water features, to the classification of land-cover, the identification of river morphology, and the monitoring of extreme flood events. The description of flood processes may particularly benefit from the integrated use of recent algorithms and monitoring techniques. In fact, flood exposure and risk over large areas and in scarce data environments have always been challenging topics due to the limited information available on river basin hydrology, basin morphology, land cover, and the resulting model uncertainty. The ability of new tools to carry out intensive analyses over huge datasets allows us to produce flood studies over large extents and with a growing level of detail. The present Special Issue aims to describe the state-of-the-art on flood assessment, monitoring, and management using new algorithms, new measurement systems and EO data. More specifically, we collected a number of contributions dealing with: (1) the impact of climate change on floods; (2) real time flood forecasting systems; (3) applications of EO data for hazard, vulnerability, risk mapping, and post-disaster recovery phase; and (4) development of tools and platforms for assessment and validation of hazard/risk models.

Keywords: hydroinformatics; flood mapping; flood monitoring; floodplains; rivers dynamics; DEM-based methods; geomorphology; data scarce environments

---

Introduction to the Special Issue

The impact of flooding is becoming increasingly pressing worldwide for several reasons [1,2]. Population growth, urbanization in alluvial areas, land use change and climate change are only some of the key factors impacting on a potential growth of floods risk. Therefore, the international community is struggling to better understand the dynamics of floods in order to provide proper planning, management and real-time forecasts.

One of the most disputed aspects is certainly climate change, whose impact is controversial and difficult to quantify. On the other hand, the steady growth of impervious surfaces and reduction of forested areas produces an undoubted increase of floods. Moreover, the exponential expansion of urban areas, frequently placed nearby rivers, makes this issue even more complex (e.g., [3,4]). Therefore, there is a huge need for new modelling applications in order to quantify and forecast floods and also to evaluate the impact of such events.

Recent studies have offered a number of innovative strategies in order to support the derivation of flood quantiles (e.g., [5–7]); to provide large scale flood mapping with simplified procedures applicable also in data scarce environments [8–16], to support flood risk management over large scales [17,18] and also to improve flood inundation monitoring with new remote sensing algorithms or exploiting social media (e.g., [19–21]). All these topics are crucial to advance our capacity to cope with floods in a changing environment.

The present special issue was promoted with the aim to provide an overview on the experiences that researchers from different parts of the world have on large scale flood monitoring, prediction and risk. The collection of papers selected introduces several these aspects, presenting novel techniques, reviews and case studies. In the following, we summarize the contents of each specific manuscript and its contribution to the topic.

The first manuscript is the one by Perera et al. [22], that explores the changes of flood impact in future climatic scenarios. The authors modelled the entire hydrological system of the Mekong basin with the TOPMODEL (BTOP) hydrological model at 20 km resolution, and the Lower Mekong Basin (LMB) area with a rainfall-runoff-inundation (RRI) model at 2 km resolution. This latter model was used to specifically analyze floods under the aforementioned climatic conditions in order to support flood management and water policy of the LMB.

The impact of climate change on rainfall statistics is introduced in the manuscript by De Paola et al. [23], which analyzed historical and projected time series of two African cities, Dar Es Salaam (TZ) and Addis Abeba (ET). The authors showed that both time series have non-stationary behavior that should be considered for engineering applications.

An important and relevant research subject is related to flood risk in ungauged basins or large scales that share the common problem of data unavailability. In this regard, an interesting contribution is presented by Ekeu-wei and Blackburn [24], who outlined a review of flood modelling and mapping processes, and the data required by these techniques. They also offer an analysis about potentials, limits and uncertainties of currently available remotely sensed datasets, highlighting how essential these open-access datasets are especially in ungauged basins of developing countries.

The problem of flood management on large metropolitan areas is also tackled also by Moufar et al. [25]. They investigate the feasible countermeasures to mitigate the floods in the Metro Colombo area providing baseline for future flood risk management and mitigation in the area. It must be mentioned that the studied area is a densely populated area of the world (with approximately 3400 inhabitants per square kilometer).

On a similar note, Papaioannou et al. [26] propose a methodological approach for implementing the EU Floods Directive in ungauged basins, with a specific focus on relatively small catchments mainly affected by fluvial flash floods. The proposed approach was applied on the Volos metropolitan area (central Greece) and validated against the flood event of 9 October 2006, using observed flood inundation data. Results highlighted that although typical engineering practices for ungauged basins introduced major uncertainties in flood risk management in urban areas, the hydrological experience may counterbalance the missing information ensuring quite realistic outcomes.

In this context, Peña and Nardi [27] investigated on a DEM-based interpolation method for upscaling flood inundation models. Results indicate the possibility of performing large scale inundation simulations in seconds maintaining a consistent representation of major flooding dynamics. The proposed method taking advantage of largely available DTMs for cost-effective parsimonious flood hazard modelling and mapping.

Besides flood extent, the inundation depth is also a key parameter for flood property damage/human losses. Hydrologic and hydraulic models are traditionally used for predicting these depth values, although their reliability depends strictly on the underlying assumptions in the model adopted. Javaheri et al. [28] propose a framework to improve flood depth estimates and reduce the error between model predictions and observations. The overall scope of the authors is to improve streamflow predictions of the National Water Model (NWM) by using a data assimilation scheme to dynamically update water level estimates in rivers.

Oddo et al. [29] introduce a simple geomorphic approach for flood depth assessment with remote sensing. This information, coupled with detailed land use data, provides rapid initial estimates of flood impacts which can provide valuable information to decision makers in the wake of extreme events.

Nowadays, the extent of inundated areas and the evolution of water expansion and regression can be effectively monitored using remotely sensed data acquired by aircraft and satellites. In this regard, Lacava et al. [30] present a sensor-independent multi-temporal approach called RST-FLOOD, where Robust Satellite Techniques (RST) are applied to detect flooded areas. In particular, the application of the RST-FLOOD methodology for the flood event that affected Basilicata and Puglia regions (southern Italy) in December 2013 is illustrated. Moderate Resolution Imaging Spectroradiometer (MODIS) and, for the first time, Suomi National Polar-orbiting Partnership (Suomi-NPP) Visible Infrared Imaging Radiometer Suite (VIIRS) imagery have been used, highlighting the great usefulness of an integrated system for a continuous monitoring of flood phenomena at large spatial scale.

In conclusion, this special issue provides a wide spectrum of results and a good overview of the research activities carried out in Large Scale Flood Monitoring and Detection.

Funding: This research received no external funding.

Conflicts of Interest: The authors declare no conflicts of interest.

References

1. Kundzewicz, Z.W.; Takeuchi, K. Flood protection and management: Quo vadimus? *Hydrol. Sci. J.* **1999**, *44*, 417–432. [CrossRef]
2. Hirabayashi, Y.; Mahendran, R.; Koirala, S.; Konoshima, L.; Yamazaki, D.; Watanabe, S.; Kim, H.; Kanae, S. Global flood risk under climate change. *Nat. Clim. Chang.* **2013**, *3*, 816–821. [CrossRef]
3. Nardi, F.; Annis, A.; Biscarini, C. On the impact of urbanization on flood hydrology of small ungauged basins: The case study of the Tiber river tributary network within the city of Rome. *J. Flood Risk Manag.* **2018**, *11*, S594–S603. [CrossRef]
4. Samela, C.; Albano, R.; Sole, A.; Manfreda, S. An open source GIS software tool for cost effective delineation of flood prone areas, Computers. *Environ. Urban Syst.* **2018**, *70*, 43–52. [CrossRef]
5. Gioia, A.; Iacobellis, V.; Manfreda, S.; Fiorentino, M Influence of infiltration and soil storage capacity on the skewness of the annual maximum flood peaks in a theoretically derived distribution. *Hydrol. Earth Syst. Sci.* **2012**, *16*, 937–951. [CrossRef]
6. Winsemius, H.C.; Van Beek, L.P.H.; Jongman, B.; Ward, P.J.; Bouwman, A. A framework for global river flood risk assessments. *Hydrol. Earth Syst. Sci.* **2013**, *17*, 1371–1892. [CrossRef]
7. Durocher, M.; Chebana, F.; Ouarda, T.B. On the prediction of extreme flood quantiles at ungauged locations with spatial copula. *J. Hydrol.* **2016**, *533*, 523–532. [CrossRef]
8. Pappenberger, F.; Dutra, E.; Wetterhall, F.; Cloke, H.L. Deriving global flood hazard maps of fluvial floods through a physical model cascade. *Hydrol. Earth Syst. Sci.* **2012**, *16*, 4143–4156. [CrossRef]
9. Herold, C.; Frédéric, M. Global flood hazard mapping using statistical peak flow estimates. *Hydrol. Earth Syst. Sci. Discuss.* **2011**, *8*, 305–363. [CrossRef]
10. Manfreda, S.; Di Leo, M.; Sole, A. Detection of Flood Prone Areas Using Digital Elevation Models. *J. Hydrol. Eng.* **2011**, *16*, 781–790. [CrossRef]
11. Manfreda, S.; Nardi, F.; Samela, C.; Grimaldi, S.; Taramasso, A.C.; Roth, G.; Sole, A. Investigation on the Use of Geomorphic Approaches for the Delineation of Flood Prone Areas. *J. Hydrol.* **2014**, *517*, 863–876. [CrossRef]

12. Manfreda, S.; Samela, C.; Gioia, A.; Consoli, G.; Iacobellis, V.; Giuzio, L.; Cantisani, A.; Sole, A. Flood-Prone Areas Assessment Using Linear Binary Classifiers based on flood maps obtained from 1D and 2D hydraulic models. *Nat. Hazards* **2015**, *79*, 735–754. [CrossRef]

13. Samela, C.; Manfreda, S.; de Paola, F.; Giugni, M.; Sole, A.; Fiorentino, M. DEM-based approaches for the delineation of flood prone areas in an ungauged basin in Africa. *J. Hydrol. Eng.* **2016**, *21*. [CrossRef]

14. Samela, C.; Troy, T.J.; Manfreda, S. Geomorphic classifiers for flood-prone areas delineation for data-scarce environments. *Adv. Water Resour.* **2017**, *102*, 13–28. [CrossRef]

15. Morrison, R.R.; Bray, E.; Nardi, F.; Annis, A.; Dong, Q. Spatial Relationships of Levees and Wetland Systems within Floodplains of the Wabash Basin, USA. *J. Am. Water Resour. Assoc.* **2018**, *54*, 934–948. [CrossRef]

16. Nardi, F.; Morrison, R.R.; Annis, A.; Grantham, T.E. Hydrologic scaling for hydrogeomorphic floodplain mapping: Insights into human-induced floodplain disconnectivity. *River Res. Appl.* **2018**. [CrossRef]

17. Voortman, H.G.; Van Gelder, P.H.A.J.M.; Vrijling, J.K. Risk-based design of large-scale flood defence systems. In *Coastal Engineering 2002: Solving Coastal Conundrums*; World Scientific: Singapore, 2003.

18. Winsemius, H.C.; Aerts, J.C.; van Beek, L.P.; Bierkens, M.F.; Bouwman, A.; Jongman, B.; Ward, P.J. Global drivers of future river flood risk. *Nat. Clim. Chang.* **2016**, *6*, 381–385. [CrossRef]

19. Giustarini, L.; Chini, M.; Hostache, R.; Pappenberger, F.; Matgen, P. Flood hazard mapping combining hydrodynamic modeling and multi annual remote sensing data. *Remote Sens.* **2015**, *7*, 14200–14226. [CrossRef]

20. D'Addabbo, A.; Refice, A.; Pasquariello, G.; Lovergine, F.; Capolongo, D.; Manfreda, S. A Bayesian Network for Flood Detection Combining SAR Imagery and Ancillary Data. *IEEE Trans. Geosci. Remote Sens.* **2016**, *54*, 3612–3625. [CrossRef]

21. Rosser, J.F.; Leibovici, D.G.; Jackson, M.J. Rapid flood inundation mapping using social media, remote sensing and topographic data. *Nat. Hazards* **2017**, *87*, 103–120. [CrossRef]

22. Perera, E.D.P.; Sayama, T.; Magome, J.; Hasegawa, A.; Iwami, Y. RCP8.5-Based Future Flood Hazard Analysis for the Lower Mekong River Basin. *Hydrology* **2017**, *4*, 55. [CrossRef]

23. De Paola, F.; Giugni, M.; Pugliese, F.; Annis, A.; Nardi, F. GEV Parameter Estimation and Stationary vs. Non-Stationary Analysis of Extreme Rainfall in African Test Cities. *Hydrology* **2018**, *5*, 28. [CrossRef]

24. Ekeu-Wei, I.T.; Blackburn, G.A. Applications of Open-Access Remotely Sensed Data for Flood Modelling and Mapping in Developing Regions. *Hydrology* **2018**, *5*, 39. [CrossRef]

25. Moufar, M.M.M.; Perera, E.D.P. Floods and Countermeasures Impact Assessment for the Metro Colombo Canal System, Sri Lanka. *Hydrology* **2018**, *5*, 11. [CrossRef]

26. Papaioannou, G.; Efstratiadis, A.; Vasiliades, L.; Loukas, A.; Papalexiou, S.M.; Koukouvinos, A.; Tsoukalas, I.; Kossieris, P. An Operational Method for Flood Directive Implementation in Ungauged Urban Areas. *Hydrology* **2018**, *5*, 24. [CrossRef]

27. Peña, F.; Nardi, F. Floodplain terrain analysis for large scale coarse 2 resolution 2D flood modelling. *Hydrology* **2018**, *5*, 52.

28. Javaheri, A.; Nabatian, M.; Omranian, E.; Babbar-Sebens, M.; Noh, S.J. Merging Real-Time Channel Sensor Networks with Continental-Scale Hydrologic Models: A Data Assimilation Approach for Improving Accuracy in Flood Depth Predictions. *Hydrology* **2018**, *5*, 9. [CrossRef]

29. Oddo, P.C.; Ahamed, A.; Bolten, J.D. Socioeconomic Impact Evaluation for Near Real-Time Flood Detection in the Lower Mekong River Basin. *Hydrology* **2018**, *5*, 23. [CrossRef]

30. Lacava, T.; Ciancia, E.; Faruolo, M.; Pergola, N.; Satriano, V.; Tramutoli, V. Analyzing the December 2013 Metaponto Plain (Southern Italy) Flood Event by Integrating Optical Sensors Satellite Data. *Hydrology* **2018**, *5*, 43. [CrossRef]

Article

Floodplain Terrain Analysis for Coarse Resolution 2D Flood Modeling

Francisco Peña [1,2,*] and Fernando Nardi [2]

[1]   Water Resources Research and Documentation Center (WARREDOC), University for Foreigners of Perugia, 06123 Perugia, Italy
[2]   Department of Civil and Environmental Engineering (DICEA), University of Florence, 50139 Florence, Italy; fernando.nardi@unistrapg.it
*   Correspondence: francisco.pena@unistrapg.it; Tel.: +39-0-755-746-401

Received: 6 August 2018; Accepted: 19 September 2018; Published: 21 September 2018

Abstract: Hydraulic modeling is a fundamental tool for managing and mitigating flood risk. Developing low resolution hydraulic models, providing consistent inundation simulations with shorter running time, as compared to high-resolution modeling, has a variety of potential applications. Rapid coarse resolution flood models can support emergency management operations as well as the coupling of hydrodynamic modeling with climate, landscape and environmental models running at the continental scale. This work sought to investigate the uncertainties of input parameters and bidimensional (2D) flood wave routing simulation results when simplifying the terrain mesh size. A procedure for fluvial channel bathymetry interpolation and floodplain terrain data resampling was investigated for developing upscaled 2D inundation models. The proposed terrain processing methodology was tested on the Tiber River basin evaluating coarse (150 m) to very coarse (up to 700 m) flood hazard modeling results. The use of synthetic rectangular cross sections, replacing surveyed fluvial channel sections, was also tested with the goal of evaluating the potential use of geomorphic laws providing channel depth, top width and flow area when surveyed data are not available. Findings from this research demonstrate that fluvial bathymetry simplification and DTM resampling is feasible when the terrain data resampling and fluvial cross section interpolation are constrained to provide consistent representation of floodplain morphology, river thalweg profile and channel flow area. Results show the performances of low-resolution inundation simulations running in seconds while maintaining a consistent representation of inundation extents and depths.

Keywords: DTM; terrain analysis; hydraulic geometry; large scale; 2D hydraulic modeling; scaling in hydrology

---

1. Introduction

Floodplain landscape morphology and roughness represent the governing factors of flood flow propagation dynamics [1–4]. Recent technological advancements, such as geomatics and remote sensing (or Earth Observation (EO)) sectors, allow for more efficient data gathering of fluvial bathymetry, floodplain topography and surface roughness. EO tools use both ground-based and airborne sensors, providing unprecedented conditions for effective inundation modeling and mapping [5].

Global flood hazard modeling is now possible [6] with hyper-resolution hydraulic modeling that are being implemented [7] taking advantage of remotely sensed data from large (i.e., satellite and aerial sensors) to small scale (i.e., drones) fluvial feature and process observation systems [8,9]. In addition, advancements of numerical hydraulic algorithms, super computational power and data rich hydrology are paving the way for hyper-resolution simulations of flood events from regional to continental domains. Nonetheless, several challenges and uncertainties impact high-resolution numerical models of flood events [10,11]. The quest for always more accurate and detailed flood models prompts the need for investigations identifying an optimal balance between hydraulic model output details and topographic input data resolution [8].

We argue that the downscaling of the flood model resolution—i.e., decreasing the size of the mesh elements that characterize the Digital Terrain Model (DTM) of the inundation domain—has to be properly developed considering the scale and properties of the flood event of interest. As a result, hyper-resolution flood models are not always strictly needed, especially when the size of the flood wave and associated dynamics (i.e., inundation depths and floodplain flow velocities) can be consistently analyzed by coarse resolution (i.e., from tens to hundreds of meters) numerical models. The upscaling of flood models represents a viable solution for flood risk assessment in several applications, especially when high resolution distributed topographic data and super-performing computers are not available or when a very fast inundation simulation is needed.

Additionally, large scale flood models are still suffering the lack of effective EO-based fluvial bathymetric data gathering technologies that are impacted by the disturbances affecting remotely sensed data in densely vegetated channels. Flood prediction and management in most remote or ungauged basins still require a significant degree of flexibility of numerical models in applying open source DTMs at $30 \neq 90$ m resolution (e.g., the NASA SRTM DTM) for terrain analysis. Even though it is foreseen that global satellite mission for DTM production will soon provide the capacity of digitizing the earth's morphology at higher resolution, from 10 m to 1 m scale, the issue of missing river bathymetry information will still impact DTM-based large scale numerical models for flood hazard simulations. While few studies attempt to investigate procedures for developing coarse resolution flood models, it is noted that there are several cases where large to continental scale flood modeling applications are needed. Notable examples are applications related to coupling hydro-modeling with climatic, landscape and environmental models or rapid inundation mapping for emergency management [12,13].

The use of synthetic cross sections, in replacing missing surveyed bathymetry, was used in previous studies, but the validity of this method was tested only in small scale or local studies [14–16]. At larger scales, we posit that the use of geomorphological laws, depicting the hydraulic geometry of stream channels [17], may provide solid means for surrogating to the lack of surveyed river cross sections and bathymetric information. Geomorphic laws to support large scale hydraulic modeling are investigated as a mean for improving flood routing model performances [18–22]. Nevertheless, DTM-based terrain analysis for implementing the use of geomorphic laws for large scale hydraulic models is still a challenging and active research issue.

This research investigated a floodplain DTM data processing procedure aiming to produce a coarse resolution bidimensional (2D) hydraulic model for fast inundation mapping. A performance assessment was developed comparing different coarse resolution inundation models in terms of simulated inundation extent, water surface levels and floodplain flow depths. The goal of this research was to test the effectiveness of a floodplain terrain and channel bathymetry data processing procedure in producing computationally efficient and consistent inundation models. A synthetic representation of channel geometry, using a rectangle morphology, was used and floodplain DTMs were resampled to derive seven coarse resolution floodplain topographic scenarios (150 m, 200 m, 300 m, 400 m, 500 m, 600 m and 700 m). The seven different scenarios were compared by evaluating the comparison between simulated inundation extents and depths as respect to a reference higher resolution flood model.

This paper is organized as follows. In Section 2, the selected data and methods are presented with specific regard to the study basin, the 2D hydraulic model and the boundary and input conditions. The proposed methodology for floodplain terrain processing for coarse resolution inundation modeling is described in Section 3. Results of the application are provided and described in Section 4, with a discussion on the presented research results inserted in Section 5.

2. Data and Methods

2.1. Study Area

The Tiber River in central Italy, selected as case study, is the second largest river basin in Italy, draining an area of 17,800 km^2. The floodplain domain is located in the downstream portion of the catchment, south of the Umbria–Lazio regional boundary. The inundation domain goes along the river channel for about 120 km between Orte (12°23′39.52″ E, 42°27′27.08″ N) and Castel Giubileo (12°29′59.34″ E; 41°59′40.29″ N), just upstream of the city of Rome, before the river flows out into the Tyrrhenian Sea (Figure 1). The Tiber River—the third longest in Italy—is the subject of frequent overflow, considering that the conveyance of the incised channel in the selected reach is approximately 900 m^3/s, as stated in the flood risk management plan issued by the Tiber River Basin Authority (TRBA) [23], and that floods with peak discharges greater than 2000 m^3/s occurred frequently in the last decade [24]. The river floodplain, which is predominantly developed, but mainly characterized by agricultural activities, is often the subject of channel overbank flows inundating the floodplain, with damages and service interruption to economic activities and minor urban settlements [25,26]. Rail and road embankments, which follow the floodplain main direction, act as levees, limiting the floodplain extent, which has an average width of 2–3 km, while the average top width of the channel is 100–120 m, as observed by aerial images.

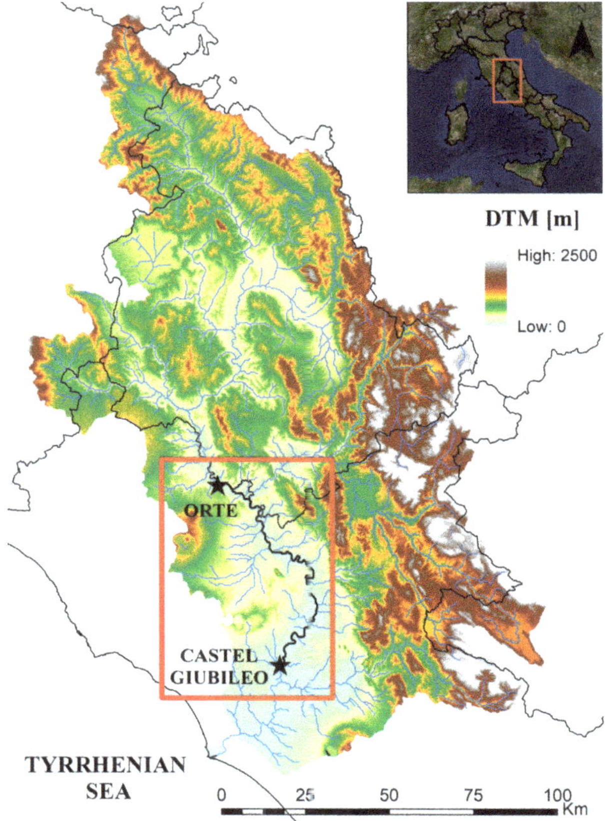

Figure 1. Map of the study basin of the Tiber River in central Italy.

2.2. Data

The data used for this study include topographic, hydrologic, and base cartographic dataset. A LIDAR aerial survey was gathered from the Italian Ministry of Environment National Cartographic Portal (PCN) and projected in the Universal Transverse Mercator (UTM) (WGS 1984, Zone 33N) coordinate system. The available LIDAR dataset included a Digital Surface Model (DSM) and a Digital Elevation Model (DEM) at 1 m resolution. The LIDAR DEM, capturing the land surface characteristics of the floodplain with the exception of river bathymetry and water bodies, was used to characterize the bare earth information needed to represent the topography of the floodplain. LIDAR data were processed to produce a Digital Terrain Model (DTM) of the study domain at 5 m resolution. The term DTM is used in this paper to distinguish the processed 5 m data from the original source LIDAR DEM. Surveyed cross sections were obtained from a validated set of fluvial bathymetry GPS surveying, provided by the Tiber River Basin Authority [25], covering the entire floodplain domain. The hydrology for the flood model was obtained by the TRBA flood risk management plan. In particular, the 200-year return period hydrograph was selected as design inflow condition for the upstream node (at Orte) of the Tiber River. Aerial imageries, gathered from the PCN and TRBA database, were used mainly for visualization purposes, as cartographic base supporting the flood modeling analysis, but also for evaluating the channel top width analysis implemented in the hydraulic model geometry construction. A 2D hydraulic model input parameter set up was gathered from the TRBA flood risk management plan that included calibrated and validated topographic data, distributed roughness parameters and rating curves for bridges and culverts that characterized the floodplain domain. This 2D flood model, calibrated and validated within the TRBA studies using real events, was used as reference model for the presented research work.

2.3. Hydraulic Model: FLO-2D

FLO-2D [27] is a Quasi-2D hydraulic model based on flood volume conservation and hydraulic routing scheme simulating channel–floodplain exchange, flood wave attenuation, and the alteration of inundation dynamics due to artificial obstructions (levees, buildings, streets, etc.) on a gridded topographic surface. The surface water propagation in the incised channel is simulated with a 1D unsteady flow model until the capacity of the channel is exceeded. The 2D floodplain flow model is coupled with the 1D channel hydrograph routing model for simulating the channel overbank flow and the initiation of shallow surface water flow propagation from channel overbank flow conditions. The 2D surface flow is routed in eight potential flow directions, four cardinal (north, east, south and west) and four diagonal directions (NE, NW, SE and SW), over the floodplain topographic model that is produced using a gridded surface. The FLO-2D numerical scheme implements the full dynamic wave momentum equation and a central finite difference routing scheme with an automatic time step adjusting algorithm, always preserving the continuity of flood volume at the domain scale. The hydraulic core algorithm is governed by topography and a roughness parameter expressed using the Manning coefficient. The channel geometry is provided associating cross sections to a predefined set of floodplain cells that are flagged as channels and sketched to represent the upstream to downstream channel link 1D domain. The fluvial bathymetry can be either schematized using natural, rectangular or trapezoidal cross sections. The FLO-2D model input data are the floodplain topographic DTM, channel geometry, inflow and outflow boundary conditions as well as grid cell parameters representing the presence of artificial features on the bare earth (levees, building, bridges, etc.) [28].

2.4. Topographic Data Processing and Floodplain Grid Resolution

The FLO-2D hydraulic model requires a grid of square cells to represent the topography of the floodplain domain. The size of the grid cell defines the resolution of the hydraulic model. The 5 m LIDAR DTM was used as source floodplain topographic information and an interpolation algorithm was implemented to produce a resampled DTM floodplain model to be used as input geometry of the hydraulic model. Nearest neighbor interpolation was selected for this study to interpolate the high resolution 5 m DTM to the desired coarser resolution [29–32]. It is noted that different interpolation methods (e.g., Inverse distance weighting and ordinary kriging) would produce different DTMs. Here, we posit that the resampling model selection is not biasing the hydraulic modeling result comparison analysis, since the different flood model resolution scenarios are all developed with the same interpolation model [33]. The resampling was performed to test the effect of morphologic information upscaling process, smoothing out microtopographic floodplain features, on inundation simulations.

Urban features such as levees, bridges or streets were also not represented in the input DTM, but inserted as additional artificial features of the hydraulic model geometry. The floodplain topographic model, used by the FLO-2D model, represents the bare earth, while the impact of artificial obstructions and features on the flood wave overbank propagation was accounted for using area and width reduction factors [28]. These reduction factors account for the decreasing of available storage, expressed by the obstructions modifying the available grid cell width and area, in the cell-to-cell surface water routing downstream.

Fluvial terrain processing for supporting large scale flood models has been investigated in several works that tested optimal methods for interpolating DTMs and cross sections. Those studies evaluate the impact of floodplain bathymetry processing on performances of hydraulic models [16,34–38]. The issue of lack or uncertainty of fluvial cross section data was the subject of several attempts to surrogate the missing bathymetric information in hydraulic modeling [16]. Investigation on different performances of flood hazard studies with changing resolution of floodplain DTMs have also highlighted the importance of identifying the optimal balance between accuracy and efficiency of inundation models. Resampling of DTMs from high to low resolution impacts the spatial and vertical accuracy of the simulated floodplain terrain, including mismatches of channel banks location and height, and underestimation (or overestimation) of the fluvial and floodplain conveyance capacity. Figure 2 illustrates a sample case of floodplain cross section processing with resampling from the high resolution 5 m DTM to produce a coarse 150 m and very coarse 400 m grid resolution floodplain topographic model. The approximation and loss of topographic information is significant, but this does not directly imply that the loss of flood modeling result accuracy is of the same order of magnitude.

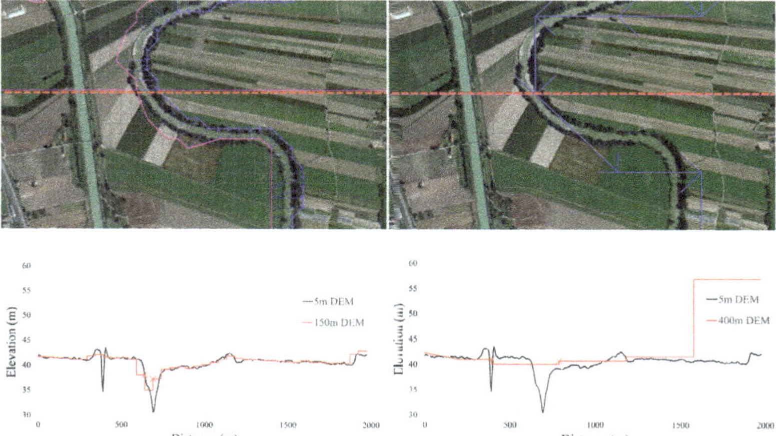

Figure 2. The impact of varying resolution represented by floodplain topographic plan view and cross sections. The 5 m floodplain terrain model is compared to interpolated 150 m (**left**) and 400 m (**right**) resolution DTMs. The upper and bottom plots show, respectively, the horizontal and vertical displacement of the interpolated floodplain grid schematization and cross section as respect to the high resolution DTM.

2.5. Boundary Conditions and Roughness Parametrization

The 200-year return time hydrograph was used as the upstream inflow condition. The peak discharge was 3600 m^3/s and simulation time was approximately 350 h corresponding to the hydrograph total duration. The downstream boundary condition was characterized by the exiting of the flood wave in undisturbed conditions by assuming uniform flow conditions at the channel and floodplain outflow nodes. Manning's roughness coefficient for channel cells was set constant to 0.04, while distributed roughness conditions for the floodplain domain were gathered from the reference TRBM 2D flood model.

2.6. Fit Index Analysis

The quantitative comparison of the different inundation extents simulated by flood models with varying resolutions is developed using a measure-to-fit index F [39–42] expressed with the following formula:

$$F = \frac{A_{ref} \cap A_{mod}}{A_{ref} \cup A_{mod}}$$

where $A_{ref} \cap A_{mod}$ represents the matching area between reference map and model results (A = true positive area) and $A_{ref} \cup A_{mod}$ is the union of the overlapping, underestimated (B = false positive) and overestimated (C = false negative) areas. The F parameter can thus be expressed as:

$$F - index = \frac{A}{(A + B + C)}$$

The F-index varies from F = 0 when the spatial intersection of models is null to a maximum of 1 (100% fit between model results and reference map).

3. Procedure

The presented methods are integrated to develop a floodplain terrain and fluvial bathymetry processing procedure that is implemented to produce a very coarse resolution (up to 700 m) flood model. The procedure aimed to test the effect of channel and floodplain topography upscaling on simulated inundation extent and dynamics comparing coarse resolution flood models with a reference higher resolution 2D hydraulic model. The original inundation model, at 150 m resolution, was used as reference case considering it was previously calibrated and validated and included surveyed natural cross sections to represent channel bathymetry. The presented tests evaluated, firstly, the possibility of replicating the reference model results at the same resolution (150 m), but substituting natural cross sections with synthetic rectangular cross section calibrated using information related to flow area and thalweg profile. Then, the upscaling of the 2D floodplain terrain model was performed producing coarser resolution inundation models (from 200 m to 700 m) evaluating differences in the hydraulic modeling results.

The implemented procedure was based on the following four steps:

(i)   Building the 2D flood reference model at 150 m resolution

The high resolution 5 m DTM was interpolated at 150 m. The simulated inundation dynamics of the 150 m simulation were consistent to the original FLO-2D model gathered from the TRBM that was originally implemented at 50 m resolution. The 1D HEC-RAS model developed by the TRBA was used to gather the available 92 fluvial cross sections. The cross sections were imported into the FLO-2D model to assign a cross section to each of the 712 river channel grid cells. The FLO-2D channel geometry was created by linear interpolation of the original Hec-Ras cross sections. The 150 m flood model was evaluated to check that the floodplain DTM, channel cross section geometry and thalweg profile accurately match the original 50 m TRBM model.

(ii)   Interpolation of natural cross sections to produce a synthetic rectangular channel bathymetric model

The 712 river channel grid cell elevations and associated cross sections were analyzed to estimate the channel cell-by-cell flow area, top width and channel maximum depth. The cross section depth characterizes the distance from top of the banks to minimum channel elevation defining the simulated thalweg profile of the fluvial channel. A rectangle was created to approximate each cross section. The channel bed elevation of the rectangle was estimated by subtracting the channel maximum depth from the DTM cell elevation to preserve the thalweg profile of the reference hydraulic model. The rectangle width was then varied to preserve the flow area estimated from the original cross section. Basically, the hypothesis was to define a rectangle that best approximates the surveyed cross section giving priority to the conservation of the channel slope (i.e., thalweg profile) and conveyance (i.e., flow area). The top width was used as calibration parameter in this channel bathymetry interpolation.

(iii)   Upscaling flood model to coarser resolutions

The reference 150 m flood model DTM was upscaled to 200 m, 300 m, 400 m, 500, 600 m and 700 m resolution. The terrain analysis process generated coarser floodplain DTM resolutions based on the following steps: (1) Use high resolution 5 m DTM as input terrain model to interpolate the digital floodplain terrain for the flood model at the desired coarser resolution. (2) Identify the channel cells by proximity analysis of the 150 m reference channel model with the new coarser channel grid. (3) Associate to each channel cell the rectangle depth and height that preserve the average slope and flow area of the channel cells intersecting the coarser channel grid cell (as in Step (ii) above).

(iv)   Inundation model runs and postprocessing of simulated water surface simulations

The coarse resolution model set up was used to perform the design event simulation representing the impact of the 200-year event on the fluvial domain. Flood modeling results were processed to depict the simulated water surface elevation, depth and velocities at the simulation resolution (200 m, 300 m, 400, 500 m, 600 m, and 700 m). The coarse resolution water surface elevation was interpolated at reference model resolution of 150 m. The simulated water surface was intersected with the high resolution DTM to produce the final inundation extent and flood depths. A postprocessing routine tool (freely available at https://github.com/antonioannis/FLO-2D-Processing-Tools) was used to interpolate and intersect the simulated coarse resolution water surface with the high resolution DTM. This analysis automatically downscaled the coarse resolution model inundating those floodplain cells that underlay the water surface elevation. Final results were plotted at 30 m resolution to allow better visualization of results. Figure 3 provides a schematic representation of the proposed procedure.

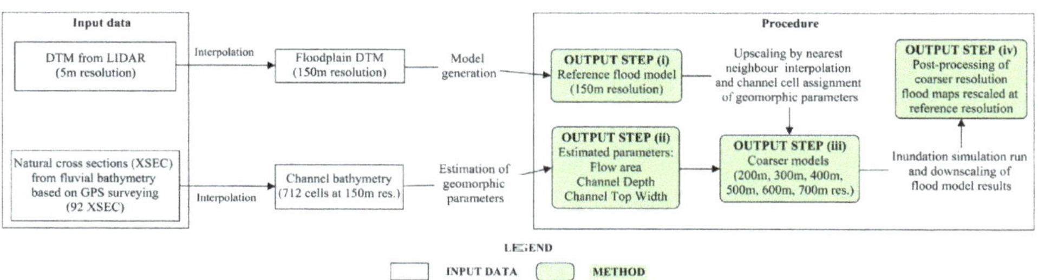

Figure 3. Flow chart describing the procedure implemented for interpolating the fluvial bathymetry and resampling floodplain terrain data to support coarse resolution flood modeling.

4. Results

The procedure is applied and results are presented in this section to depict the effect of the proposed floodplain terrain processing on flood modeling results with varying performances when upscaling the model from coarse to very coarse resolutions.

Figure 4 presents the results of Steps (i) and (ii) of the procedure by showing the cross section width, depth and flow area associated to interpolated synthetic rectangular cross sections at varying resolution with respect to surveyed cross sections of the reference model. It is shown how the channel bed elevation and flow area are preserved in all models for adjusted channel top width.

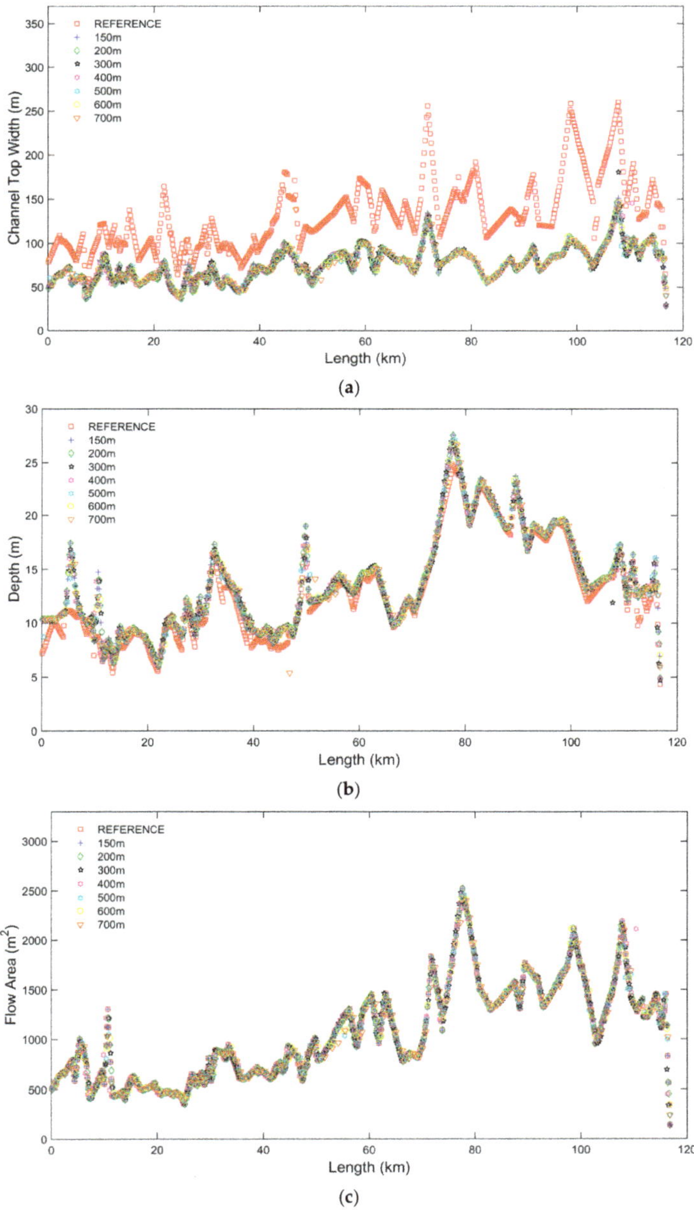

Figure 4. Plots of channel top width (**a**), depth (**b**) and flow area (**c**) using the different model set ups: from the 150 m reference model using the naturally surveyed cross sections to the seven realizations of the synthetic rectangular channel flood model for 150 m, 200, 300 m, 400 m, 500, 600 m and 700 m resolutions.

The simulated flood profile along the river channel is represented in Figure 5. The visual comparison shows that the six upscaled models have a consistent Water Surface Elevation (WSE) with discrepancies increasing with decreasing resolution. It is noted that the simulated surface seems always consistent even in some segments of the river reach where the spatial adjustment determines notable differences in the location of the bed profile. An underestimation of the water surface elevation is also noted when the channel slope is steep, while in the flat downstream valleys the differences decrease.

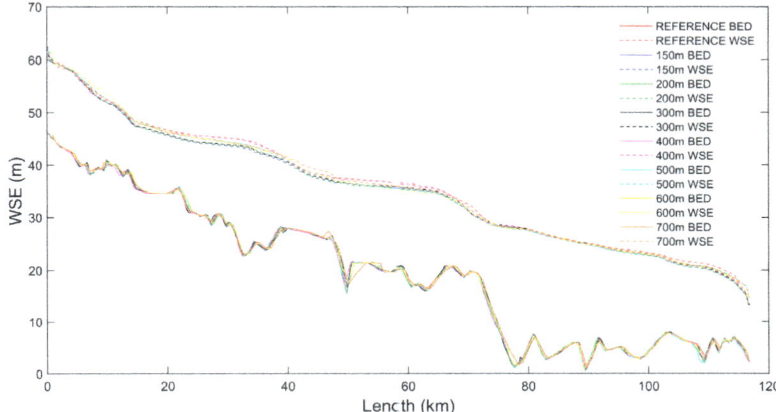

Figure 5. River channel flow profile with simulated maximum water surface level at 150 m and coarser resolutions.

The simulated channel maximum discharge is reported in Figure 6. Here, the impact of the resolution on simulated floodplain flow dynamics is evident, considering that the expected impact of the different floodplain DTM realizations on the flood wave attenuation process. The use of synthetic cross section provides notable differences as well as the approximation of the channel bank elevation that plays here a major role, considering the importance of accurate river–floodplain exchange simulations. This result shows the significant impact of the resampling procedure on the simulation of flood wave propagation dynamics (i.e., discharge, velocities).

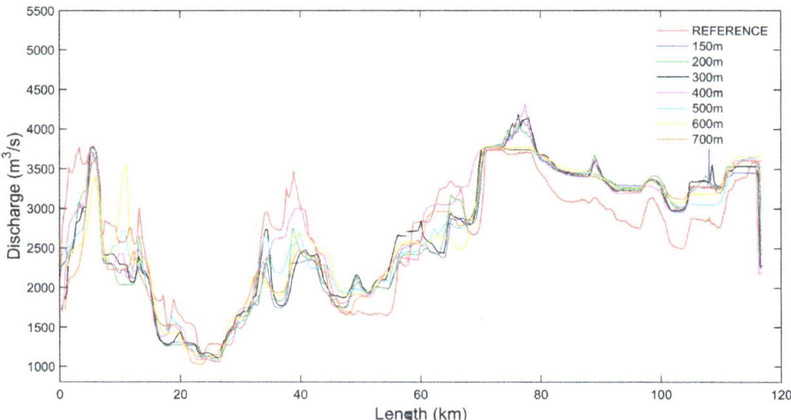

Figure 6. River channel flow profile with simulated maximum peak discharge at 150 m and coarser resolutions.

Results presented in Figures 5 and 6 show that inundation simulations for the Tiber case study are not driven by peak discharge, but are mainly governed by flood volumes transiting along the valley. The maximum conveyance of the Tiber is not greater than 1000 m^3/s, while the 200-year peak discharge is greater than 3000 m^3/s for the entire domain. The consistency of flood profiles is motivated by the fact that, once the channel has reached its maximum conveyance capacity, the floodplain of the Tiber River acts as a large storage. The presented tests support the validity of consistent simulations of coarse to very coarse resolution models even in cases where the topography is represented using only 1–3 grid cells per floodplain cross sections. While it is expected that results of very coarse resolution will not match the accuracy of high resolution models for the reconstruction of flood wave dynamics, we posit that inundation extent and depths can be reasonably captured.

A significant advantage of the coarse resolution is the speed of the simulation and the computational efficiency. Table 1 shows the varying simulation computational performance with varying DTM resolutions. A summary of the main specifications of the different model set ups is provided with results also evaluating the total inundated area. The run time decreased from 37 min to 2.5 min using the same computer power (a regular workstation). Table 1 also includes the results of the measure-to-fit analysis with the F-index results.

Table 1. Performance metrics of running time, maximum simulated inundated area and F-index comparing reference model with upscaled flood models at different DTM resolutions.

Grid Size (m)	Number of Cells	Running Time (min)	Inundated Area (m^2)	F-Index (-)
REF	11,191	36.97	125,370,000	
150	11,191	34.25	115,267,500	0.917
200	11,169	23.17	116,880,000	0.868
300	2838	6.53	122,400,000	0.825
400	1588	3.46	108,000,000	0.744
500	838	3.99	120,250,000	0.796
600	561	2.05	119,880,000	0.761
700	498	2.49	118,580,000	0.720

The visual comparison of simulated inundation depths is represented in Figures 7–9. Figure 7 shows the differences between the simulated maximum floodplain flow depths at coarse (150 m) to very coarse (700 m) resolution compared to the reference model. The flood model resolution is maintained with no postprocessing here developed to show the visual effect of the varying resolutions. Figure 8 presents the spatial distribution of differences between floodplain maximum flow levels of the reference model compared to the upscaled flood simulation at different resolutions. Figure 9 shows the results of the simulated maximum flow depths from the reference 2D hydraulic model simulation for the entire study domain. Three insets are inserted into Figure 9 for zooming in on an upstream, central and downstream area of the model visualizing the comparison of simulated inundation depths for the different flood model resolutions. Note in Figures 8 and 9 the use of the postprocessing tool of Step (iv) of the presented procedure for plotting high resolution simulated inundation maps.

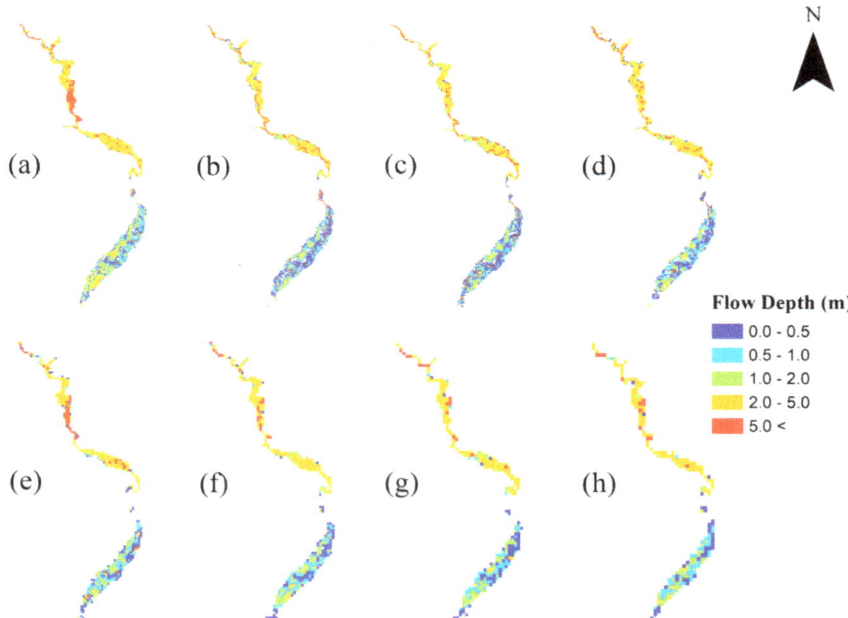

Figure 7. Distribution of inundation flow depths for: (**a**) reference model at 150 m and natural cross sections; as compared to upscaled flood models synthetic rectangular cross section at resolution of: (**b**) 150 m; (**c**) 200 m; (**d**) 300 m; (**e**) 400 m, (**f**) 500 m (**g**) 600 m; and (**h**) 700 m.

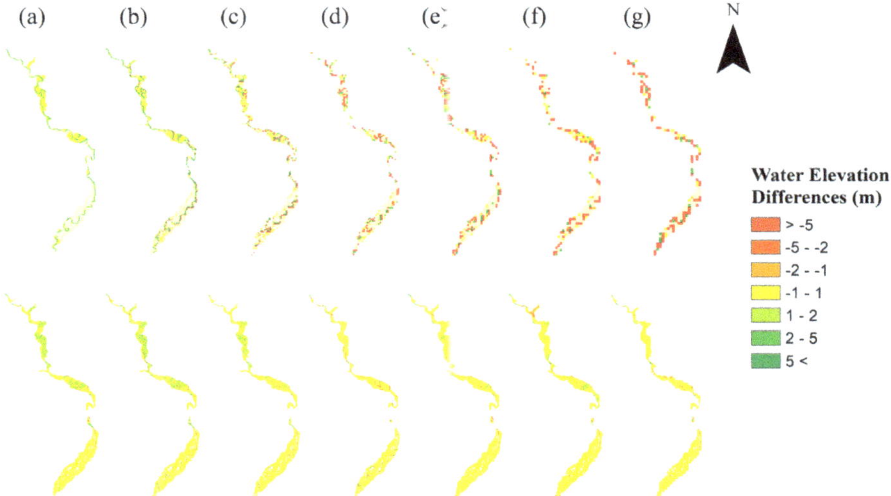

Figure 8. Distribution of simulated surface water elevation differences for coarser resolution models against the reference model using the model resolution (**top**) and the postprocessed 30 m resolution (**bottom**) with: (**a**) 150 m; (**b**) 200 m; (**c**) 300 m; (**d**) 400 m, (**e**) 500 m; (**f**) 600 m; and (**g**) 700 m.

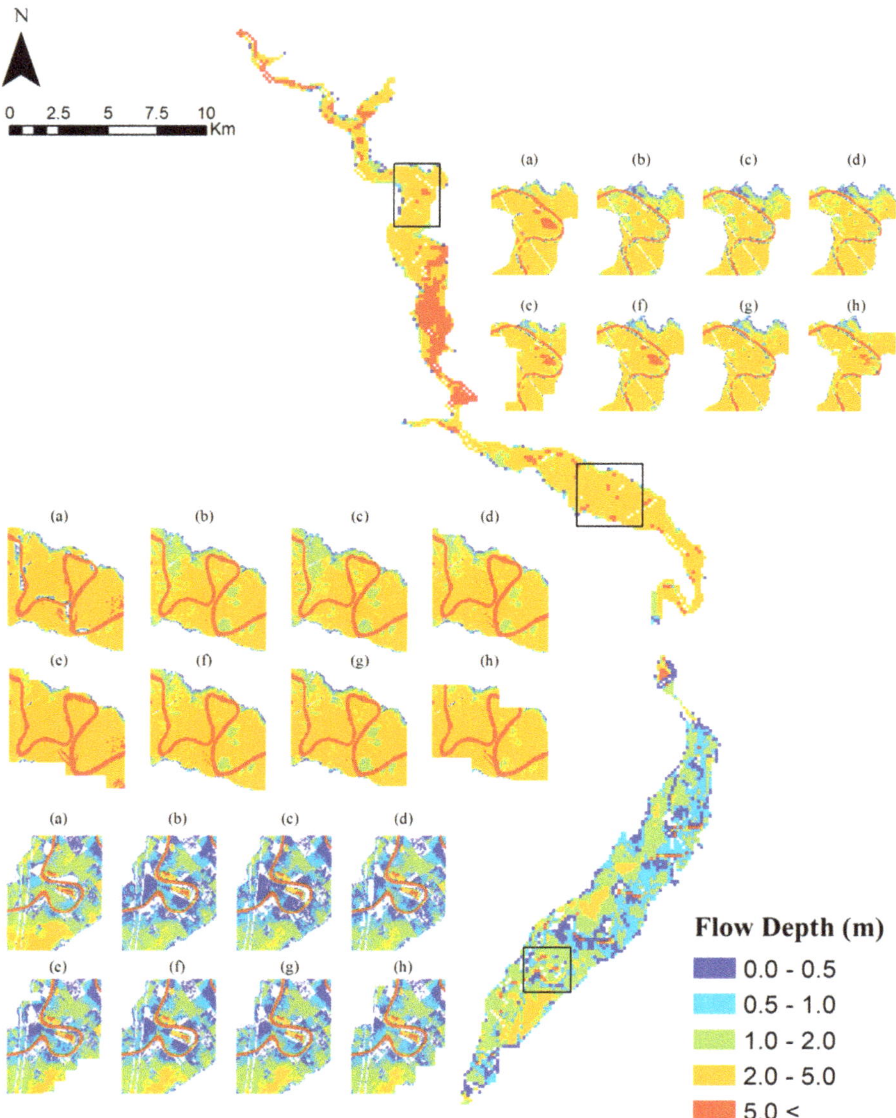

Figure 9. Distribution of inundation flow depths of the reference model at 150 m resolution (full scale map) and three selected subdomains where inundation depths are depicted using the postprocessing geospatial algorithm for visualizing the different flood modelling scenario at 30 m model resolution comparing (**a**) reference model; (**b**) 150 m; (**c**) 200 m; (**d**) 300 m; (**e**) 400 m; (**f**) 500 m; (**g**) 600 m; and (**h**) 700 m.

5. Discussion

This research investigated the development of very coarse resolution 2D inundation numerical models supporting large scale flood mapping. A procedure for interpolating floodplain terrain and bathymetric data is proposed and tested with the goal of producing consistent inundation models using coarse resolution floodplain topography and synthetic representation of channel morphology. Results were evaluated compared to validated reference 2D flood models quantifying performance metrics of running time and distributed inundation extents and depths of coarse (150 m) to very coarse (700 m) hydraulic model simulations.

The proposed floodplain terrain analysis procedure for creating coarse resolution 2D inundation models, that are able to consistently simulate the inundation extent at large scale, may be beneficial for flood risk management in multiple ways. Fast and reasonably accurate 2D flood simulations, running in seconds or minutes at large scales, pave the way for the use of 2D models in early warning systems that commonly use 1D models. Coarse inundation models are also needed for implementing cross-disciplinary applications where hydrodynamic models are coupled with climatic (e.g., Global Circulation Models (GCM)), landscape evolution or ecologic models.

This research contributes to the development of alternative cost-effective parsimonious flood modeling approaches that can produce consistent results using largely available data, considering that floodplain morphology, and channel bathymetry specifically, is in constant change and measurements cannot be afforded periodically, becoming a challenge for both gauged and ungauged basins [43]. Results show that naturally surveyed cross sections can be replaced by synthetic geometry cross sections of rectangular shape maintaining a correct representation of channel slope (i.e., thalweg profile) and channel conveyance (i.e., channel flow area). This result links this research with large scale flood hazard modeling based on the use of geomorphic laws or literature values for providing valid estimation of morphologic parameters (river/floodplain width, depth and flow area) associated to varying climatic, geomorphic and hydrologic regimes [20–22].

This study also highlights the value of floodplain terrain analysis for processing high resolution DTMs to produce upscaled geometry for 2D inundation mapping. Flood model upscaling follows scaling principles that characterize inundation dynamics. The DTM scale (i.e., grid size resolution) shall be defined according to the scale of the simulated flood processes. A major river defining a large floodplain domain with flooding dynamics characterized by inundation widths and flow depths that are in the order of, respectively, kilometer (1–3 km) and meter (1–10 m) scale requires a proportional scale for the topographic and bathymetric representation. If this scaling principle is correctly applied, we argue that the loss of details in floodplain DTMs at coarser resolutions does not necessarily imply that simulated inundations will be affected by major errors. Fluvial bathymetry interpolation and floodplain terrain resampling, also using synthetic channel geometry, allow fast and accurate flood models (Figures 5 and 6). Nevertheless, we cannot neglect that floodplain interpolation are difficult to be automated, but semi-automatic procedures are needed that require flood modeler interpretation, calibration and validation, especially when converting natural to rectangular shape cross sections.

The proposed procedure still requires, in fact, manual processing work by the flood analyst. Therefore, results can be sensitive to errors, time consuming activities and to subjectivity impacting the replicability and generalization of results. The full automatization of this procedure, integrating an objective analysis of the proper scale and resolution, for creating upscaled inundation models require further investigations using geomorphological laws for estimating rectangular cross sections. Moreover, the behavior of different numerical scheme of 2D flood modeling and the development of an extended set of case studies in other climatic and morphologic settings characterize future work needed for the generalization of the proposed procedure.

6. Conclusions

In this paper, a procedure of floodplain terrain analysis was introduced for supporting the development of coarse resolution inundation models. The procedure was tested for the 120 km domain of the Tiber River in central Italy to evaluate the consistency of simulated inundation extent and depth distributions, derived from 2D hydraulic models applying different floodplain DTM resolutions, as compared to a reference flood model that was validated in previous studies. The main conclusive remarks of this research are as follows:

(1)   The proposed floodplain terrain analysis procedure does not aim to replicate the performances of accurate high-resolution flood modeling that requires surveying of fluvial bathymetry and floodplain morphology. Flood hazard management for safe urban planning and risk mitigation requires detailed modeling of floodplain and urban features exposed to inundation risk. The aim of this research was to produce consistent coarse resolution flood models, running in seconds, for investigating applications where 2D inundation modeling is actually not used for its computational and data need burden. For example, this research may pave the way to the use of 2D inundation modeling in large scale real time emergency management operations or in the coupling of hydrodynamic models with climatic global circulation models, landscape evolution models or in continental ecologic modeling applications.

(2)   Results show the tests related to the substitution of surveyed natural fluvial cross sections with synthetic rectangular cross sections. The use of a synthetic channel geometry is a valid working hypothesis if the channel thalweg profile and flow area are preserved. The impact of channel geometry simplification on the inundation dynamics is evident, considering the importance of channel morphology on flood wave routing, with significant misinterpretation of simulated discharge and velocities. Nevertheless, simulated inundation extents and depths are modeled in the range of flood modeler expectations.

(3)   Visual representation of water depths and quantitative analyses, based on the use of a measure-to-fit F index, for evaluating the potential overestimation or underestimation of the flooding extent, show that the upscaling procedure produces consistent results of decreasing accuracy with increasing floodplain DTM grid size. Consistent inundation simulation results are obtained interpolating the floodplain DTM resolution from 150 m to 700 m. The running time of simulations is reduced by one order of magnitude from approximately 30 min to 3 min for the 700 m simulation.

(4)   The proposed procedure for flood model upscaling still requires expert knowledge and subjective manual processing of reference data for correctly interpolating DTM floodplain topographic and bathymetric data. Further research is needed for flood model calibration and validation towards the replicability and generality of the proposed procedure in other case studies.

Author Contributions: Francisco Peña (FP) and Fernando Nardi (FN) jointly conceptualized the research experiment, from the design of the procedure, to the presentation of results. FP gathered and processed the case study data, developed the floodplain terrain analysis and 2D flood modelling, wrote initial version of manuscript with tables and figures. FP and FN jointly produced the final version of the manuscript, figures and tables.

Funding: This work was funded by Regione Lazio Grant No. A11598 (Research grant "Media Valle del fiume Tevere").

Acknowledgments: Reviewers and editors are acknowledged for their valuable comments that helped authors in significantly improving this manuscript. Antonio Annis is also acknowledged for providing technical assistance in the use of the FLO-2D postprocessing tool and for his support in the development of the Fit index analysis.

Conflicts of Interest: The authors declare no conflict of interest.

References

1.   Bates, P.D.; Horritt, M.S.; Aronica, G.; Beven, K. Bayesian updating of flood inundation likelihoods conditioned on flood extent data. *Hydrol. Process.* **2004**, *18*, 3347–3370. [CrossRef]

2.   Pappenberger, F.; Beven, K.; Horritt, M.; Blazkova, S. Uncertainty in the calibration of effective roughness parameters in HEC-RAS using inundation and downstream level observations. *J. Hydrol.* **2005**, *302*, 46–69. [CrossRef]

3.   Merwade, V.; Olivera, F.; Arabi, M.; Edleman, S. Uncertainty in flood inundation mapping: Current issues and future directions. *J. Hydrol. Eng.* **2008**, *13*, 608–620. [CrossRef]

4.   Di Baldassarre, G.; Schumann, G.; Bates, P.D.; Freer, J.E.; Beven, K.J. Flood-plain mapping: A critical discussion of deterministic and probabilistic approaches. *Hydrol. Sci. J.* **2010**, *55*, 364–376. [CrossRef]

5. Domeneghetti, A.; Schumann, G.J.P.; Frasson, R.P.M.; Wei, R.; Pavelsky, T.M.; Castellarin, A.; Brath, A.; Durand, M.T. Characterizing water surface elevation under different flow conditions for the upcoming SWOT mission. *J. Hydrol.* **2018**, *561*, 848–861. [CrossRef]

6. Ward, P.J.; Jongman, B.; Salamon, P.; Simpson, A.; Eates, P.; De Groeve, T.; Muis, S.; De Perez, E.C.; Rudari, R.; Trigg, M.A.; et al. Usefulness and limitations of global flood risk models. *Nat. Clim. Chang.* **2015**, *5*, 712–715. [CrossRef]

7. Neal, J.; Dunne, T.; Sampson, C.; Smith, A.; Bates, P. Optimisation of the two-dimensional hydraulic model LISFOOD-FP for CPU architecture. *Environ. Mode . Soft.* **2018**, *107*, 148–157. [CrossRef]

8. Schumann, G.J.-P.; Bates, P.D.; Apel, H.; Aronica, G.T. *Global Flood Hazard: Applications in Modeling, Mapping and Forecasting*; American Geophysical Union and John Wiley & Sons: Hoboken, NJ, USA, 2018.

9. Tauro, F.; Selker, J.; Van De Giesen, N.; Abrate, T.; Uijlenhoet, R.; Porfiri, M.; Manfreda, S.; Caylor, K.; Moramarco, T.; Benveniste, J.; et al. Measurements and observations in the XXI century (MOXXI): Innovation and multi-disciplinarity to sense the hydrological cycle. *Hydrol. Sci. J.* **2018**, *63*, 169–196. [CrossRef]

10. Trigg, M.A.; Birch, C.E.; Neal, J.C.; Bates, P.D.; Smith, A.; Sampson, C.C.; Yamazaki, D.; Hirabayashi, Y.; Pappenberger, F.; Dutra, E.; et al. The credibility challenge for global fluvial flood risk analysis. *Environ. Res. Lett.* **2016**, *11*, 094014. [CrossRef]

11. Jung, Y.; Merwade, V. Uncertainty quantification in flood inundation mapping using generalized likelihood uncertainty estimate and sensitivity analysis. *J. Hydrol. Eng.* **2012**, *17*, 507–520. [CrossRef]

12. De Frasson, R.P.M.; Wei, R.; Durand, M.; Minear, J.T.; Domeneghetti, A.; Schumann, G.; Williams, B.A.; Rodriguez, E.; Picamilh, C.; Lion, C.; et al. Automated river reach definition strategies: Applications for the surface water and ocean topography mission. *Water Resour. Res.* **2017**, *53*, 8164–8186. [CrossRef]

13. Vorogushyn, S.; Bates, P.D.; de Bruijn, K.; Castellarin, A.; Kreibich, H.; Priest, S.; Schröter, K.; Bagli, S.; Blöschl, G.; Domeneghetti, A.; et al. Evolutionary leap in large-scale flood risk assessment needed. *WIREs Water* **2017**. [CrossRef]

14. Price, R.K. An optimized routing model for flood forecasting. *Water Resour. Res.* **2009**, *45*, 1–15. [CrossRef]

15. Gichamo, T.Z.; Popescu, I.; Jonoski, A.; Solomatine, D. River cross-section extraction from the ASTER global DEM for flood modeling. *Environ. Model. Soft.* **2012**, *31*, 37–46. [CrossRef]

16. Bhuyian, M.N.M.; Kalyanapu, A.J.; Nardi, F. Approach to digital elevation model correction by improving channel conveyance. *J. Hydrol. Eng.* **2014**, *20*. [CrossRef]

17. Leopold, L.; Maddock, T. *The Hydraulic Geometry of Stream Channels and Some Physiographic Implications*; Geological Survey Professional Paper; U.S. Department of the Interior: Washington, DC, USA, 1953.

18. Lehner, B.; Verdin, K.; Jarvis, A. New global hydrography derived from spaceborne elevation data. *Eos* **2008**, *89*, 93–94. [CrossRef]

19. Pappenberger, F.; Dutra, E.; Wetterhall, F.; Cloke, H.L. Deriving global flood hazard maps of fluvial floods through a physical model cascade. *Hydrol. Earth Syst. Sci.* **2012**, *16*, 4143–4156. [CrossRef]

20. Andreadis, K.M.; Schumann, G.J.P.; Pavelsky, T. A simple global river bankfull width and depth database. *Water Resour. Res.* **2013**, *49*, 7164–7168. [CrossRef]

21. Sampson, C.C.; Smith, A.M.; Bates, P.D.; Neal, J.C.; Alfieri, L.; Freer, J.E. A high-resolution global flood hazard model. *Water Resour. Res.* **2015**, *51*, 7358–7381. [CrossRef] [PubMed]

22. Domeneghetti, A. On the use of SRTM and altimetry data for flood modeling in data-sparse regions. *Water Resour. Res.* **2016**, *52*, 2901–2918. [CrossRef]

23. Tiber River Basin Authority. *Piano Direttore dell'Autorità di Bacino del fiume Tevere (Flood Risk Management Plan)*; Autorità di Bacino del fiume Tevere: Rome, Italy, 2010. (In Italian)

24. Spada, E.; Sinagra, M.; Tucciarelli, T.; Barbetta, S.; Moramarco, T.; Corato, G. Assessment of river flow with significant lateral inflow through reverse routing modeling. *Hydrol. Process.* **2017**, *31*, 1539–1557. [CrossRef]

25. Manfreda, S.; Nardi, F.; Samela, C.; Grimaldi, S.; Taramasso, A.; Roth, G.; Sole, A. Investigation on the use of geomorphic approaches for the delineation of flood prone areas. *J. Hydrol.* **2014**, *517*, 863–876. [CrossRef]

26. Tauro, F.; Olivieri, G.; Petroselli, A.; Porfiri, M.; Grimaldi, S. Flow monitoring with a camera: A case study on a flood event in the Tiber River. *Environ. Monit. Assess.* **2016**, *188*, 1–11. [CrossRef] [PubMed]

27. O'Brien, J.S.; Julien, P.Y.; Fullerton, W.T. Two-dimensional water flood and mudflow simulation. *J. Hydrol. Eng.* **1993**, *119*, 244–261. [CrossRef]

28. O'Brien, J.S. *FLO-2D Users Manual*; FLO-2D Software, Inc.: Nutrioso, AZ, USA, 2011.

29. Sibson, R. *A Brief Description of Natural Neighbor Interpolation*; John Wiley & Sons: New York, NY, USA, 1981.

30. ESRI. *ArcGIS Desktop: Release 10*; Environmental Systems Research Institute: Redlands, CA, USA, 2011.
31. Grimaldi, S.; Teles, V.; Bras, R.L. Sensitivity of a physically based method for terrain interpolation to initial conditions and its conditioning on stream location. *Earth Surf. Proc. Land.* **2004**, *29*, 587–597. [CrossRef]
32. Grimaldi, S.; Teles, V.; Bras, R.L. Preserving first and second moments of the slope area relationship during the interpolation of digital elevation models. *Adv. Water Resour.* **2005**, *28*, 583–588. [CrossRef]
33. Merwade, V.; Cook, A.; Coonrod, J. GIS techniques for creating river terrain models for hydrodynamic modeling and flood inundation mapping. *Environ. Model. Soft.* **2008**, *23*, 1300–1311. [CrossRef]
34. Brandt, S.A. Resolution issues of elevation data during inundation modeling of river floods. In Proceedings of the XXXI International Association of Hydraulic Engineering and Research Congress, Seoul, Korea, 11–16 September 2005.
35. Da Paz, A.R.; Collischonn, W.; Tucci, C.E.M.; Padovani, C.R. Large-scale modelling of channel flow and floodplain inundation dynamics and its application to the Pantanal (Brazil). *Hydrol. Process.* **2011**, *25*, 1498–1516. [CrossRef]
36. Jung, H.C.; Hamski, J.; Durand, M.; Alsdorf, D.; Hossain, F.; Lee, H.; Azad Hossain, A.K.M.; Hasan, K.; Khan, A.S.; Zeaul Hoque, A.K.M. Characterization of complex fluvial systems using remote sensing of spatial and temporal water level variations in the Amazon, Congo, and Brahmaputra rivers. *Earth Surf. Proc. Land.* **2010**, *35*, 294–304. [CrossRef]
37. Biancamaria, S.; Bates, P.D.; Boone, A.; Mognard, N.M. Large-scale coupled hydrologic and hydraulic modelling of the Ob river in Siberia. *J. Hydrol.* **2009**, *379*, 136–150. [CrossRef]
38. Merwade, V.; Saksena, S. Incorporating the effect of DEM resolution and accuracy for improved flood inundation mapping. *J. Hydrol.* **2015**, *530*, 180–194.
39. Pappenberger, F.; Frodsham, K.; Beven, K.J.; Romanovicz, R.; Matgen, P. Fuzzy set approach to calibrating distributed flood inundation models using remote sensing observations. *Hydrol. Earth Syst. Sci.* **2007**, *11*, 739–752. [CrossRef]
40. Nardi, F.; Morrison, R.R.; Annis, A.; Grantham, T.E. Hydrologic scaling for hydrogeomorphic floodplain mapping: Insights into human-induced floodplain disconnectivity. *River Res. Appl.* **2018**. [CrossRef]
41. Horritt, M.S.; Bates, P.D. Effects of spatial resolution on a raster based model of flood flow. *J. Hydrol.* **2001**. [CrossRef]
42. Aronica, G.; Bates, P.D.; Horritt, M.S. Assessing the uncertainty in distributed model predictions using observed binary pattern information within GLUE. *Hydrol. Process.* **2002**, *16*, 2001–2016. [CrossRef]
43. Nardi, F.; Annis, A.; Biscarini, C. On the impact of urbanization on flood hydrology of small ungauged basins: The case study of the Tiber river tributary network within the city of Rome. *J. Flood Risk Manag.* **2018**, *11*, S594–S603. [CrossRef]

Article

Analyzing the December 2013 Metaponto Plain (Southern Italy) Flood Event by Integrating Optical Sensors Satellite Data

Teodosio Lacava [1,*], Emanuele Ciancia [2], Mariapia Faruolo [1], Nicola Pergola [1], Valeria Satriano [1] and Valerio Tramutoli [2]

[1] National Research Council, Institute of Methodologies for Environmental Analysis, C. da S. Loja, 85050 Tito Scalo (PZ), Italy; mariapia.faruolo@imaa.cnr.it (M.F.); nicola.pergola@imaa.cnr.it (N.P.); valeria.satriano@imaa.cnr.it (V.S.)

[2] School of Engineering, University of Basilicata, Via dell'Ateneo Lucano, 10, 85100 Potenza, Italy; emanuele.ciancia@imaa.cnr.it (E.C.); valerio.tramutoli@unibas.it (V.T.)

* Correspondence: teodosio.lacava@imaa.cnr.it; Tel.: +39-0971-427242

Received: 5 July 2018; Accepted: 5 August 2018; Published: 7 August 2018

Abstract: Timely and continuous information about flood dynamics are fundamental to ensure an effective implementation of the relief and rescue operations. Satellite data provided by optical sensors onboard meteorological satellites could have great potential in this framework, offering an adequate trade-off between spatial and temporal resolution. The latest would benefit from the integration of observations coming from different satellite systems, also helping to increase the probability of finding cloud free images over the investigated region. The Robust Satellite Techniques for detecting flooded areas (RST-FLOOD) is a sensor-independent multi-temporal approach aimed at detecting flooded areas which has already been applied with good results on different polar orbiting optical sensors. In this work, it has been implemented on both the 250 m Moderate Resolution Imaging Spectroradiometer (MODIS) and the 375 m Suomi National Polar-orbiting Partnership (SNPP) Visible Infrared Imaging Radiometer Suite (VIIRS). The flooding event affecting the Basilicata and Puglia regions (southern Italy) in December 2013 has been selected as a test case. The achieved results confirm the RST-FLOOD potential in reliably detecting, in case of small basins, flooded areas regardless of the sensor used. Flooded areas have indeed been detected with similar performance by the two sensors, allowing for their continuous and near-real time monitoring.

Keywords: flood; remote sensing; data integration; RST-FLOOD; MODIS; VIIRS; optical data

1. Introduction

Among natural disasters, floods are more easily investigable using satellite data. In 2016, among the 36 activations of the International Charter "Space & Major Disasters", 17 (i.e., 44%) of them were related to floods [1], demonstrating the effectiveness of the information that satellite remote sensing can provide in this context. Several indications can be furnished in all the phases (i.e., mitigation, preparedness, response, recovery) of the flood risk management cycle. In particular, among the different possible contributions, information about flood mapping and monitoring activities can be useful for all the above-mentioned phases [2–4]. A flood map is prepared during/after a flood occurrence in order to delineate the inundated areas, while flood maps of different times are suitable for monitoring water expansion and regression [2,3]. Moreover, multi-temporal maps can aid in detecting critical spatial changes in flood hazards and vulnerability over time. These products can be used for flood prone area delineation in order to prevent future floods, providing crucial information to identify appropriate protection measures and strategies for risk mitigation and producing efficient response plans [5]. Once a flood event occurs, emergency planners and rescuers can use inundation maps to detect the most affected areas, then identify evacuation routes and plan the assistance and aid. During the recovery stage, flood maps can support the identification of places for reconstruction and multi-date maps aid in the monitoring of community rebuilding [6].

Microwave and optical-band data have been widely used to produce flooding maps, exploiting the specific advantages of each spectral band and technology used [7,8]. Microwave sensors enable an all-day (i.e., 24 h) and all-weather detection capability, with spatial resolution ranging from few meters to dozens of kilometers when moving from active to passive technologies, respectively [5]. On the other hand, active sensors (e.g., Synthetic Aperture Radar (SAR)) allow for medium-long temporal frequency (up to five to six days), while passive radiometers have sub-daily temporal resolution [5]. Therefore, SAR data can provide infrequent detailed information about small-scale flooded areas, while passive microwave data could represent the most suitable solution if timely information is required for investigating large-scale flooding events [9]. Optical sensors onboard polar satellites, usually deployed in satellite constellation, can assure the better trade-off among spectral/spatial/temporal resolutions useful for a near real-time and continuous monitoring of flooded areas [10,11]. Obviously, cloud cover can fully hamper any kind of acquisition in this spectral region, limiting the applicability of this data. Hence, the integration of data acquired by sensors operating at different wavelengths is preferred to avoid such an issue, as well as to make the most of their potential [12].

The Robust Satellite Techniques (RST) [13] is a general multi-temporal satellite data analysis methodology that has already been applied for detecting flooded areas (RST-FLOOD [12,14]) on Advanced Very High Resolution Radiometer (AVHRR) and Moderate Resolution Imaging Spectroradiometer (MODIS) data. Both of these sensors acquire data in the Visible (VIS) and Near Infrared (NIR) regions, allowing for flooded area detection thanks to the particular spectral behavior of water at these wavelengths. Compared to other common land cover and features (like bare or vegetated soils), water generally shows a reflectance (R) in the NIR lower than in the VIS region, with the latter that corresponds for both sensors to the Red (RED) portion of the electromagnetic spectrum. Therefore, in the presence of water bodies or flooded areas, values lower than the surroundings have to be expected for the combinations of spectral reflectances acquired in the VNIR (VIS and NIR) region, like the ratio R_{NIR}/R_{VIS} [15,16] or the difference $R_{NIR}-R_{VIS}$ [17].

In this paper, we further assess the RST-FLOOD performance in detecting flooded areas by investigating a few days of the flood event that affected the Basilicata and Puglia regions (southern Italy—Figure 1) in December 2013 [18–23]. In order to investigate this event, we firstly implemented the MODIS-based approach using data acquired in its first two channels at 250 m of spatial resolution. In the previous work [12], the feasibility of this data providing reliable information about flooded areas was only preliminarily explored, referring to a much larger event than the one analyzed here in terms of flood extent. Then, exploiting the RST-FLOOD inherent characteristics, we exported the proposed approach on Suomi National Polar-orbiting Partnership (SNPP) Visible Infrared Imaging Radiometer Suite (VIIRS) imagery. The VIIRS instrument extends and improves upon a series of measurements initiated by its operational and research predecessors, the AVHRR flown on board multiple National Oceanic and Atmospheric Administration (NOAA) and Meteorological Operational (Metop) satellites, and the National Aeronautics and Space Administration (NASA) MODIS, aboard the Terra and Aqua satellites, due to its better spatial and spectral resolution as well as radiometric accuracy and stability. The accuracy of the achieved results has been evaluated through a comparison analysis with a Landsat Enhanced Thematic Mapper Plus image (ETM+ on Landsat 7) acquired concurrently with MODIS and VIIRS data.

The main aim of this study is to assess the potential of RST-FLOOD, when implemented on medium spatial resolution images, in effectively analyzing floods occurring within small catchments.

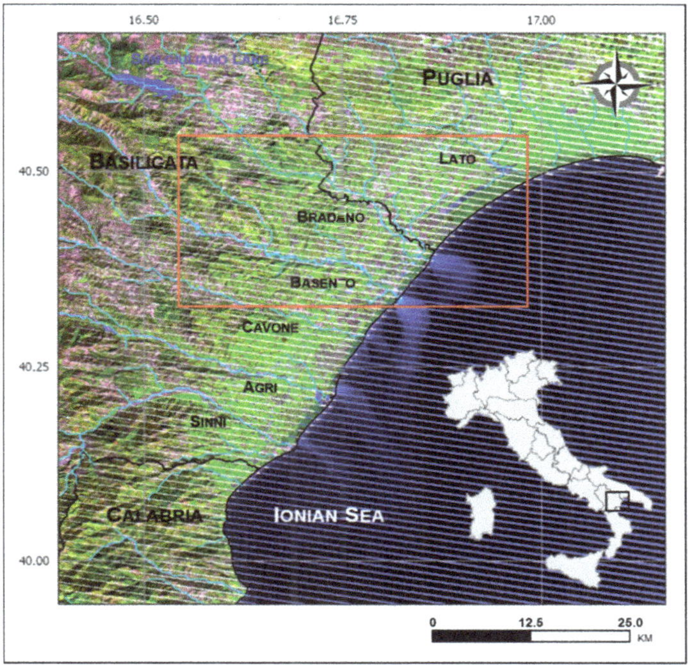

Figure 1. Localization of the Region of Interest (ROI). In the background, the Landsat Enhanced Thematic Mapper Plus (Landsat 7 ETM+) false-color (R = Short Wave InfraRed-SWIR; G = NIR; B = RED) composite image of 5 December 2013 at 9:31 GMT. The red box is the area used in the text.

2. Study Area

A flood event affected a large portion of the Metaponto plain, in the southeastern part of the Basilicata region (southern Italy), including a subset of the Puglia region (Figure 1), in the first week of December 2013 [18,19,23]. This event was caused by a significant amount of rainfall due to the "Ciclone Nettuno" storm that occurred in the Basilicata region between 30 November and 3 December 2013 [24]. A medium cumulative precipitation value of 150 mm was registered for the whole Basilicata Region between 1 and 2 December 2013, with peaks above 200 mm in the Metaponto plain [18,21]. The Sinni, Agri, Cavone, Basento and Bradano rivers in Basilicata and the Lato River in Puglia flooded in several points along their path, as well as in correspondence of the freeway "Strada Statale 106", causing its closure and serious damage to farms and agricultural crops [18]. All the rivers involved in the event are strongly seasonally dependent, with maximum hydrometric levels usually reached between the late fall and the early spring, and minimum levels reached in summer at cross-sections lower than 100 m [25]. Therefore, the studied event represents a suitable test case for evaluating the potential of medium-resolution optical sensors in investigating small-scale floods.

3. Data and Methods

3.1. Satellite Data

MODIS data acquired in the first two bands (i.e., channel 1, VIS, at 0.62–0.67 μm and channel 2, NIR, at 0.841–0.876 μm) at 250 m of spatial resolution were used in this work. In more detail, imagery acquired by the sensor onboard the Aqua satellite in the 12:00–14:00 GMT temporal range, for the month of December in the 2002–2016 period have been investigated. MODIS Level 1B (MYD02QKM) and geolocation (MYD03) data directly produced at the satellite receiving station of the Institute of Methodologies for Environmental Analysis (IMAA), located in Tito Scalo (Basilicata region, southern Italy), have been processed. These data have been produced by running the Community Satellite Processing Package (CSPP) software with antenna data directly acquired. Level 1B and geolocation data downloaded from the Level-1 and Atmosphere Archive & Distribution System (LAADS) Distributed Active Archive Center (DAAC) archive [26] have been also used to fill any gaps in the analyzed historical series.

Similarly, for VIIRS, data collected in the first two imagery bands (i.e., I1, VIS, at 0.60–0.68 µm and I2, NIR, at 0.85–0.88 µm) and the thermal infrared (TIR) imagery one (I5, TIR, at 10.5–12.4 µm) at 375 m of spatial resolution were investigated. In particular, acquisitions coming from the SNPP satellite in the same daily temporal range of Aqua were considered for the month of December in the 2012–2016 period. Also for VIIRS, the SNPP Sensor Data Record (SDR) and the I-band terrain-corrected geolocation (GITCO) data directly produced at the Institute of Methodologies for Environmental Analysis (IMAA) receiving station, which was downloaded from the NOAA Comprehensive Large Array-data Stewardship System (CLASS) archive [27], have been exploited to populate the historical series.

Finally, in absence of in situ ground truth data, the Landsat 7 ETM+ image acquired on 5 December 2013 at 9:31 GMT over the ROI and downloaded from the U.S. Geological Survey (USGS) portal [28] (Figure 1), was used to assess the accuracy of the achieved results. The RGB false color (R = SWIR, 1.55–1.75 µm; G = NIR, 0.775–0.90 µm; B = RED, 0.63–0.69 µm) of such an image, related to the ROI, is shown in Figure 1. In order to better highlight the area most affected by the water presence, the Normalized Difference Vegetation Index (NDVI) has been computed, looking for values below zero, which should correspond to water affected areas [29], including both flooded pixels and permanent water. A Boolean (water/no water) mask was produced and shown in Figure 2.

An area of about 105 km^2 has been recognized as affected by water presence, including permanent waters related both to rivers and the San Giuliano Lake included within the ROI.

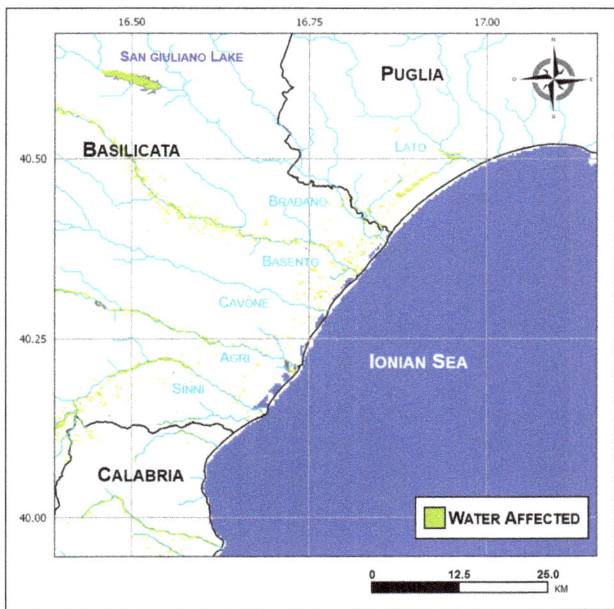

Figure 2. In green, the pixels of the Landsat 7 ETM+ image, shown in Figure 1, recognized as water affected are depicted (see text).

3.2. RST-FLOOD

The RST-FLOOD approach has already been implemented on AVHRR and MODIS data exploiting the above cited particular spectral behaviour of water in the VNIR (VIS and NIR) region of the electromagnetic spectrum [12,14]. This has led to the development of the two following Absolutely Local Index of Change of the Environment (ALICE—[13]) indices:

$$ALICE_{NIR-VIS}(x,y,t) \; = \; \frac{R_{NIR-VIS}(x,y,t) - \mu_{NIR-VIS}(x,y)}{\sigma_{NIR-VIS}(x,y)}, \tag{1}$$

$$ALICE_{NIR/VIS}(x,y,t) \; = \; \frac{R_{NIR/VIS}(x,y,t) - \mu_{NIR/VIS}(x,y)}{\sigma_{NIR/VIS}(x,y)}, \tag{2}$$

where $R_{NIR-VIS}(x,y,t)$ (or $R_{NIR/VIS}(x,y,t)$) is the reflectance difference (ratio) signal measured at time t for each pixel (x,y) of the analyzed satellite scene, $\mu_{NIR-VIS}(x,y)$ (or $\mu_{NIR/VIS}(x,y)$) and $\sigma_{NIR-VIS}(x,y)$ (or $\sigma_{NIR/VIS}(x,y)$), the named reference fields, are, respectively, the "normal" value expected for the signal and its natural variability. They are both computed by processing a multi-year dataset of co-located cloud-free imagery, collected under homogeneous observational conditions (e.g., around the same time of day and during the same month of the year). For its inherent formulation, each ALICE provides, at the pixel level, a measure of the deviation of the recorded signal from its expected (in unperturbed or normal conditions) value and automatically compares this deviation with its normal variability, which includes all the possible noise sources not related to the event being monitored. For example, the normal signal variability (i.e., the standard deviation reference field) is high for those pixels characterized both by the presence of water and land, because they are likely to be affected by a high signal fluctuation due to both residual geo-location errors and the natural cross-section changes. Therefore, for those areas, anomalous ALICE values will only be detected when high signal deviation from the expected one will be measured. In any case, in correspondence to flooded areas, negative $ALICE_{NIR-VIS}(x,y,t)$ (and $ALICE_{NIR/VIS}(x,y,t)$) values should be observed.

Moreover, for their construction, both ALICE indices are standardized variables that, as the number (N) of the records increases, tend toward a Gaussian like distribution. Under this hypothesis, values of $ALICE_{NIR/VIS}$ (or $ALICE_{NIR-VIS}$) <-2 can be associated to rare events (probability of occurrence less than 2.5%) and values <-3 to very rare events (probability of occurrence less than 0.13%). Hence, statistically significant signal anomalies are expected for $ALICE_{NIR/VIS}$ (or $ALICE_{NIR-VIS}$) <-2 at least, with an increasing level of confidence when moving to the lowest ones (i.e., -3, -4) [12,14,24].

In order to investigate the selected flood event, the above-cited MODIS and VIIRS historical temporal series were separately processed, firstly generating their corresponding reference fields (i.e., temporal mean and standard deviation) for both difference and ratio signals, and then looking for signal anomalies in the event images.

During the generation of the reference fields and during the change detection step, cloudy pixels were identified and discarded from the detection step by implementing the One Channel Algorithm (OCA) method [30,31]. Such an approach, still based on the RST prescriptions, analyzes the historical series of MODIS channel 2 to identify clouds as statistically high reflectivity objects, or VIIRS I5 data to identify them as statistically cold bodies. With regards to cloud shadows, after a visual inspection of images within the used dataset, a simple 2 km (i.e., five-and six-pixel for MODIS and VIIRS, respectively) buffer around the detected clouds has been applied, thus ensuring a good trade-off between the possibility to produce false positives and omission errors.

4. Results

Two almost concurrently acquired imagery for the first two cloud-free days over the ROI by MODIS and VIIRS have been analyzed in terms of ALICE indices computation (Equations (1) and (2)). In detail, the results investigating the MODIS imagery of 4 December 2013 at 12:25 GMT and 5 December at 11:30 GMT are presented and discussed in Section 4.1, while those referring to the VIIRS sensor,

4 December at 12:10 GMT and 5 December at 11:50, are shown in Section 4.2. Finally, Section 4.3 is focused on the comparison analysis among results achieved in the previous two sections with the Landsat 7 ETM+ image shown in Figure 1.

4.1. MODIS RST-FLOOD

The RST-FLOOD maps produced by analyzing the two above-mentioned MODIS data are shown in Figure 3, where in the top panels (Figure 3a,b), the areas flagged as anomalous using the ALICE$_{NIR-VIS}$ index (i.e., Equation (1)) are shown, while those related to the ratio index (i.e., Equation (2)) are plotted in Figure 3c,d. In all the maps, indices values less than -2 are depicted in violet and orange colors when the difference and ratio indices have been applied, respectively.

Several anomalous pixels have been detected by the RST-FLOOD indicators in all the output maps. These pixels are mostly located along the Basento and Lato rivers and are in good agreement with the local information about the flood localization [18,21,22,32], thus indicating that it can be associated to the inundated zones. Moreover, a difference between the two indicators can be observed, with a higher sensitivity of the ALICE$_{NIR-VIS}$ to the flood inundated area (see areas along the Basento and Lato rivers), as well as to the effect of turbid waters in the San Giuliano lake (see Table 1). Concerning flood dynamics, a lesser number of anomalous pixels has been identified in the maps of 5 December 2013 by both indices (Table 1). In more detail, an area of about 16 km^2 was detected as flooded on 4 December 2013, decreasing to 12 km^2 the next day.

In order to provide a deeper view of the achieved results, a magnification of the area within the red box of Figure 1 is plotted in Figure 4, where for each of the analyzed MODIS images, the results achieved by combining both the indices are highlighted. This aggregation allowed for the better definition of the potentially flooded areas, while the common detections (namely the pixels flagged as anomalous by both indices—green pixels in Figure 4), may be associated to the definitely flooded areas. Furthermore, in these maps, the temporal persistence of the anomalous area is evident, as well as the better sensitivity of the ALICE$_{NIR-VIS}$ index than the ALICE$_{NIR/VIS}$ one.

Table 1. Number of anomalous pixels identified by the two ALICE indices in the two analyzed MODIS imagery.

	ALICE$_{NIR-VIS}$	ALICE$_{NIR/VIS}$
4 December 2013	219	98
5 December 2013	185	58

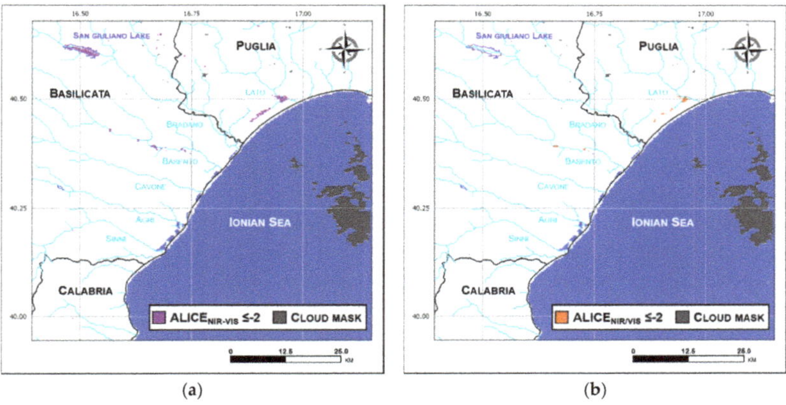

(a)   (b)

Figure 3. *Cont.*

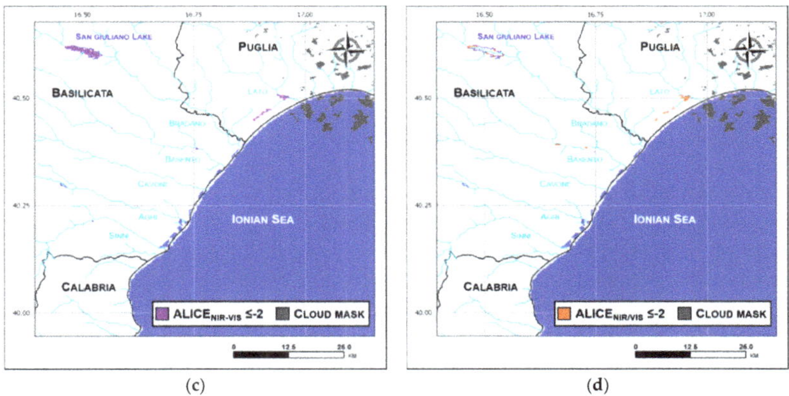

(c) (d)

Figure 3. Anomalous pixels detected by Moderate Resolution Imaging Spectroradiometer (MODIS)-based Robust Satellite Techniques for detecting flooded areas (RST-FLOOD) using: (**a**) Absolutely Local Index of Change of the Environment Near Infrared-Visible (ALICE$_{\text{NIR-VIS}}$), (**b**) ALICE$_{\text{NIR/VIS}}$ on 4 December 2013 at 12:25 GMT and (**c**) ALICE$_{\text{NIR-VIS}}$, (**d**) ALICE$_{\text{NIR/VIS}}$ on 5 December 2013 at 11:30 GMT.

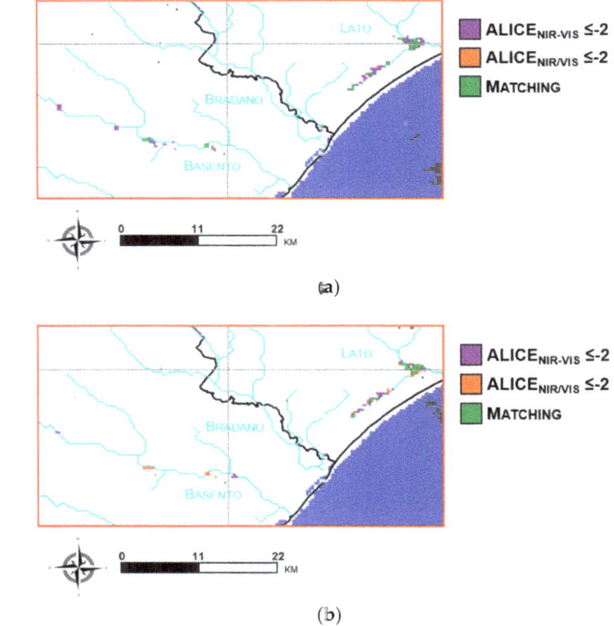

(a)

(b)

Figure 4. Anomalous pixels detected within the red box shown in Figure 1 by MODIS-based RST-FLOOD on (**a**) 4 December 2013 at 12:25 GMT and (**b**) 5 December 2013 at 11:30 GMT using ALICE$_{\text{NIR-VIS}}$ (violet pixels) and ALICE$_{\text{NIR/VIS}}$ (orange pixels). In green, the pixels detected as anomalous by both indices are highlighted.

4.2. VIIRS RST-FLOOD

Figure 5 shows the anomalous pixels identified within the ROI by implementing RST-FLOOD on VIIRS data. First, it is interesting to note that, when compared to MODIS, the two RST-FLOOD indices seem to show a higher sensitivity to flooding and to not significantly differ in terms of detected anomalies, as revealed by the analysis of the numbers reported in Table 2. In detail, an averaged area of approximately 52 km^2 was detected as flooded, almost three times higher than the one previously identified using MODIS. Although the VIIRS VNIR bands had lower spatial resolution than the MODIS one, the implementation of RST-FLOOD on VIIRS allowed for the detection of a larger number of anomalous pixels, not only along the Basento and Lato rivers, but also close to Cavone, Agri and Sinni, most likely due to the flooding. The almost equivalent number of detected pixels in two days confirm this high sensitivity, indicating that the effect of the flood was still detectable on 5 December. Concerning the San Giuliano Lake, the effect of suspended sediments is still present, even with a reduced impact. Finally, some spurious effects are also observable in the correspondence of a few pixels not close to riverbeds, disappearing when higher confidence levels of RST-FLOOD indices are used.

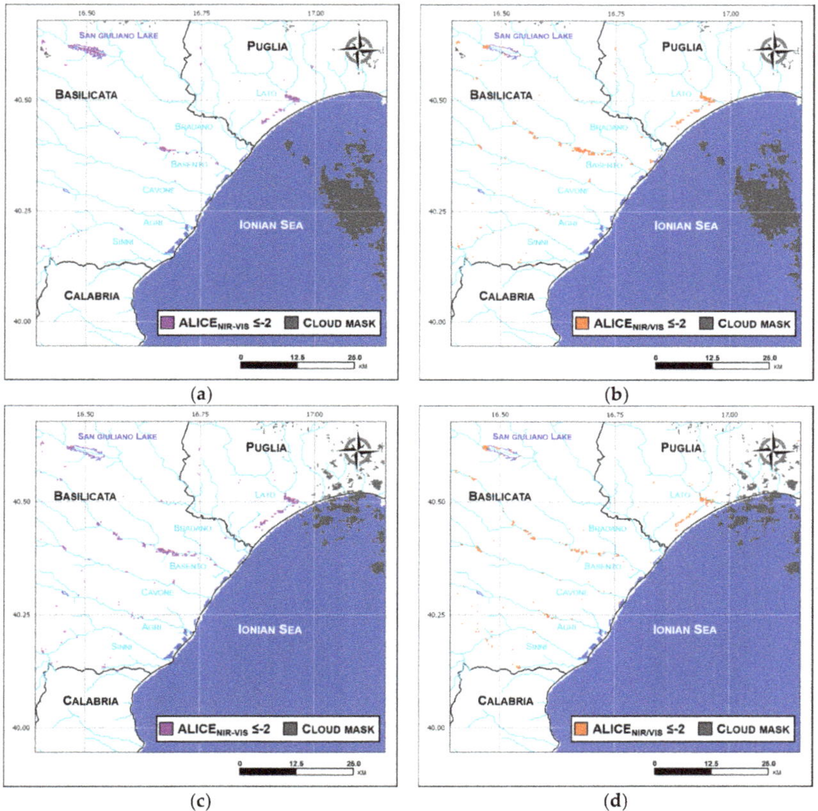

Figure 5. Anomalous pixels detected by Visible Infrared Imaging Radiometer Suite (VIIRS)-based RST-FLOOD using: (**a**) ALICE$_{NIR\text{-}VIS}$, (**b**) ALICE$_{NIR/VIS}$ on 4 December 2013 at 12:10 GMT and (**c**) ALICE$_{NIR\text{-}VIS}$, (**d**) ALICE$_{NIR/VIS}$ on 5 December 2013 at 11:50 GMT.

Table 2. Number of anomalous pixels identified by the two ALICE indices in the two analyzed VIIRS imagery.

	ALICE$_{NIR-VIS}$	ALICE$_{NIR/VIS}$
4 December 2013	214	272
5 December 2013	262	243

The magnification of the areas mainly affected by the event as detected by VIIRS have been taken into account and reported in Figure 6. The maps shown in this figure provide a clear confirmation of the considerations discussed above for MODIS, emphasizing the advantage in coupling the detections provided by the two RST-FLOOD indices, both in detecting the definitely flooding affected areas (green pixels in Figure 6) and improving the delimitation of their extent.

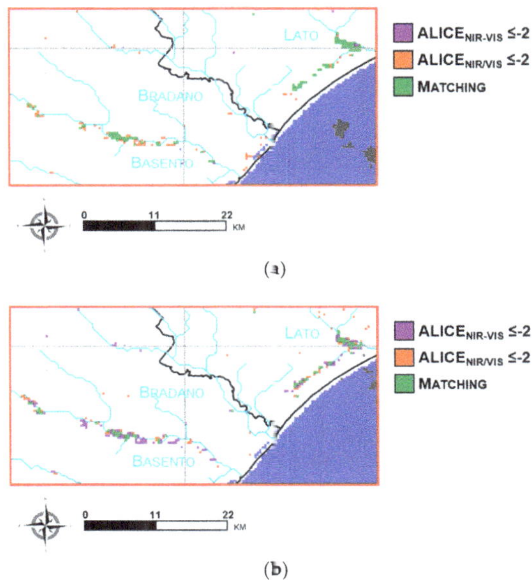

Figure 6. Anomalous pixels detected within the red box shown in Figure 1 by VIIRS-based RST-FLOOD on (**a**) 4 December 2013 at 12:10 GMT and (**b**) 5 December 2013 at 11:52 GMT using ALICE$_{NIR-VIS}$ (violet pixels) and ALICE$_{NIR/VIS}$ (orange pixels). In green, the pixels detected as anomalous by both indices are highlighted.

4.3. Comparison with Landsat

To assess the reliability of the proposed approach, the Landsat 7 ETM+ image of 5 December 2013, at 9:31 GMT, was exploited (Figure 1). In Figure 7a, a magnification of the area within the red box in Figure 1 is plotted, using the Boolean mask already plotted in Figure 2 in the background, where water affected areas are clearly visible along the Basento and Lato rivers, notwithstanding no data acquisition due to the Landsat 7's Scan Line Corrector (SLC) failure. An area of about 28 km^2 was identified as water affected within the investigated box (green pixels in Figure 7a).

The RST-FLOOD results carried out for 5 December 2013 have been superimposed on those areas aggregating MODIS and VIIRS detections (at 11:30 GMT and 11:52 GMT, respectively) achieved with the same ALICE index (Figure 7b,c), estimating an area ranging between 15 and 18 km^2 when the ratio (Figure 7c) and the difference (Figure 7b) index is used, respectively. Considering all the Landsat

7 water affected pixels depicted in Figure 7a as a benchmark for assessing the extension of flooded areas, a maximum underestimation of about 46% was computed for RST-FLOOD. The large difference in terms of spatial resolution among ETM+ and the two optical sensors used in this work forms the basis of this result [12]. Spurious and isolated ETM+ water affected pixels cannot be detected at the spatial resolution allowed by MODIS and VIIRS. Moreover, even when those pixels are aggregated, their total contribution in terms of sub-pixel effect can provide a result lower than the expected value as defined by RST-FLOOD, and therefore they are not identified as anomalous. On the other hand, such an outcome is quite relevant, because it confirms the potential of the proposed approach in effectively detecting the presence of flooded areas notwithstanding the medium spatial resolution of the used data. Furthermore, concerning Landsat 7 water affected pixels, it is worth mentioning that they also take into account permanent water, as well as other effects not directly ascribable to water presence.

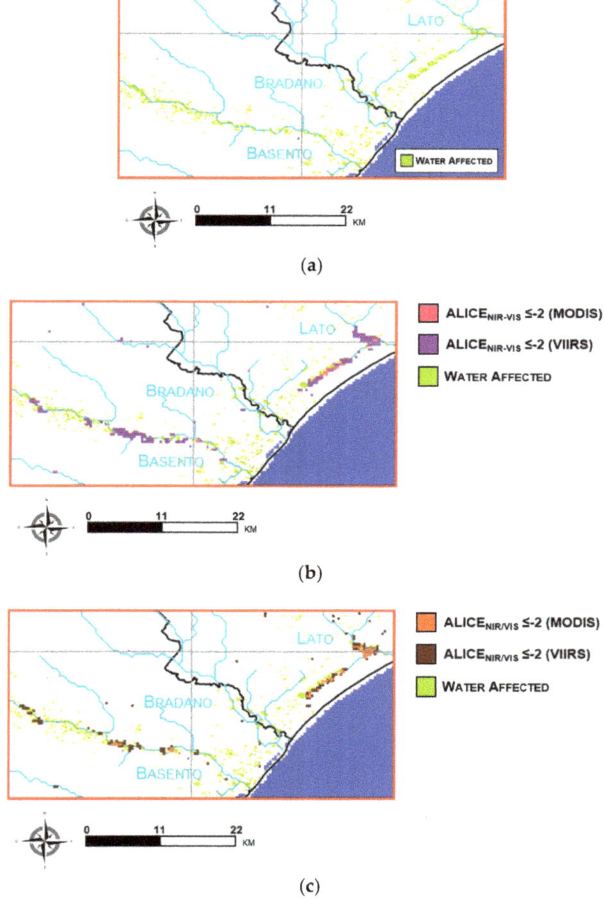

Figure 7. (**a**) Subset of the image shown in Figure 2 highlighting water affected pixels; (**b**) anomalous pixels detected by $ALICE_{NIR\text{-}VIS}$ on 5 December 2013 at 11:30 GMT for MODIS (pink) and at 11:52 GMT for VIIRS (violet); (**c**) anomalous pixels detected by $ALICE_{NIR/VIS}$ on 5 December 2013 at 11:30 GMT for MODIS (orange) and at 11:50 GMT for VIIRS (brown).

In any case, such a comparison shows the satisfactory capability of the proposed approach in detecting flooded areas regardless of the sensors used, and the different sensitivities of RST-FLOOD indicators when implemented on MODIS and VIIRS imagery. These aspects assume great relevance in detecting flooded areas by optical satellite data, reinforcing the great usefulness of implementing an integrated satellite system within the flood hazard management cycle, allowing for a more detailed identification of flooded areas and a continuous monitoring of the ongoing phenomenon on a large spatial scale. Advantages arising from such an integration are clearly observable when analyzing Figure 8, where the outputs produced by applying the same index on both of the sensors for the two available images are shown. An area of about 80 km^2 has been recognized as flooded in the two days examined by both MODIS and VIIRS sensors through the ALICE$_{NIR-VIS}$ index implementation, while 72 km^2 has been detected by the ALICE$_{NIR/VIS}$ index, with the main difference due to the pixel issues related to the San Giuliano turbid waters.

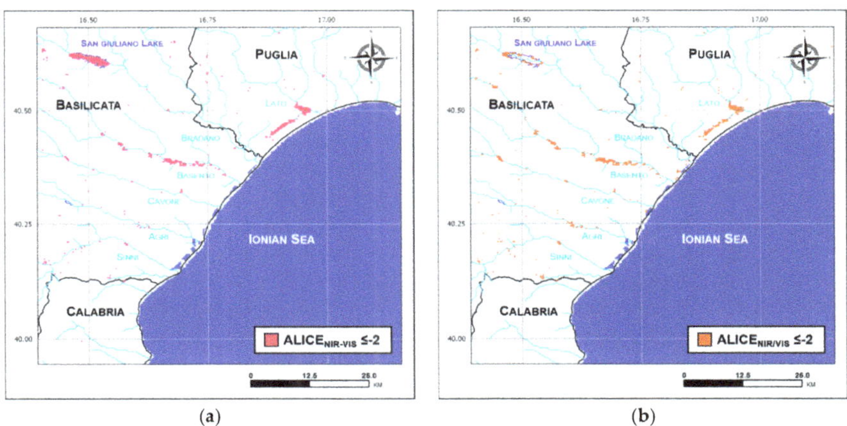

Figure 8. (**a**) Flooded areas detected within the Region of Interest (ROI) by integrating results achieved by ALICE$_{NIR-VIS}$ implemented on MODIS and VIIRS data for the two investigated days; (**b**) as in (**a**) using the ALICE$_{NIR/VIS}$.

5. Discussion

Timely and continuous information about flood dynamics is fundamental to ensure an effective implementation of relief and rescue operations. Using data acquired by optical sensors onboard meteorological satellites for flood detection and monitoring may be detrimental due to cloud cover that can hamper any kind of acquisition. Despite this limitation, there are several advantages that make optical data a good complement to other satellite-based systems, such as: (i) the large swath that allows for both large area coverage and high temporal resolution; (ii) a medium spatial resolution (in the order of hundreds of meters) useful for detecting medium-major flood events; and (iii) the deployment in satellites constellation that further enables an increase in revisiting time. In addition, sensors operating in different spectral regions that can complement the information acquired using optical data are often present aboard meteorological satellites [33]. Both MODIS and VIIRS, the optical sensors considered in this work, show almost all of the above-cited features, making them suitable for flooded area detection. While the MODIS capability in this framework has been largely demonstrated [11], there are only a few works based on VIIRS I-band data [9,34,35].

In this work, we implemented the RST-FLOOD approach on MODIS and VIIRS daytime data to analyze a few days of the flooding event that occurred in the Metaponto plain (southern Italy) in the first week of December 2013. RST-FLOOD has been already applied using 1 km MODIS visible and near infrared data, with only a preliminarily feasibility analysis of the potential of the same data when acquired at 250 m of spatial resolution [12]. In this work, we further assess the capability of this latter specific configuration of RST-FLOOD in detecting flooded areas, and also test its performance when implemented on VIIRS imagery data at 375 m of spatial resolution.

The main advantages of RST-FLOOD with respect to traditional techniques are: (i) the use of local (i.e., at the pixel scale) adaptive and dynamic thresholds; and (ii) no dependence on any kind of auxiliary/ancillary information [12]. Ancillary datasets or fixed thresholds on the signal under investigation are often used to face the main challenges, such as cloud and terrain shadows and the discrimination of flooded areas [34,35], limiting flood detection reliability. The accuracy/availability of auxiliary information can directly affect the quality of the achieved results, while fixed thresholds may suffer from sensitivity/accuracy limits because of the signal variability due to the specific site/local setting of the scene under investigation [14].

The only constraint for RST-FLOOD implementation is the availability of a satellite historical series long enough to guarantee a consistent identification of the expected values in terms of temporal mean and standard deviation. An independent work focusing on RST [36] found that such a historical series should consist of at least of 80 images to produce reliable reference fields, corresponding to at least three years of data, considering a monthly temporal window and a daily frequency of observation.

Results achieved by applying the RST-FLOOD indices on the two sensors suggest the benefit of their integration for a continuous monitoring of the ongoing phenomenon at large spatial scale. Such an integration enables a clear increase in the observational frequency, improving the flood evolution monitoring capability of each single sensor. This result is fundamental for decision makers, especially during the crisis, in order to better identify critical situations and ensure an effective implementation of relief and rescue operations. Furthermore, hydrological models would also benefit from the integration of continuous and updated information about the real-time situation, having the opportunity to check the quality of their setup as well as improving the quality of their outputs. Finally, the integration of multi-sensor datasets will generally increase the probability of clear sky acquisitions, reducing the impact of clouds, which are the main limitations of optical satellite observations, especially for this kind of application. It is worth mentioning that cumulated flooded maps of the event, allowed by high temporal resolution weather satellites, could help for a better delineation of the involved area, enabling an assessment and relative updating of the flooded risk map.

Basilicata rivers are representative of small hydrological basins, with small-size cross-sections (<100 m); therefore, the achieved results indicate that data acquired by medium-resolution optical sensors, if adequately analyzed, can also be profitable for flood detection monitoring of watersheds. In the near future, other test cases should be studied to further confirm the quality of the results achieved here by investigating a different scenario from the one considered here. For example, flood events of a smaller size (in terms of both channel width and floodplain extension) than the ones analyzed here should be investigated to better understand the sensitivity limit of the proposed approach when applied to data at 375 m of spatial resolution. Events occurring in urban areas should also be analyzed to confirm the feasibility of the methodology presented here. Furthermore, it is worth mentioning that while the proposed indices are almost "mandatory" for MODIS data, considering the spatial resolution constraints, VIIRS has other imagery bands, such as the I3 (SWIR: 1.58–1.64 μm), which will enable other combinations currently under investigation.

6. Conclusions

In this paper, RST-FLOOD has been applied to analyze the flood event that affected the Basilicata and Puglia regions in the first week of December 2013. Two different indicators, based on the difference and ratio between NIR and VIS reflectance (i.e., channel 2 and channel 1 for MODIS, I2 and I1 for VIIRS), respectively, have been used to detect flooded areas.

When implemented on MODIS data, the RST-FLOOD indices showed a behavior similar to the one observed when studying a different test case. A similar behavior between the two ALICE indices has been recognized, with the one based on the difference being more exposed to a few false positives due to suspended sediments flown into the San Giuliano lake. A better sensitivity of the VIIRS-based indices was observed. This aspect, which needs more investigation, can be preliminarily justified by considering the different specific spectral and radiometric accuracies of the two sensors and the longer historical series of MODIS data.

The flood maps provided by both RST-FLOOD indices, aggregating the MODIS and VIIRS detections, are in good geographical agreement with the one derived using Landsat 7 ETM+ data of 5 December 2013. Such an integration allowed for the discrimination of a flood area extent up to 80 km^2, lower than that potentially detectable by using a high spatial resolution sensor like the ETM+ (about 24% greater). Despite this, the double advantage of an integrated system in effectively supporting flood risk management is clearly demonstrated. The high temporal revisiting capability offered by optical sensors aboard weather satellites allows for the improvement of the observational capability of an operational monitoring system, while the combined use of different indices improves the accuracy in flood extent mapping.

Author Contributions: Conceptualization, T.L., M.F., N.P. and V.T.; Methodology, T.L., N.P. and V.T.; Validation, M.F., T.L. and V.S.; Data Curation, E.C., M.F. and V.S., Writing—Original Draft Preparation, T.L., M.F. and E.C.; Writing—Review & Editing, N.P. and V.T.

Funding: This research received no external funding.

Conflicts of Interest: The authors declare no conflict of interest.

References

1. International Charter "Space & Major Disasters". Annual Report. 2016. Available online: https://disasterscharter.org/documents/10180/66908/16thAnnualReport (accessed on 29 June 2018).
2. Sanyal, J.; Lu, X.X. Application of Remote Sensing in Flood Management with Special Reference to Monsoon Asia: A Review. *Nat. Hazards* **2004**, *33*, 283–301. [CrossRef]
3. Franci, F.; Mandanici, E.; Bitelli, G. Remote sensing analysis for flood risk management in urban sprawl contexts. *Geomat. Nat. Hazards Risk* **2015**, *6*, 583–599. [CrossRef]
4. Ward, P.J.; de Perez, E.C.; Dottori, F.; Jongman, B.; Luo, T.; Safaie, S.; Uhlemann-Elmer, S. The need for mapping, modeling, and predicting flood hazard and risk at the global scale. In *Global Flood Hazard: Applications in Modeling, Mapping, and Forecasting*, 1st ed.; Geophysical Monograph 233; Schumann, G.J.-P., Bates, P.D., Apel, H., Aronica, G.T., Eds.; John Wiley & Sons, Inc.: Hoboken, NJ, USA; American Geophysical Union: Washington, DC, USA, 2018; ISBN 978-1-119-21786-2.
5. Dasgupta, A.; Grimaldi, S.; Ramsankaran, R.; Pauwels, V.R.N.; Walker, J.P.; Chini, M.; Hostache, R.; Matgen, P. Flood mapping using synthetic aperture radar sensors from local to global scales. In *Global Flood Hazard: Applications in Modeling, Mapping, and Forecasting*, 1st ed.; Geophysical Monograph 233; Schumann, G.J.-P., Bates, P.D., Apel, H., Aronica, G.T., Eds.; John Wiley & Sons, Inc.: Hoboken, NJ, USA; American Geophysical Union: Washington, DC, USA, 2018; ISBN 978-1-119-21786-2.
6. Franci, F. The Use of Satellite Remote Sensing for Flood Risk Management. Ph.D. Thesis, Alma Mater Studiorum Università di Bologna, Bologna, Italy, 2015. [CrossRef]
7. Fayne, J.; Bolten, J.; Lakshmi, V.; Ahamed, A. Optical and physical methods for mapping flooding with satellite imagery. In *Remote Sensing of Hydrological Extremes*; Lakshmi, V., Ed.; Springer: Cham, Switzerland, 2017.
8. Markert, K.L.; Chishtie, F.; Anderson, E.R.; Saah, D.; Griffin, R.E. On the merging of optical and SAR satellite imagery for surface water mapping applications. *Results Phys.* **2018**, *9*, 275–277. [CrossRef]
9. Sun, D.; Li, S.; Zheng, W.; Croitoru, A.; Stefanidis, A.; Goldberg, M. Mapping floods due to Hurricane Sandy using NPP VIIRS and ATMS data and geotagged Flickr imagery. *Int. J. Digit. Earth* **2016**, *9*, 427–441. [CrossRef]
10. Lacava, T.; Brocca, L.; Coviello, I.; Faruolo, M.; Pergola, N.; Tramutoli, V. Integration of optical and passive microwave satellite data for flooded area detection and monitoring. In *Engineering Geology for Society and Territory*; Springer: New York, NY, USA, 2014; Volume 3, pp. 631–635.

11. Brakenridge, G.R. Flood risk mapping from orbital remote sensing. In *Global Flood Hazard: Applications in Modeling, Mapping, and Forecasting*, 1st ed.; Geophysical Monograph 233; Schumann, G.J.-P., Bates, P.D., Apel, H., Aronica, G.T., Eds.; John Wiley & Sons, Inc.: Hoboken, NJ, USA; American Geophysical Union: Washington, DC, USA, 2018; ISBN 978-1-119-21786-2.

12. Faruolo, M.; Coviello, I.; Lacava, T.; Pergola, N.; Tramutoli, V. A multi-sensor exportable approach for automatic flooded areas detection and monitoring by a composite satellite constellation. *IEEE Trans. Geosci. Remote Sens.* **2013**, *51*, 2136–2149. [CrossRef]

13. Tramutoli, V. Robust Satellite Techniques (RST) for Natural and Environmental Hazards Monitoring and Mitigation: Theory and Applications. In Proceedings of the Fourth International Workshop on the Analysis of Multitemporal Remote Sensing Images, Louven, Belgium, 18–20 July 2007.

14. Lacava, T.; Filizzola, C.; Pergola, N.; Sannazzaro, F.; Tramutoli, V. Improving flood monitoring by the Robust AVHRR Technique (RAT) approach: The case of the April 2000 Hungary flood. *Int. J. Remote Sens.* **2010**, *31*, 2043–2062. [CrossRef]

15. Sheng, Y.; Su, Y.; Xiao, Q. Challenging the cloud-contamination problem in flood monitoring with NOAA/AVHRR imagery. *Photogramm. Eng. Remote Sens.* **1998**, *64*, 191–198.

16. Sheng, Y.; Gong, P.; Xiao, Q. Quantitative dynamic flood monitoring with NOAA AVHRR. *Int. J. Remote Sens.* **2001**, *22*, 1709–1724.

17. Xiao, Q.; Chen, W. Songhua River flood monitoring with meteorological satellite imagery. *Remote Sens. Inf.* **1987**, *4*, 37–41.

18. Centro Funzionale Decentrato delle Protezione Civile Basilicata: Eventi Metereologici Eccezionali dei Giorni 1,2 e 3 Dicembre 2013 nel Territorio della Regione Basilicata. 2013. Available online: http://www.centrofunzionalebasilicata.it/ew/ew_pdf/r/Report%20evento%20dicembre%202013.pdf (accessed on 29 June 2018).

19. Autorità di Bacino della Puglia: Valutazione Globale Provvisoria del Piano di Gestione del Rischio di Alluvioni. 2015. Available online: http://www.adb.puglia.it/public/files/downloads/20151104_PGRA/VGP.pdf (accessed on 29 June 2018).

20. Dal Sasso, S.F.; Cantisani, A.; Lanorte, V.; Pacifico, G.; Manfreda, S. Gli eventi storici della Basilicata. In *Le Precipitazioni Estreme in Basilicata*, 1st ed.; Manfreda, S., Sole, A., De Costanzo, G., Eds.; Universosud Società Cooperativa: Potenza, Italy, 2015; pp. 6–24, ISBN 978-88-99432-03-4. Available online: http://www.centrofunzionalebasilicata.it/it/pdf/pioggia_download.pdf (accessed on 29 June 2018).

21. D'addabbo, A.; Refice, A.; Pasquariello, G.; Lovergine, F.P.; Capolongo, D.; Manfreda, S. A Bayesian network for flood detection combining SAR imagery and ancillary data. *IEEE Trans. Geosci. Remote Sens.* **2016**, *54*, 3612–3625. [CrossRef]

22. De Musso, N.M.; Capolongo, D.; Refice, A.; Lovergine, F.P.; D'Addabbo, A.; Pennetta, L. Spatial evolution of the December 2013 Metaponto plain (Basilicata, Italy) flood event using multi-source and high-resolution remotely sensed data. *J. Maps* **2018**, *14*, 219–229. [CrossRef]

23. Il Giornale della Protezione Civile: Rassegna Stampa del 4/12/2013. 2013. Available online: https://www.ilgiornaledellaprotezionecivile.it/html/download.html?id=7315738794M (accessed on 29 June 2018).

24. Di Polito, C.; Ciancia, E.; Coviello, I.; Doxaran, D.; Lacava, T.; Pergola, N.; Satriano, V.; Tramutoli, V. On the Potential of Robust Satellite Techniques Approach for SPM Monitoring in Coastal Waters: Implementation and Application over the Basilicata Ionian Coastal Waters Using MODIS-Aqua. *Remote Sens.* **2016**, *8*, 922. [CrossRef]

25. Autorità di Bacino della Basilicata: Mappe della Pericolosità e Mappe del Rischio Idraulico, Relazione. 2014. Available online: http://www.adb.basilicata.it/adb/Pstralcio/pianoacque/Relazione_ottobre_2014.pdf (accessed on 29 June 2018).

26. Level-1 and Atmosphere Archive & Distribution System (LAADS) Distributed Active Archive Center (DAAC) Archive. 2018. Available online: https://ladsweb.modaps.eosdis.nasa.gov/ (accessed on 29 June 2018).

27. NOAA Comprehensive Large Array-Data Stewardship System (CLASS). 2018. Available online: https://www.avl.class.noaa.gov/saa/products/welcome (accessed on 29 June 2018).

28. U.S. Geological Survey (USGS). EarthExplorer. 2018. Available online: https://earthexplorer.usgs.gov/ (accessed on 29 June 2018).

29. Szabó, S.; Gácsi, Z.; Balázs, B. Specific features of NDVI, NDWI and MNDWI as reflected in land cover categories. *Landsc. Environ.* **2016**, *10*, 194–202. [CrossRef]

30. Cuomo, V.; Filizzola, C.; Pergola, N.; Pietrapertosa, C.; Tramutoli, V. A self-sufficient approach for GERB cloudy radiance detection. *Atmos. Res.* **2004**, *72*, 39–56. [CrossRef]

31. Pietrapertosa, C.; Pergola, N.; Lanorte, V.; Tramutoli, V. Self Adaptive Algorithms for Change Detection: OCA (the One-channel Cloud-detection Approach) an adjustable method for cloudy and clear radiances detection. In Proceedings of the Technical Proceedings of the Eleventh International (A)TOVS Study Conference (ITSC-XI), Budapest, Hungary, 20–26 September 2000; pp. 281–291.

32. Vivi Castelleneta: Alluvione a Castellaneta Marina. Case in Pericolo. 2013. Available online: http://www.vivicastellaneta.it/notizie/item/2140-alluvione-a-castellaneta-marina-case-in-pericolo (accessed on 29 June 2018).

33. Lacava, T.; Ciancia, E.; Coviello, I.; Di Polito, C.; Faruolo, M.; Pergola, N.; Satriano, V.; Tramutoli, V. A satellite multi-sensor approach for flooded areas detection and monitoring. In *Advances in Watershed Hydrology*; Moramarco, T., Barbetta, S., Brocca, L., Eds.; Water Resources Publications, LLC.: Highlands Ranch, CO, USA, 2015; Chapter 5, pp. 83–96, ISBN 13-978-1-887201-85-8.

34. Huang, C.; Chen, Y.; Wu, J.; Li, L.; Liu, R. An evaluation of Suomi NPP-VIIRS data for surface water detection. *Remote Sens. Lett.* **2015**, *6*, 155–164. [CrossRef]

35. Li, S.; Sun, D.; Goldberg, M.D.; Sjoberg, B.; Santek, D.; Hoffman, J.P.; DeWeese, M.; Restrepo, P. Lindsey, S., Holloway, E., Automatic near real-time flood detection using Suomi-NPP/VIIRS data. *Remote Sens. Environ.* **2018**, *204*, 672–689. [CrossRef]

36. Koeppen, W.C.; Pilger, E.; Wright, R. Time series analysis of infrared satellite data for detecting thermal anomalies: A hybrid approach. *Bull. Volcanol.* **2011**, *73*, 577–593. [CrossRef]

Article

GEV Parameter Estimation and Stationary vs. Non-Stationary Analysis of Extreme Rainfall in African Test Cities

Francesco De Paola [1,*], Maurizio Giugni [1], Francesco Pugliese [1], Antonio Annis [2] and Fernando Nardi [2]

[1] Department of Civil Architectural and Environmental Engineering (DICEA), University of Naples Federico II, 80125 Naples, Italy; maurizio.giugni@unina.it (M.G.); francesco.pugliese2@unina.it (F.P.)

[2] Water Resources Research and Documentation Centre (WARREDOC), University for Foreigners of Perugia, 06123 Perugia, Italy; antonio.annis@unistrapg.it (A A.); fernando.nardi@unistrapg.it (F.N.)

* Correspondence: francesco.depaola@unina.it or depaola@unina.it; Tel.: +39-081-768-3420

Received: 13 March 2018; Accepted: 16 May 2018; Published: 18 May 2018

Abstract: Nowadays, increased flood risk is recognized as one of the most significant threats in most parts of the world, with recurring severe flooding events causing significant property and human life losses. This has entailed public debates on both the apparent increased frequency of extreme events and the perceived increases in rainfall intensities within climate changing scenarios. In this work, a stationary vs. Non-Stationary Analysis of annual extreme rainfall was performed with reference to the case studies of the African cities of Dar Es Salaam (TZ) and Addis Ababa (ET). For Dar Es Salaam (TZ) a dataset of 53 years (1958–2010) of maximum daily rainfall records (24 h) was analysed, whereas a 47-year time series (1964–2010) was taken into account for Addis Ababa (ET). Both gauge stations rainfall data were suitably fitted by Extreme Value Distribution (EVD) models. Inference models using the Maximum Likelihood Estimation (MLE) and the Bayesian approach were applied on EVD considering their impact on the shape parameter and the confidence interval width. A comparison between a Non-Stationary regression and a Stationary model was also performed. On this matter, the two time series did not show any Non-Stationary effect. The results achieved under the CLUVA (Climatic Change and Urban Vulnerability in Africa) EU project by the Euro-Mediterranean Centre for Climate Change (CMCC) (with 1 km downscaling) for the IPCC RCP8.5 climatological scenario were also applied to forecast the analysis until 2050 (93 years for Dar Es Salaam TZ and 86 years for Addis Ababa ET). Over the long term, the process seemed to be Non-Stationary for both series. Moreover with reference to a 100-year return period, the IDF (Intensity-Duration-Frequency) curves of the two case-studies were estimated by applying the Maximum Likelihood Estimation (MLE) approach, as a function of confidence intervals of 2.5% and 97.5% quantiles. The results showed the dependence of Non-Stationary effects of climate change to be conveniently accounted for engineering design and management.

Keywords: climate change; IDF curves; Bayesian analysis; Non-Stationary process

---

1. Introduction

Nowadays, the increased flood risk is recognized as one of the most important threats, from both the actual and, more importantly, the climate change scenario, with frequent severe flooding at the global scale causing significant loss of property and life. In 2009 (Messina, Sicily Region), 2010 (Atrani, Campania Region) and 2011 (Genova, Liguria Region), the Italian territory was forced to face widespread flooding events, highlighting the significant vulnerability of the territory and the inadequacy of early warning and flood protection systems. According to the official report of the

Italian Environmental Ministry, over 1000 buildings and businesses were flooded, causing 40 deaths. Nevertheless, worldwide flooding events caused by extreme rainfall events are causing the degradation of water quality, damages and potential loss of life [1,2].

These events prompted public debates on the apparent causes of the increased frequency of extreme events. Indeed, scientists are querying the changing statistics of rainfall intensity, especially considering the climate model predictions of increasing frequency and intensity of heavy rainfall in the high latitudes under enhanced greenhouse conditions. In 2007, the United Nations Intergovernmental Panel on Climate Change (IPCC) issued a report [3] stating that "It is very likely that hot extremes, heat waves, and heavy precipitation events will continue to become more frequent", with the "very likely" assertion indicating an occurrence probability greater than 90%. Thus, the actual and recent observations and future predictions are pointing out the importance of suitable assessments of the probability occurrence of extreme events related to long return periods T.

A comprehensive review of trend analysis and climate change projections of extreme precipitations in Europe is given by [4]. Many studies demonstrate changes in seasonal extreme precipitations in some parts of Europe [5–8].

To estimate extreme value statistics for frequency analysis of extreme precipitation, the assumption of stationarity is valid if we do not consider changes in climate that affect the hydrological regime. In this matter, during the last decade, several studies adopted Non-Stationary frequency analysis for extreme precipitation in different parts of the world [9–11]. On the other hand, simplified indexes [12,13] or Innovative Trend Analysis [14] have also been applied and tested for identifying potential trends in extreme precipitations.

Extreme Value Distributions (EVDs) are usually able to accurately represent the frequency of hydrologic over-threshold physical processes. Parameter estimation through Likelihood-based methods allows extreme quantiles to be calculated, providing the appraisal of the parameter and the associated dynamic for given return period T (e.g., 100 years). However, EVD parameters are typically estimated from the extremes of a dataset (e.g., annual maxima), which may result in characterizing unrealistic values. A possible solution to improve the reliability of statistical analysis concerns the application of a Bayesian framework, which allows us to constrain the estimates as a function of predefined information, which could realistically reproduce the hydrologic processes governing the available data. Further essential points regard the potential Non-Stationarity that could involve meaningful variation of the mean value of the distribution, with corresponding variation of the return levels related to a return period T.

This work first investigated the behaviour of the GEV distribution to estimate the annual maxima of rainfall depths adopting both MLE and Bayesian methods, evaluating their relative impacts on the confidence interval widths. Moreover, stationary and Non-Stationary analyses of extreme rainfall were applied to historical time series (with a relative limited extension), and the extended ones used climate models, thus evaluating potential Non-Stationary effects induced by the climate models.

Maximum daily (24 h) rainfall data series for the cities of Dar Es Salaam (TZ) with a 53-year dataset (1958–2010), and Addis Ababa (ET) with a 47-year dataset (1964–2010) were analysed to evaluate the Goodness of Fit (GoF) of EVD models. Specifically, a comparison was realized between Maximum Likelihood Estimation (MLE) and Bayesian models. Moreover, results gathered from the CLUVA (Climatic Change and Urban Vulnerability in Africa) EU project by the Euro-Mediterranean Centre for Climate Change (CMCC) (with 1 km downscaled resolution) for the IPCC RCP8.5 climatological scenario were also considered, extending the rainfall dataset to 2050 (93 years for Dar Es Salaam TZ and 87 years for Addis Ababa ET).

The impact of the actual versus changing hydrologic forcing scenarios was evaluated, with reference to a 100-year return period, estimating the Intensity-Duration-Frequency (IDF) curves through the Maximum Likelihood Estimation (MLE) as a function of confidence intervals of 16% and 84% quantiles, respectively.

2. Methodology

2.1. Extreme Value Theory

The Generalized Extreme Value (GEV) distribution was proposed by [15] following the original formulation of [16]. The GEV approach is widely applied to model extremes of hydrologic processes such as floods [17], rainfall [18] and sea waves [19] Further applications also include different subjects such as financial market risk modelling [20].

By supposing $X_1, X_2, \ldots, X_n$ as a sequence of independent random variables from a common distribution function F, the statistic order of interest is $M_n = max\{X_1, X_2, \ldots, X_n\}$, namely, the maximal value of the independent identically distributed (i.i.d.) random variables. Here, X_i represents the daily rainfall amount (mm) and n the number of observations in one year (365 or 366), so that M_n is the annual maximum. When the parent distribution function F is known, the distribution of M_n can be derived from statistical theory such that $P(M_n \leq z) = [F(z)]^n$.

With F unknown, a limit distribution for F_n as $n \to \infty$ is searched for, in a way similar to how the application of the Central Limit Theorem (CLT) allows approximation of the distribution of samples by means of the Normal distribution. Considering the linearly renormalized variable M^*_n:

$$M^*_n = \frac{M_n - b_n}{a_n} \tag{1}$$

for sequences $a_n > 0$ and b_n, M_n should be normalized because $F_n(z) \to 0$ as $n \to \infty$ for fixed z lower than the upper end-point of F. This causes the degeneration of the distribution to a point mass on the upper end-point of F (i.e., the smallest z such that $F(z) = 1$).

The whole range of limit distributions for M^*_n is given by the Extremal Types Theorem [21]. If a series of constants $\{a_n > 0\}$ and $\{b_n\}$ exists, such that $P(M^*_n \leq z) \to G(z)$ as $n \to \infty$ (with G a non-degenerate distribution function), then G must be included into one of the following three families: (I) Gumbel; (II) Fréchet and (III) Weibull distributions.

Families I, II and III can be combined into a single family named the Generalized Extreme Value (GEV) distribution [4], given by the following Equation (2):

$$G(z) = \exp\left\{-\left[1 + \xi\left(\frac{z - \mu}{\sigma}\right)\right]^{-\frac{1}{\xi}}\right\}, \quad -\infty < \mu < \infty, \ \sigma > 0, \ -\infty < \xi < \infty \tag{2}$$

defined as $\{z: 1 + \xi(z - \mu)/\sigma > 0\}$, where μ, σ, ξ are the location, scale and shape parameters, respectively.

The shape parameter ξ governs the tail behaviour of the distribution at its upper end. The Weibull class has a finite upper endpoint, whereas the Gumbel and Fréchet classes provide relatively different rates of decay in the tail. The Fréchet is a more heavy tailed distribution because it decays polynomially with respect to the Gumbel class, which decays exponentially instead. The tail behaviour is strongly significant, as it corresponds to quite different characteristics of extreme value behaviour. To perform the unification to a single GEV distribution, making a choice about the best model before parameter estimation is required. The key advantage of the GEV distribution over the three EV types derives from the assertion that the estimation of the ξ parameter via inference methods allows the data to select which family (and tail behaviour) to adopt without any prior decision. Moreover, the uncertainty in the estimated value of ξ measures the uncertainty correlated with the effectiveness of the three types for the available data set.

By inverting Equation (2), an estimation of extreme quantiles can be obtained, being the *p*th-upper quantile of the *z* distribution given by $G(z_p) = 1 - p$. A z_p estimation is achieved by substituting the estimates of μ, σ and ξ (for $0 < p < 1$):

$$\widehat{z}_p = \begin{cases} \widehat{\mu} - \dfrac{\widehat{\sigma}}{\widehat{\xi}}\left[1 - (-\log(1-p))^{-\widehat{\xi}}\right] & for\,\widehat{\xi} \neq 0 \\[2mm] \widehat{\mu} - \widehat{\sigma}\log(-\log(1-p)) & for\,\widehat{\xi} = 0 \end{cases} \tag{3}$$

where z_p is the return level correlated to the return period $T = 1/p$. Return levels represent usual estimation parameters for extreme events, resulting in the "100-year flood" widely applied in hydrological applications for civil engineering. Loosely speaking, z_p is the level expected to be meanly exceeded once every $1/p$ years (assuming G as the annual maximum). More formally, for extreme rainfall, z_p is the daily rainfall amount exceeded by the annual maximum for any year with a probability equal to p.

The plot of z_p against $1/p$ is known as the return level plot, resulting in an effective tool to graphically observe the return levels. The sign of the shape parameter takes on an important role when extrapolating to long return periods. Indeed, a small error in estimating the ξ parameter can lead to a much larger error in the return level estimation. Thus, accurate evaluation of the GEV shape parameter is essential to plan and develop flood protection models [22].

2.2. Bayesian Approach

The Bayesian approach allows us to make inferences from the Likelihood function, overcoming the limitation of the usually small size of the time series characterized by the annual maxima.

According to this approach, as opposed to the MLE, a parameter θ of a distribution is not an unknown constant but it is treated as a random variable related to a prior normal *pdf* $f(\theta)$ with zero mean and a certain variance v_θ.

If we consider $x = (x_1, \ldots, x_n)$ as independent realizations of a random variable so that $\{f(x;\theta) : \theta \in \Theta\}$, according to the Bayes Theorem:

$$f(\theta/x) = \frac{f(\theta)f(x/\theta)}{\int_\theta f(\theta)f(x/\theta)d\theta} \tag{4}$$

where $f(\theta|x)$ is the posterior distribution, $f(\theta)$ is the abovementioned prior distribution and $f(x|\theta) = \prod_{i=1}^{n} f(x_i|\theta)$ is the Likelihood. In case of many parameters such as the GEV distribution, the denominator of Equation (4) can be computationally complex. To overcome this issue, Markov Chain Monte Carlo (MCMC) techniques, based on multiple simulations, are usually adopted [21].

The sequences of simulated values of the parameters can be generated with the Metropolis-Hastings algorithm, e.g., applying the evdbayes package within R software [23]. The algorithm generates a sequence of parameters adopting a random-walk characterized by a normal distribution with a mean equal to the previous value of the parameter in the chain and a given variance. After a burn-in period, the sequence of the parameters is approximately stationary and the estimate of the parameter is computed as the mean of the sequence excluding the values within the burn-in period.

2.3. Non-Stationary Frequency Model

Under the climate change context, the intensity and/or the frequency of extreme rainfall events can change with time, so that the hypothesis of stationarity of the series of annual maxima is not satisfied [24].

The Non-Stationary behavior of an extreme value distribution can be expressed in different ways, such as changing its average values, the variability of its variables or both of them simultaneously [25].

More complex cases of Non-Stationarity could involve changes in distribution asymmetry or even in the parametric form [24].

Most studies look at trends induced by average variables [26] or extreme values indicators [27] without considering the variation of the shape of the distribution or of its variability. Following this simplest hypothesis, in this work, the GEV distribution for a Non-Stationary frequency model is expressed as GEV(μ_i, σ, ζ), where $\mu_i = \mu_0 + \mu_{trend}\, t_i$ with t_i the counter of the year from the analysed time series (see Section 3.1.1).

3. Case Studies

The extreme value statistics for frequency analysis of extreme precipitation were performed for Dar Es Salaam (TZ) and Addis Ababa (ET) case studies for both cases of historical data and extended time series using the COSMO CLM model by Euro-Mediterranean Centre for Climate Change (CMCC). Table 1 shows the basic statistics of the mentioned time series. Kolmogorov-Smirnov, Anderson-Darling and Chi-Square GoF tests have been performed on the GEV distribution application for both case studies related to the historical and extended time series. Results of the GoF were positive for each test, considering the rejection of the null hypothesis at the level a = 0.01.

Table 1. Basic statistics of the times series for both cases studies of Dar Es Salaam and Ababa.

Basic Statistics	Dar Es Salaam		Addis Ababa	
	Historical Data	Hist. + CMCC Data	Historical Data	Hist. + CMCC Data
Mean [mm]	78.8	71.3	52.1	54.2
Standard deviation [mm]	23.6	26.8	18.4	20.2
Skewness	1.20	0.97	1.20	1.19
Kurtosis	0.98	0.59	0.94	0.96

3.1. Dar Es Salaam (TZ) Rainfall Data

3.1.1. Historical Data

The first analysis was based on the data series of daily rainfall observations recorded at Dar Es Salaam (TZ) Airport, Tanzania (Latitude: -6.87 N; Longitude: 39.20 E; Elevation: 53 m a.s.l.) with reference to the time span 1 January 1958–31 December 2010. Annual maxima are plotted in the following Figure 1.

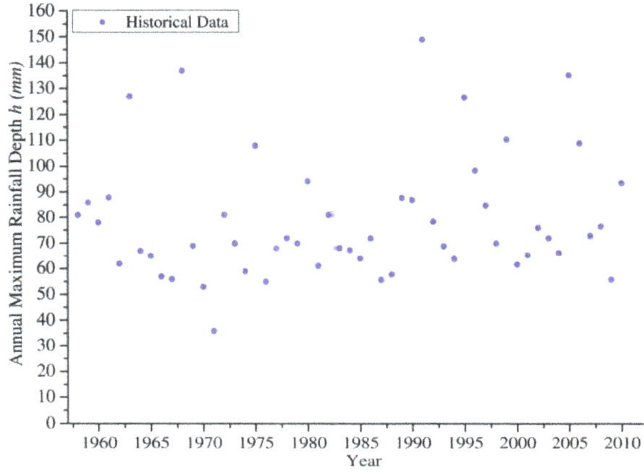

Figure 1. Annual maximum 24 h rainfalls recorded at Dar Es Salaam TZ (1958–2010).

The assumption of data as independent observations from the GEV distribution was applied. Based on the Maximum Likelihood Estimation (MLE) method (by using the *eXtremes* [28] and *evd* [29] packages in *R* software [30]), the following estimation of μ, σ and ξ parameters was obtained $(\hat{\mu}, \hat{\sigma}, \hat{\xi})$ = (68.25, 16.93, 0.039), with standard errors equal to 2.55, 1.83 and 0.083, respectively. Approximate 95% confidence intervals for each parameter were [63.25, 73.25], [13.35, 20.52] and [−0.124, 0.202] for μ, σ and ξ, respectively. This showed that the 95% confidence interval is well extended for values lower than zero, although the estimation of the shape parameter was positive, pointing out the uncertainty of the performed evaluation.

The survey for the 100-year return level was $\hat{z}_{0.01}$ = 153.6 mm, with a 95% confidence interval of [129, 208] mm, and the return level plot in Figure 2 shows the linear trend of the function as a consequence of the estimation of the ξ parameter tending towards 0. Diagnostic plots (not shown for the sake of brevity), such as probability plot and quantile plots, showed that each set of plotted points are roughly linear, validating the use of the GEV model.

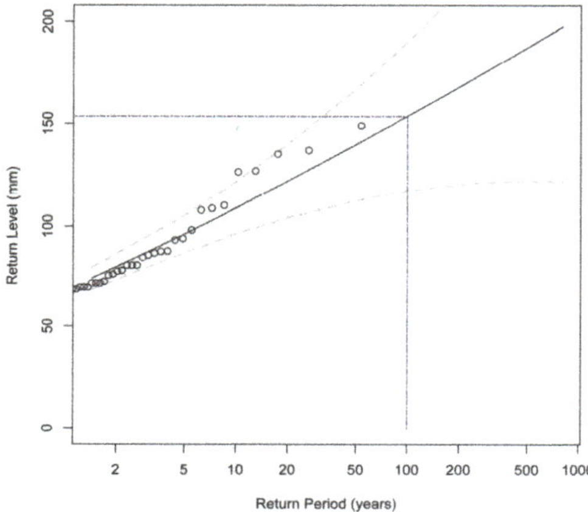

Figure 2. Return Level (in mm) plot using maximum likelihood surveys.

A more reliable survey using a Bayesian approach was implemented by applying the *evdbayes* package in *R* software [23]. The algorithm provides functions for the Bayesian analysis of extreme value models, using the Markov Chain Monte Carlo (MCMC) method. The solely genuine prior information available referred to the GEV shape parameter; thus, prior information about μ and σ parameters were not-informative normal distributions with variance 10^4. In the Bayesian analysis, more specific empirical evidence provided by Koutsoyiannis [31,32] was applied, as a function of $\xi \approx 0.15$ for Europe.

A normal distribution around 0.15 with variance 0.2 was formed, restricting the ξ variation to a physically reasonable range [22]. By applying the MLEs as the initial vector θ_0 = $(\hat{\mu}, \hat{\sigma}, \hat{\xi})$ = (68.25, 16.93, 0.039) and the proposal standard deviations *psd* = (6.191, 0.230, 0.216) (identified with some pilot runs), a Markov Chain Monte Carlo (MCMC method) was generated with a length of 100,000, satisfying mixing properties (Figure 3). By graphically examining the chain (Figure 3) and using the Geweke diagnostic [33], a burn-in period of 10 iterations was found.

The sample means and standard deviations of each marginal component of the chain were:

$$\hat{\mu} = 68.22\ (2.65);\ \hat{\sigma} = 17.74\ (1.98);\ \hat{\xi} = 0.04716\ (0.0834) \tag{5}$$

whereas the 95% reasonable intervals were [63.10, 73.52], [14.32, 22.12] and [−0.11, 0.22] for μ, σ, and ξ, respectively.

The sequence of simulated (μ_i, σ_i, ξ_i) values was transformed, leading to a sample from the corresponding posterior distribution of the 100-year return level (Figure 4). This gave a $\hat{z}_{0.01}$ estimation equal to 161.8 mm with 95% reasonable interval of [131.6, 219.1] mm. The plot of the posterior return level given in Figure 5 shows the upper 95% interval to be more remote than the lower interval from the median level.

This was due to the heavier upper tail of the posterior distribution (Figure 4), achieved for the non-negative prior on ξ. The summary of Dar Es Salaam (TZ) data is given in Table 2.

Moreover, μ was estimated to be 68 mm; nevertheless, MLE returned a lower estimation of the scale parameter σ, with respect to that derived from the Bayesian method. In terms of credibility intervals, the estimation of the shape parameter was more precise using the Bayesian method.

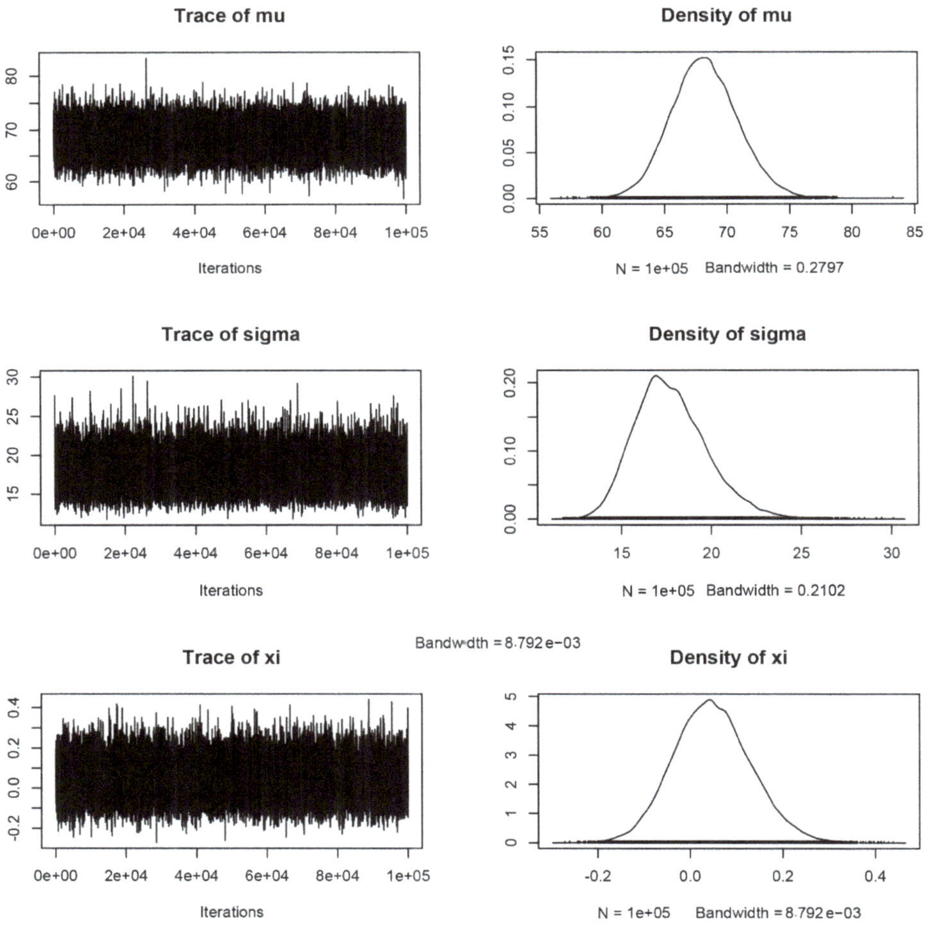

Figure 3. MCMC realizations of the GEV parameters with a Bayesian analysis of the Dar Es Salaam (TZ) rainfall data.

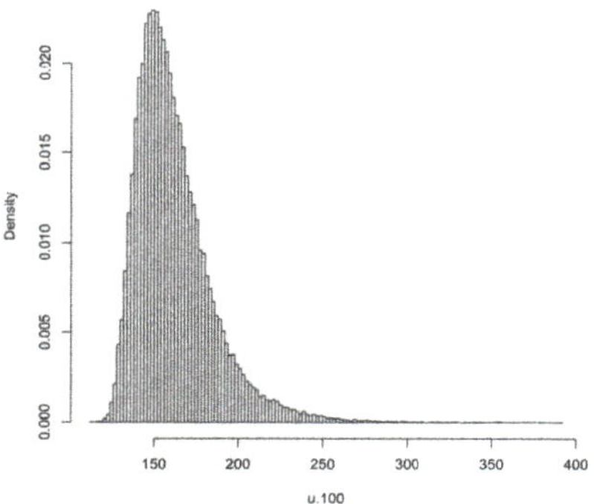

Figure 4. Estimated posterior density of the 100-year return level.

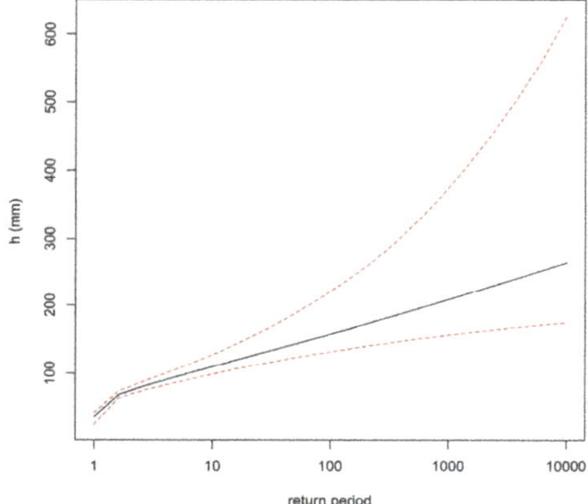

Figure 5. Posterior return level plot in Bayesian analysis of Dar Es Salaam (TZ) rainfall data: the median (solid line) and the 95% intervals of the posterior probability (dashed lines).

The Bayesian estimates were relatively insensitive to the prior distributions, as shown by the similar parameter and quantile estimates. The computational efforts of the Bayesian approach by using the *R* software (*evdbayes* package) were significantly short, requiring very short extra processing time with respect to the MLE. The sole prejudices, in terms of required computational time, regarded both the prior setting up and ensuring that the Markov Chain Monte Carlo had desirable properties. The inclusion of genuine prior information was a compelling factor in favour of the Bayesian inference. This, together with the limited amount of historical data available for the Dar Es Salaam (TZ) analysis, provided significant evidence to prefer the Bayesian analysis instead of the MLE one.

In Figure 6, the autocorrelations for all three parameters after a 5 lag period decreased rapidly. Therefore, it is shown that the result has good mixing.

Table 3 shows the results of the 3 diagnostics according to the Gelman–Rubin, Geweke and Raftery–Lewis methods, as reported in [34] for checking the convergence of the algorithm. The Gelman–Rubin diagnostic is equal to 1.000 for both μ, σ, and ξ. Therefore, it is known that the chains could be accepted, and this indicates the estimates come from a state space of the parameter, as depicted in Figure 7. In Table 3, Geweke's test statistics are 0.4307, 0.6353 and 0.9895 for μ, σ and ξ, respectively. Therefore, also in this case, the chain is acceptable, as shown in Figure 8. The last quantitative diagnostic is the Raftery–Lewis method. In Table 3, the dependence factors I are 4.320,

3.860 and 3.85 for μ, σ and ξ, respectively. According to this method, high dependence factors (>5) show significant correlations between estimates, indicating poor mixing. Therefore, the estimated values have good mixing.

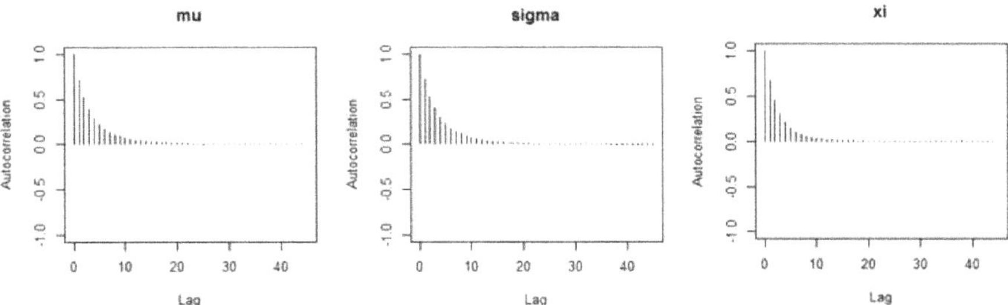

Figure 6. Behaviour of autocorrelations for all three parameters of the GEV distribution with lag effects.

Table 2. Summary of results obtained from different methods of estimation (Dar Es Salaam TZ data).

Method		MLE	Bayesian $\xi \sim N(0.15, 0.20)$
	μ (mm)	68.25	68.22
Estimates	σ (mm)	16.93	17.74
	ξ	0.039	0.047
	μ (mm)	[63.25, 73.25]	[63.10, 73.52]
95% intervals	σ (mm)	[13.35, 20.52]	[14.32, 22.12]
	ξ	[−0.124, 0.202]	[−0.110, 0.220]
100-year return level			
Estimates (mm)		153.6	161.8
95% intervals (mm)		[129, 208]	[132, 219]

Table 3. Diagnostics by Gelman–Rubin method Geweke method, and Raftery–Lewis method.

Parameter	Gelman–Rubin (R)	Geweke ($Z_{0.025} = \pm 1.96$)	Raftery–Lewis (I)
μ	1	0.4307	4.320
σ	1	0.6353	3.860
ξ	1	0.9895	3.850
Parameter	Gelman–Rubin (R)	Geweke ($Z_{0.025} = \pm 1.96$)	Raftery–Lewis (I)
μ	1	0.4307	4.320
σ	1	0.6353	3.860
ξ	1	0.9895	3.850

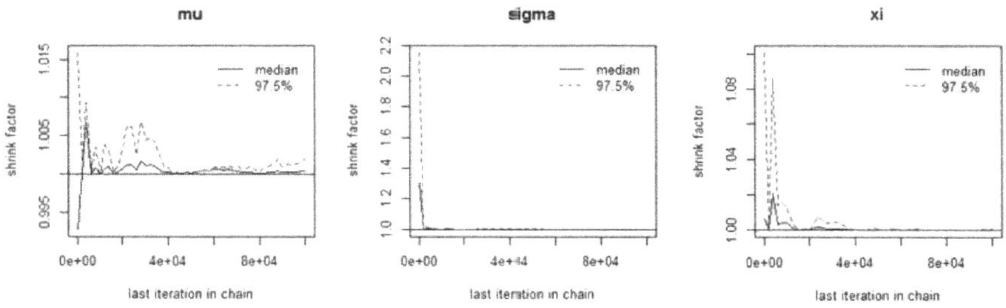

Figure 7. Gelman plot diagnostic for the three parameters μ, σ, and ξ of Dar Es Salaam (TZ) historical data.

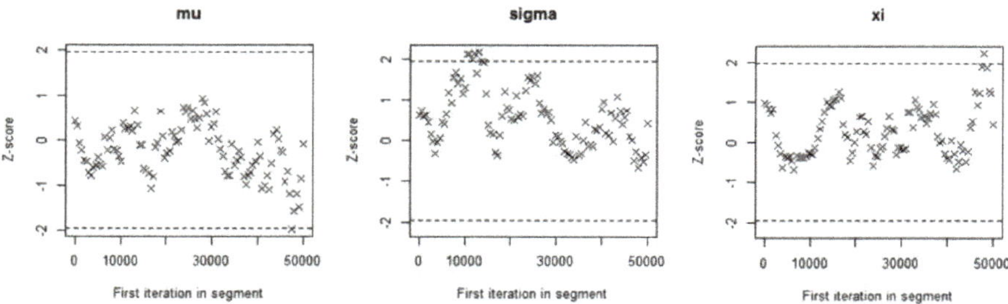

Figure 8. Geweke plot diagnostic for the three parameters μ, σ, and ξ of Dar Es Salaam (TZ) historical data.

After analyzing the historical data with a stationary approach, the analysis was performed to verify the feasibility of Non-Stationarity data by applying both the GEV and assuming a linear trend for the location parameter.

The GEV log-Likelihood is based on the assumption that the data to be fitted are the observed values of independent random variables $X_1, \ldots, X_n$, where $X_i \sim \text{GEV}(\mu, \sigma, \xi)$ for each $i = 1, \ldots, n$. This assumption can be extended to $X_i \sim \text{GEV}(\mu_i, \sigma, \xi)$, where $\mu_i = \mu_0 + \mu_{trend}\, t_i$. The parameters (μ, μ_{trend}) are estimated, and the vectors of covariates $t = (t_1, \ldots, t_n)$ are specified by the user.

In this case study, the MLE fit for the location parameter was $\hat{\mu} = 61.34 + 0.25 \cdot t_i$ (where $t_i = 0$, 1, 2, $\ldots$, 52 years) and associated standard errors were 4.92 and 0.15 for μ_0 and μ_{trend} (Figure 9), respectively. The $\hat{\sigma}$ and $\hat{\xi}$ estimates were 16.22 and 0.071 with standard errors of 1.81 and 0.091, respectively. As shown in Figure 9, this resulted once again in a satisfactory fit. The observed trend of the fitted data with the GEV distribution can be considered a relevant result and suggests further analysis be performed on extended time series with climatic models. An analytic approach to determine the better fit between stationary and Non-Stationary approaches is the Likelihood-Ratio test (*eXtremes* package [28]). In this test case, the Likelihood-Ratio was equal to about 2.4897, i.e., lower than the 95% quantile of the $X_1{}^2$ distribution of 3.8415, suggesting that the covariate t_i model did not provide a significant improvement to the model without a covariate. This assumption was also supported by the estimation of the *p*-value, equal to 0.114.

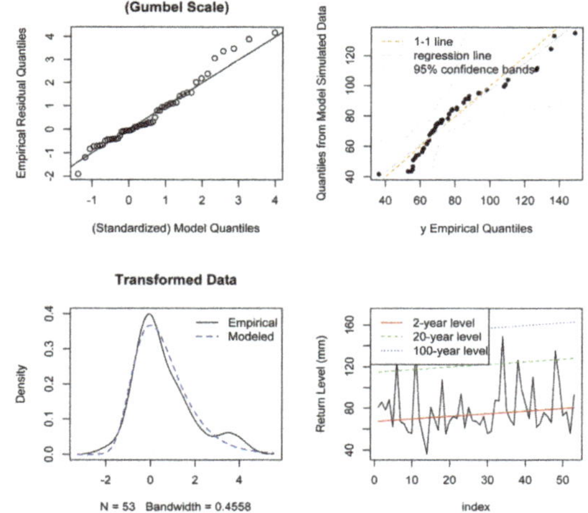

Figure 9. Observed data fitted with the GEV distribution with a trend.

3.1.2. Historical Data with CMCC Simulation Data

In this section, the analysis is extended by integrating the rainfall observations recorded at Dar Es Salaam (TZ) Airport with the simulated data until year 2050, derived from the climatic forecasting simulations performed by the Euro-Mediterranean Centre for Climate Change (CMCC) for the IPCC scenario RCP8.5, using the COSMO CLM model Data were downscaled to 1 km spatial precision. Thus, the total dataset was composed of 93 annual maximum rainfall events, 53 observed and 40 simulated data points (Figure 10).

A similar analysis was performed for the historical data, and the results are summarized in Table 4. For the Bayesian analysis, the MLE was applied as the initial vector $\theta_0 = (\hat{\mu}, \hat{\sigma}, \hat{\xi}) = (59.62, 20.96, -0.00645)$ by using proposal standard deviations $psd = (5.931, 0.193, 0.194)$. A Markov Chain Monte Carlo (MCMC method) was generated, with a length of 100,000 and good mixing properties. By examining the chain graphically and using the Geweke diagnostic, a burn-in period of very few iterations (about 50) was found to be satisfactory. As in the previous analysis, once a stationary approach was applied, the verification of possible Non-Stationarity of the data was done by using the GEV, as a function of a linear trend for the location parameter. In this case, the MLE fit for the location parameter was $\hat{\mu} = 72.46 - 0.273 \cdot t_i$ (where $t_i = 0, 1, 2, \dots , 92$ years), and associated standard deviations were 3.88 and 0.0683 for μ_0 and μ_{trend}, respectively. The $\hat{\sigma}$ and $\hat{\xi}$ estimations were 18.83 and 0.0542 with associated standard deviations of 1.60 and 0 074, respectively. The Likelihood-ratio was about 13.3543, resulting in a greater 95% quantile of the χ_1^2 distribution of 3.8415. The latter suggested that the covariate t_i model was a significant improvement over the model without a covariate, obtaining a small p-value of 0.000258.

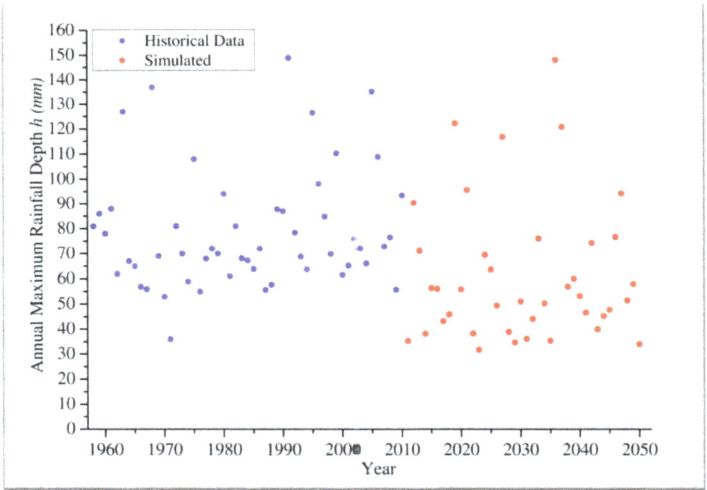

Figure 10. Annual maximum 24 h rainfall recorded at Dar Es Salaam TZ (1958–2010, blue circles) and simulated by CMCC (2011–2050, red circles).

The posterior return level plot represented in Figure 11 shows once again how the upper 95% interval was farther from the median than the lower one.

A naive Bayesian analysis was thus performed, taking near-flat priors that reflected the absence of external information. Indeed, prior on μ_{trend} was a non-informative normal distribution, in a way similar to μ and σ parameters, with a standard deviation of 100. Using MLEs as the initial vector $\theta_0 = (\hat{\mu}_0, \hat{\sigma}, \hat{\xi}, \hat{\mu}_{trend}) = (72.46, 18.83, 0.0542, -0.27)$, and using proposal standard deviations $psd = (5.679, 0.202, 0.187, 0.095)$, a Markov Chain Monte Carlo (MCMC method) was generated with a length of

100,000 and good mixing properties. As usual, the proposal standard deviation was determined by pilot runs. By examining the chain graphically (Figure 12) and using the Geweke diagnostic plot through the coda package in *R* [35] (Figure 13), a burn-in period of only 100 iterations was estimated. The sample means and standard deviations of each marginal component of the chain were:

$$\hat{\mu}_0 = 71.81\ (3.93);\ \hat{\sigma} = 19.46\ (1.72);\ \hat{\xi} = 0.0592\ (0.0743);\ \hat{\mu}_{trend} = -0.266\ (0.0701) \tag{6}$$

whereas the 95% reasonable intervals were [64.14, 79.58], [16.42, 23.15], [−0.0792, 0.211] and [−0.405, −0.129] for μ_0, σ, ξ and μ_{trend}, respectively.

Table 4. Summary of results for the different methods of estimation (Dar Es Salaam, TZ).

Method		MLE	Bayesian $\xi \sim N(0.15, 0.20)$
	μ (mm)	59.62	59.46
Estimates	σ (mm)	20.96	21.41
	ξ	−0.00645	0.00756
	μ (mm)	[54.82, 64.42]	[54.61, 64.41]
95% intervals	σ (mm)	[17.49, 24.43]	[18.12, 25.39]
	ξ	[−0.158, 0.145]	[−0.136, 0.171]
100-year return level			
Estimates (mm)		154.6	161.6
95% intervals (mm)		[133, 202]	[135, 211]

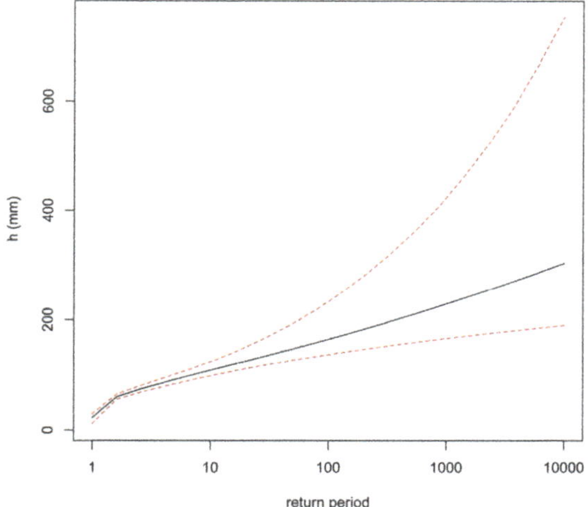

Figure 11. Posterior return level plot in Bayesian analysis of the Dar Es Salaam (TZ) rainfall data: the median (solid line) and the 95% intervals of the posterior probability (dashed lines).

In this case, the estimate for the 100-year return level $\hat{z}_{0.01}$ was not feasible because of the linear trend of the location parameter. In Figure 14, the $\hat{z}_{0.1}$, $\hat{z}_{0.01}$ and $\hat{z}_{0.001}$ return levels are plotted against time from year 1958 to year 2050, as a function of the applied Bayesian analysis. Moreover, the 95% credible intervals are plotted with dashed and dotted lines.

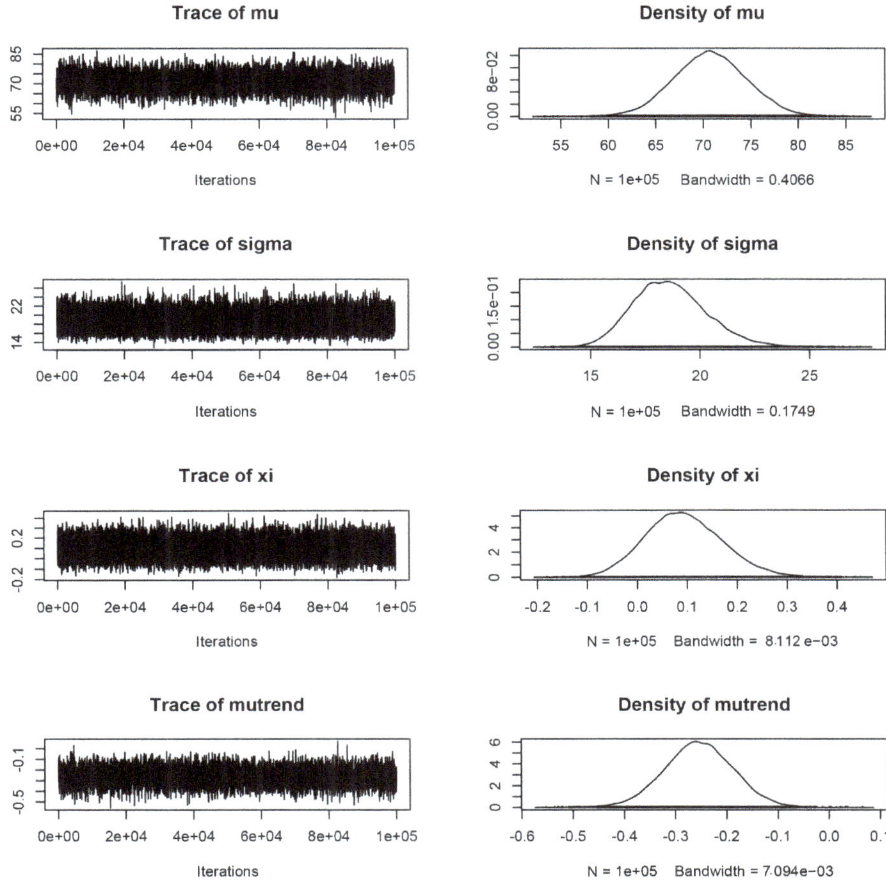

Figure 12. MCMC realizations of the GEV parameters in a Bayesian Non-Stationary Analysis of Dar Es Salaam (TZ) rainfall data.

In Figure 14, the linear variation over time (although improving the distribution pattern) of the mean parameter was observed, leading towards return level estimations being insignificant over time. As an example, for the 100-year return level, in 1958, 175 mm was achieved with 95% reasonable intervals of [127, 259] mm, becoming 150 mm in 2050 with 95% reasonable intervals of [90, 248] mm. A reduction of only 25 mm was thus observed. In Table 5, the whole set of results for the Non-Stationary Analysis is summarized.

Considering the two methods, the μ_0 parameter was estimated to be approximately 72 mm. Nevertheless, as shown in the Maximum Likelihood simulations, it returned a lower estimate of the σ scale parameter than that from the Bayesian method. The estimation of the shape parameter was more precise for the Bayesian method, whereas the estimation of the μ_{trend} parameter was equally precise with both approaches.

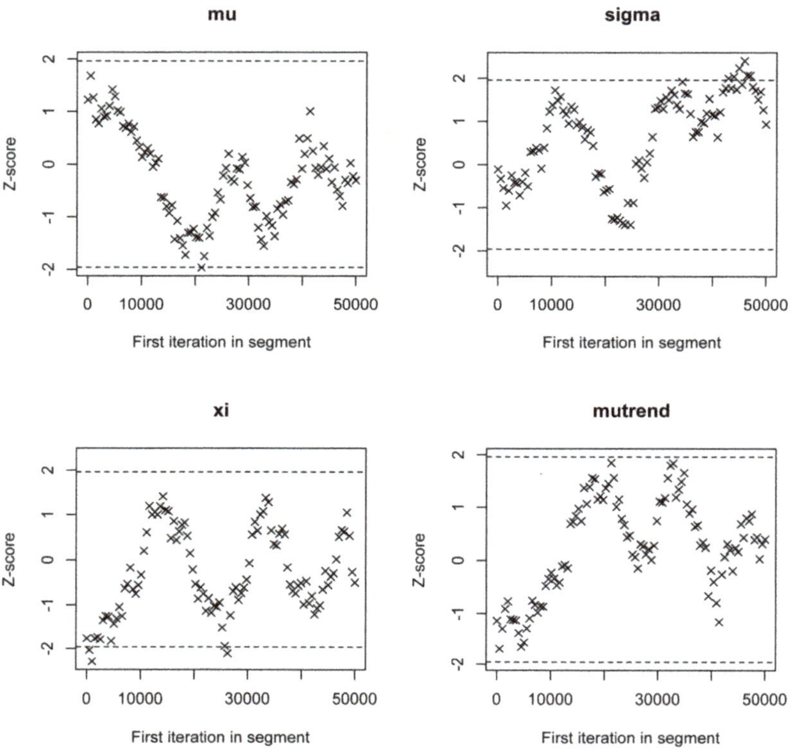

Figure 13. Geweke plot showing the good properties of the performed chain.

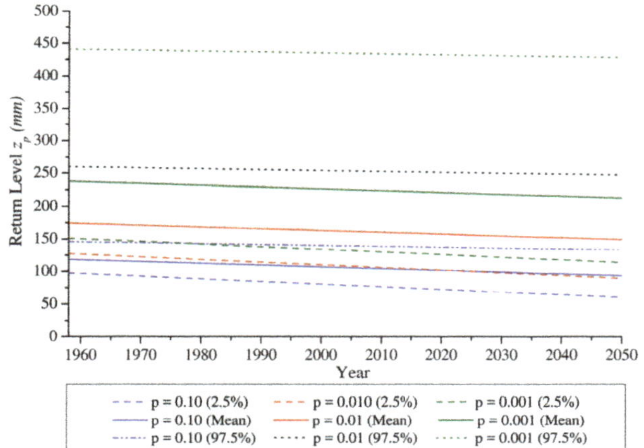

Figure 14. $\hat{z}_{0.1}$, $\hat{z}_{0.01}$ and $\hat{z}_{0.001}$ return levels plotted with 95% credible intervals.

Table 5. Summary of results for the different methods of estimation (Non-Stationary Analysis, Dar Es Salaam, TZ).

Method		MLE	Bayesian $\zeta \sim N(0.15, 0.20)$
	μ_0 (mm)	72.46	71.81
Estimates	σ (mm)	18.83	19.46
	ζ	0.0542	0.0592
	μ_{trend}	-0.273	-0.266
	μ_0 (mm)	[64.71, 79.69]	[64.14, 79.58]
95% intervals	σ (mm)	[15.70, 21.97]	[16.42, 23.15]
	ζ	[$-0.0911, 0.199$]	[$-0.0792, 0.211$]
	μ_{trend}	[$-0.407, -0.139$]	[$-0.405, -0.129$]

3.2. Addis Ababa (ET) Rainfall Data

3.2.1. Historical Data

The second case study regarded the data series of daily rainfall observations recorded at Addis Ababa (ET) Bole, Ethiopia (Latitude: 9.03 N; Longitude: 38.75 E; Elevation: 2354 m a.s.l.), referring to the time span from 1 January 1964 to 31 October 2010. The annual maxima rainfalls are plotted in Figure 15.

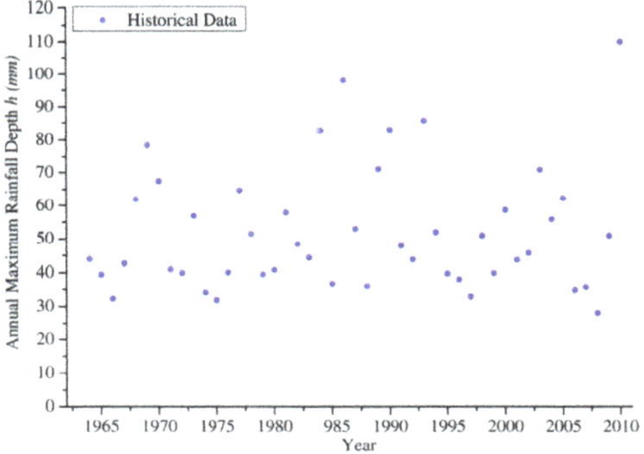

Figure 15. Annual maximum 24 h rainfalls recorded at Addis Ababa ET (1964–2010).

The same assumptions of the Dar Es Salaam (TZ) case study were applied. Table 6 summarizes the results of the stationary analysis for Addis Ababa (ET).

Similarly to the Bayesian analysis, the same hypotheses were assumed. By using MLEs as initial vector $\theta_0 = (\hat{\mu}_0, \hat{\sigma}, \hat{\zeta}) = (42.76, 11.11, 0.234)$ and the proposal standard deviations $psd = (3.373, 0.323, 0.365)$, an MCMC method was generated with a length of 100,000, showing satisfactory mixing properties. By examining the chain graphically and using the Geweke diagnostic, a burn-in period of insignificant iterations was found.

The sequence of simulated values $(\mu_i, \sigma_i, \zeta_i)$ was transformed, leading towards a sample from the corresponding posterior distribution of the 100-year return level. This gave an estimate $\hat{z}_{0.01} = 147.3$ mm with 95% credible interval [97, 268] mm. The posterior return level plot represented in Figure 16 shows how the upper 95% interval was farther than the lower interval with respect to the median trend.

Table 6. Summary of results for the different methods of estimation (Addis Ababa, ET).

Method		MLE	Bayesian $\xi \sim N(0.15, 0.20)$
	μ (mm)	42.76	42.92
Estimates	σ (mm)	11.11	11.79
	ξ	0.234	0.229
	μ (mm)	[39.08, 46.44]	[39.27, 47.05]
95% intervals	σ (mm)	[8.09, 14.12]	[8.92, 15.57]
	ξ	[−0.039, 0.506]	[−0.014, 0.501]
100-year return level			
Estimates (mm)		134.43	147.30
95% intervals (mm)		[95, 237]	[97, 268]

Considering the two methods, μ was estimated to be approximately equal to 43 mm; nevertheless, as confirmed by the Maximum Likelihood simulations, it returned a lower estimate of the scale parameter σ than that from the Bayesian method. The estimation of the shape parameter was more precise for the Bayesian method in terms of the credibility intervals, whilst the opposite was true for the quantile intervals, as a consequence of the greater estimation of ξ parameter.

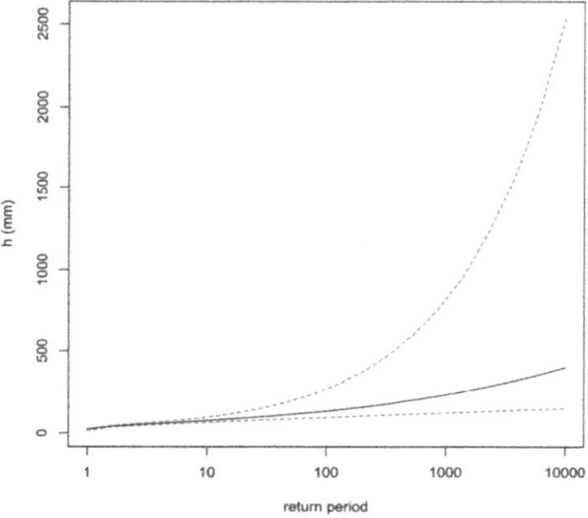

Figure 16. Posterior return level plot in Bayesian analysis of the Addis Ababa (ET) rainfall data: the median (solid line) and the intervals containing 95% of the posterior probability (dashed lines).

Once the historical data were analysed with a stationary approach, analysis was also performed to verify a feasible Non-Stationarity of the data, by applying both the GEV and assuming a linear trend for the location parameter. The parameters (μ, μ_{trend}) were estimated, and the vector of covariates $t = (t_1, \dots, t_n)$ was specified by the user. In this case, the MLE fit for the location parameter was $\hat{\mu} = 43.67 - 0.047 \cdot t_i$ (where $t_i = 0, 1, 2, \dots, 52$ years) and associated standard errors were 2.58 and 0.088 for μ_0 and μ_{trend}, respectively. The $\hat{\sigma}$ and $\hat{\xi}$ estimate parameters were 10.93 and 0.258, corresponding to associated standard errors of 1.56 and 0.151, respectively. Here again, the Likelihood-ratio was about 0.2575, lower than the 95% quantile of the X_1^2 distribution of 3.8415. This suggested the covariate t_i model is not a significant improvement with respect to the model without a covariate. The *p*-value of 0.612 was in fact estimated.

3.2.2. Historical Data with CMCC Simulation Data

As for the Dar Er Salaam (TZ) case study, the analysis was improved by considering the rainfall observations recorded at Addis Ababa (ET) Bole joined with the simulated data until 2050 performed by Euro-Mediterranean Centre for Climate Change (CMCC) for the IPCC scenario RCP8.5, by using the COSMO CLM model. Data were downscaled to 1 km spatial precision again. Thus, the total dataset contained 87 annual maximum rainfall events: 47 observed and 40 simulated data points (Figure 17).

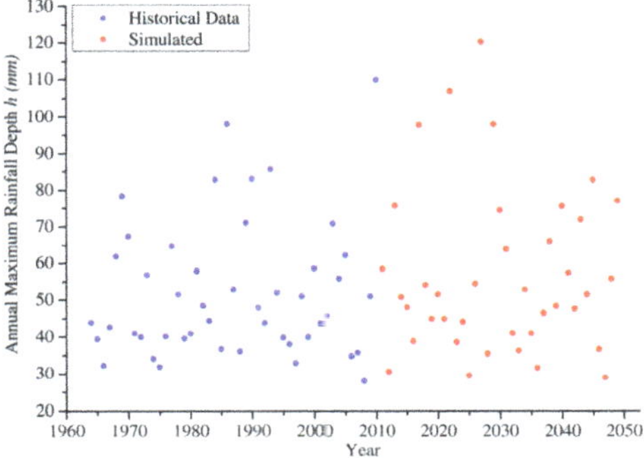

Figure 17. Annual maximum 24 h rainfalls recorded at Addis Ababa ET (1964–2010, blue circles) and simulated by CMCC (2011–2049, red circles).

The same analysis, already performed for the historical data, was developed, reaching results summarized in Table 7. For the Bayesian analysis, again using MLEs as the initial vector $\theta_0 = (\hat{\mu}_0, \hat{\sigma}, \hat{\xi}) = (43.93, 12.58, 0.214)$, and the proposal standard deviations $psd = (3.11, 0.206, 0.234)$, the MCMC method with length 100,000 provided good mixing properties.

The posterior return level plot given in Figure 18 shows that, also in this case, the upper 95% interval was more remote than the lower interval from the median.

Once the data were analysed with a stationary approach, the analysis was devoted to verify a feasible Non-Stationarity of the data, using the GEV as a function of the linear trend for the location parameter. Here, the MLE fit for the location parameter was $\hat{\mu} = 44.02 - 0.00236 \cdot t_i$ (where $t_i = 0$, 1, 2, ... , 85 years) and associated standard errors were 2.46 and 0.496 for μ_0 and μ_{trend}, respectively. The estimates of parameters $\hat{\sigma}$, and $\hat{\xi}$. were 12.57 and 0.216 with associated standard errors of 1.31 and 0.112, respectively. The Likelihood-ratio was about 0.002, i.e., lower than the 95% quantile of the X_1^2 distribution of 3.8415. Thus, the covariate t_i model was not able to significantly improve the model without a covariate. The *p*-value of 0.965 was in fact estimated.

Table 7. A summary of results for the different methods of estimation (Addis Ababa, ET).

Method		MLE	Bayesian $\xi \sim N(0.15, 0.20)$
	μ (mm)	43.94	43.97
Estimates	σ (mm)	12.58	12.97
	ξ	0.214	0.218
	μ (mm)	[40.82, 47.05]	[40.91, 47.22]
95% intervals	σ (mm)	[10.06, 15.10]	[10.54, 15.85]
	ξ	[0.0050, 0.423]	[0.0287, 0.435]
100-year return level			
Estimates (mm)		142.5	151.6
95% intervals (mm)		[108, 224]	[109, 238]

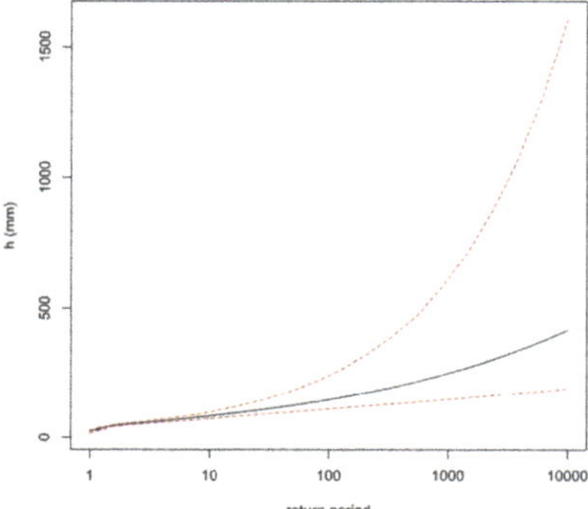

Figure 18. Posterior return level plot in Bayesian analysis of the Addis Ababa rainfall data: the median (solid line) and the intervals containing 95% of the posterior probability (dashed lines).

4. Evaluation of Intensity-Duration-Frequency (IDF) Curves

The Intensity-Duration-Frequency (IDF) curves were introduced in hydrology to synthetically define, for a fixed return period T, a generic duration d of a rainfall event and for a given location, information about the maximum rainfall height h and the maximum rainfall intensity i. Through knowledge of these characteristics, synthetic rainfall graphs can be represented, which are useful to reconstruct flood hydrographs.

The analysis was carried out considering the rainfall observations recorded at Dar Es Salaam (TZ) and Addis Ababa (ET), joined with the data until 2050 provided by CMCC for the IPCC Scenario RCP8.5, referring to a downscaling of 1 km spatial precision. The total dataset defined in the previous paragraphs was taken into account.

Generally, IDF curves can be characterized by two or three parameters expressions:

$$h(d, T) = a(T)d^n \tag{7}$$

where $a(T)$ and n are the parameters to be estimated through a probabilistic approach.

To define the extreme values in a smaller time steps ($10'$, $30'$, 1 h, 3 h, 6 h, 12 h), the generation of a synthetic sequence of rainfall was required, with statistical properties equal to those for the observed rainfall data. To calculate the extreme values in a smaller time window, the daily rainfall was successively disaggregated by using two models:

- *cascade-based disaggregation model*
- *short-time intensity disaggregation method*

These methods have already been validated in Africa [36].

Assuming that daily rainfalls derive from a marked Poisson process, rainfall lag and depths are drawn from exponential Probability Distribution Functions (PDFs) (whose parameters are calculated from the observed rainfall series), using a simple stochastic model, able to describe the occurrence of rainfall as a compound Poisson process with frequency of events λ. The distribution of times τ between precipitation events is an exponential function with mean $1/\lambda$, and exponentially distributed rainfall amounts h with mean γ. This model satisfactory fitted the observed daily rainfall data for individual seasons. Specifically:

- *in a cascade-based disaggregation model* [37], precipitation data of daily resolution are converted into either 12-hourly, 6-hourly, or 3-hourly values, based on the principles of multiplicative cascade processes. For each year, known γ, λ, it is possible to generate some years of disaggregated values and from these is taken the maximum value for each time window (3 h, 6 h, 12 h);

- *in a short-time intensity disaggregation model* [33], three fine-resolution time intervals of 1-h, 1/2-h and 10-min are considered.

The $\hat{\mu}$, $\hat{\sigma}$ and $\hat{\xi}$ parameters for the different time steps (10′, 30′, 1 h, 3 h, 6 h, 12 h and 24 h) were evaluated, also considering the 2.5% and 97.5% percentiles, applying a GEV-Bayesian analysis and the z_p related to the 100-year return period, calculated through Equation (3).

The IDF curves for T = 100 years for historical data for Dar Es Salaam (TZ) and Addis Ababa (ET) are shown in Figure 19a,b, respectively. In Figure 20a,b IDF curves by combining the historical data and the projected ones are plotted instead. The a and n are summarized in Table 8, with reference to 5, 50, 100 and 500-year return periods.

From a comparison between Tables 8 and 9, it was observed that, for the Dar Es Salaam (TZ) case study, similar uncertainties were evaluated from both historical and historical plus projected data. Nevertheless, the IDF derived from the projected data provides greater rainfall intensities values than the historical ones for the whole set of considered durations.

For Addis Ababa (ET), Tables 6 and 7 show that the uncertainties estimated from historical data are significantly greater than those derived from historical plus projected ones. This implied less uncertainty in the simulated data. In this case, the IDF derived from the projected data provided greater rainfall intensities than the historical only for durations longer than 16 h.

The Non-Stationary effect, considered in the evaluation with a linear trend of the GEV location parameter, in all the considered situations, i.e., both historical and historical plus projected, did not show significant improvement with respect to the model without a covariate (stationary case).

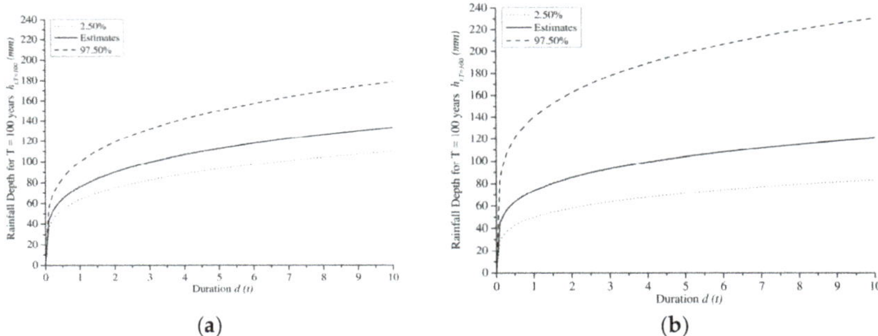

Figure 19. IDF curves for historical data of: (**a**) Dar Es Salaam (TZ) and (**b**) Addis Ababa (ET) cities, with reference to the estimated data (continuous lines), the 2.5% (dotted lines) and 97.5% quantiles (dashed lines).

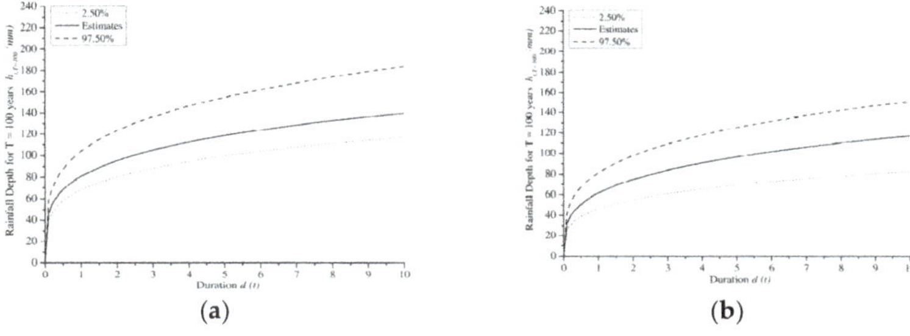

Figure 20. IDF curves for historical data combined with projected ones of: (**a**) Dar Es Salaam (TZ) and (**b**) Addis Ababa (ET) cities, with reference to the estimated data (continuous lines), the 2.5% (dotted lines) and 97.5% quantiles (dashed lines).

Table 8. IDF parameters from the historical data of Dar Es Salaam (TZ) and Addis Ababa (ET).

| T (Years) | Dar Es Salaam (TZ) | | | | | | Addis Ababa (ET) | | | | | |
| | 2.5° Quant. | | Estimates | | 97.5° Quant. | | 2.5° Quant. | | Estimates | | 97.5° Quant. | |
	a (mm/h^n)	n (-)	a (mm/h^n)	n (-)	a (mm/h^n)	n (-)	a (mm/h^n)	n (-)	a (mm/h^n)	n (-)	a (mm/h^n)	n (-)
5	42.41	0.24	46.81	0.24	52.17	0.24	28.14	0.22	32.77	0.22	38.68	0.21
50	59.81	0.23	69.71	0.24	87.48	0.25	45.65	0.22	62.92	0.22	103.41	0.21
100	63.73	0.24	76.12	0.24	100.04	0.25	49.83	0.22	73.79	0.21	139.89	0.22
500	71.00	0.24	90.89	0.26	133.47	0.28	58.23	0.24	114.01	0.24	283.32	0.23

Table 9. IDF parameters from the historical data combined with projected ones of Dar Es Salaam (TZ) and Addis Ababa (ET).

| T (Years) | Dar Es Salaam (TZ) | | | | | | Addis Ababa (ET) | | | | | |
| | 2.5° Quant. | | Estimates | | 97.5° Quant. | | 2.5° Quant. | | Estimates | | 97.5° Quant. | |
	a (mm/h^n)	n (-)	a (mm/h^n)	n (-)	a (mm/h^n)	n (-)	a (mm/h^n)	n (-)	a (mm/h^n)	n (-)	a (mm/h^n)	n (-)
5	41.53	0.24	45.99	0.23	51.06	0.23	26.91	0.26	29.73	0.26	33.04	0.26
50	63.87	0.23	73.82	0.24	91.08	0.25	42.33	0.25	51.13	0.26	66.68	0.26
100	68.41	0.23	80.81	0.24	103.89	0.25	46.30	0.25	61.63	0.28	81.25	0.27
500	76.30	0.24	96.72	0.26	138.65	0.29	54.59	0.25	81.32	0.29	126.48	0.29

5. Conclusions

In this paper the effectiveness of the GEV distribution to estimate the annual maxima of rainfall depths is assessed. The shape parameter ξ of such a distribution resulted in the essential parameter to evaluate the properties of extreme value behaviour, namely, the tail of the distribution at its upper end-point.

Estimation of GEV parameters by using methods, such as the Maximum Likelihood, can be unreliable, due to the small length of rainfall records, typically not longer than 30–50 years.

In this study, rainfall data from both Dar Es Salaam (TZ) and Addis Ababa (ET) were analysed, by performing statistical analysis as a function of the 100-year return level.

With regard to Dar Es Salaam (TZ), the confidence intervals for such long return periods were significantly large due to the uncertainty of the extrapolation procedure. With reference to the Addis Ababa (ET) case study, the results were quite different, with a higher shape parameter. In this case, the use of a Bayesian framework to incorporate prior knowledge could improve the estimation reliability by restricting the ξ variation to a physically reasonable range. The Bayesian method showed similar values for the 100-year return level, in spite of wider reasonable intervals at the 95% observed level.

The analysis of the historical data joined with those yielded by the climate model of the CMCC allowed us to estimate the shape parameter with the related standard errors for both test cases.

By applying the Likelihood-ratio test, the Non-Stationary effect was observed for Dar Es Salaam (TZ).

Considering the MLE and Bayesian methods, the μ_0 parameter estimated by using the Bayesian method was more precise, whereas the estimation of the μ_{trend} parameter for Dar Es Salaam (TZ) was comparable with both methods.

Results achieved for the two test cases showed that the time series, even though not very numerous, were well represented by stationary GEV. Taking into account additional data from climate models, as a consequence of the sample size increase, the distribution model was forced to be represented by a Non-Stationary GEV.

From these results, it was shown that the Non-Stationary effects are frequently induced by climate models. Indeed, using this trend to analyse the return levels z_p, the influence of linear trends in location parameters was observed.

Finally, aiming to answer possible questions about the climate models capability to induce any possible Non-Stationarity in the rainfall series, in future work, further case-studies will be analysed to assess the effectiveness of observed results for different forcing rainfall conditions.

Author Contributions: F.D.P., M.G. and F.N. conceived and performed the statistical analysis; M.G. and A.A. analysed the data; F.P. contributed materials; F.D.P., F.N. and F.P. wrote the paper.

Funding: APC was sponsored by MDPI.

Acknowledgments: In this section you can acknowledge any support given which is not covered by the author contribution or funding sections. This may include administrative and technical support, or donations in kind (e.g., materials used for experiments).

Conflicts of Interest: The authors declare no conflict of interest.

References

1. De Paola, F.; Giugni, M.; Topa, M.E.; Bucchignani, E. Intensity-Duration-Frequency (IDF) rainfall curves, for data series and climate projection in African cities. *SpringerPlus* **2014**, *3*, 1–18. [CrossRef] [PubMed]
2. De Risi, R.; Jalayer, F.; De Paola, F.; Iervolino, I.; Giugni, M.; Topa, M.E.; Mbuya, E.; Kyessi, A.; Manfredi, G.; Gasparini, P. Flood Risk Assessment for Informal Settlements. *Nat. Hazards* **2013**, *69*, 1003–1032. [CrossRef]
3. Intergovernmental Panel on Climate Change (IPCC) *Climate Change 2007—The Physical Science Basis: Working Group I Contribution to the Fourth Assessment Report of the IPCC*; Cambridge University Press: New York, NY, USA, 2007.

4.   Madsen, H.; Lawrence, D.; Lang, M.; Martinkova, M.; Kjeldsen, T.R. Review of trend analysis and climate change projections of extreme precipitation and floods in Europe. *J. Hydrol.* **2014**, *519*, 3634–3650. [CrossRef]

5.   Zolina, O. Changes in Intense Precipitation in Europe, Chapter 6. In *Changes in Flood Risk in Europe*; Kundzewicz, Z.W., Ed.; European Environment Agency Special Publication; IAHS Press: Wallingford, UK, 2012; Volume 10, pp. 97–120.

6.   Gregersen, I.B.; Sørup, H.J.D.; Madsen, H.; Rosbjerg, D.; Mikkelsen, P.S.; Arnbjerg-Nielsen, K. Assessing future climatic changes of rainfall extremes at small spatio-temporal scales. *Clim. Chang.* **2013**, *118*, 783–797. [CrossRef]

7.   Jones, M.R.; Fowler, H.J.; Kilsby, C.G.; Blenkinsop, S. An assessment of changes in seasonal and annual extreme rainfall in the UK between 1961 and 2009. *Int. J. Climatol.* **2013**, *33*, 1178–1194. [CrossRef]

8.   Van den Bessalar, E.J.M.; Klen-Tank, A.M.G.; Buishand, T.A. Trends in European precipitation extremes over 1951–2010. *Int. J. Climatol.* **2015**, *33*, 2682–2689. [CrossRef]

9.   Vasiliades, L.; Galiatsatou, P.; Loukas, A. Nonstationary frequency analysis of annual maximum rainfall using climate covariates. *Water Resour. Manag.* **2015**, *29*, 339–358. [CrossRef]

10.   Cheng, L.; AghaKouchak, A. Nonstationary precipitation intensity-duration-frequency curves for infrastructure design in a changing climate. *Sci. Rep.* **2014**, *4*, 7093. [CrossRef] [PubMed]

11.   Thiombiano, A.N.; El Adlouni, S.; St-Hilaire, A.; Ouarda, T.B.; El-Jabi, N. Nonstationary frequency analysis of extreme daily precipitation amounts in Southeastern Canada using a peaks-over-threshold approach. *Theor. Appl. Climatol.* **2017**, *129*, 413–426. [CrossRef]

12.   Sheikh, M.M.; Manzoor, N.; Ashraf, J.; Adnan, M.; Collins, D.; Hameed, S.; Manton, M.J.; Ahmed, A.U.; Baidya, S.K.; Borgaonkar, H.P. Trends in extreme daily rainfall and temperature indices over South Asia. *Int. J. Climatol.* **2015**, *35*, 1625–1637. [CrossRef]

13.   Zilli, M.T.; Carvalho, L.; Liebmann, B.; Silva Dias, M.A. A comprehensive analysis of trends in extreme precipitation over southeastern coast of Brazil. *Int. J. Climatol.* **2017**, *37*, 2269–2279. [CrossRef]

14.   Wu, H.; Qian, H. Innovative trend analysis of annual and seasonal rainfall and extreme values in Shaanxi, China, since the 1950s. *Int. J. Climatol.* **2017**, *37*, 2582–2592. [CrossRef]

15.   Jenkinson, A.F. The frequency distribution of the annual maximum (or minimum) values of meteorological elements. *Q. J. R. Meteorol. Soc.* **1955**, *81*, 158–171. [CrossRef]

16.   Fisher, R.A.; Tippet, L.H.C. Limiting forms of the frequency distribution of the largest or smallest member of a sample. *Math. Proc. Camb. Philos. Soc.* **1928**, *24*, 180–190. [CrossRef]

17.   Martins, E.S.; Stedinger, J.R. Generalized maximum-likelihood generalized extreme-value quantile estimators for hydrologic data. *Water Resour. Res.* **2000**, *36*, 737–744. [CrossRef]

18.   Coles, S.; Tawn, J.A. A Bayesian Analysis of Extreme Rainfall Data. *Appl. Stat.* **1996**, *45*, 463–478. [CrossRef]

19.   De Haan, L.; de Ronde, J. Sea and Wind: Multivariate extremes at work. *Extremes* **1998**, *1*, 7–45. [CrossRef]

20.   Geluk, J.L.; de Haan, L.; de Vries, C.G. *Weak and Strong Financial Fragility*; Tinbergen Institute: Amsterdam, The Netherlands, 2007.

21.   Coles, S. *An Introduction to Statistical Modeling of Extreme Values*; Springer Series in Statistics; Springer: London, UK, 2001; ISBN 978-1-4471-3675-0.

22.   Cotton, J. *Bayesian Priors and Estimation of Extreme Rainfall in Camborne*; University of Exeter: Cornwall, UK, 2008.

23.   R Development Core Team. *R: A Language and Environment for Statistical Computing*; R Foundation for Statistical Computing: Vienna, Austria, 2012; ISBN 3-900051-07-0.

24.   Mailhot, A.; Duchesne, S. Design criteria of urban drainage infrastructures under climate change. *J. Water Resour. Plan. Manag.* **2009**, *136*, 201–208. [CrossRef]

25.   Lemmen, D.S.; Warren, F.J.; Lacroix, J.; Bush, E. *From Impacts to Adaptation: Canada in a Changing Climate 2007*; Government of Canada: Ottawa, ON, Canada, 2008.

26.   Vincent, L.A.; Mekis, E. Changes in daily and extreme temperature and precipitation indices for Canada over the 20th century. *Atmos. Ocean* **2006**, *44*, 177–193. [CrossRef]

27.   Frei, C.; Schöll, R.; Fukutome, S.; Schidli, J.; Vidale, P.L. Future change of precipitation extremes in Europe: Intercomparison of scenarios from regional climate models. *J. Geophys. Res.* **2006**, *111*, D06105. [CrossRef]

28.   Gilleland, E.; Katz, R.W. New software to analyse how extremes change over time. *Eos Trans. Am. Geophys. Union* **2011**, *92*, 13–14. [CrossRef]

29.   Stephenson, A.G. evd: Extreme Value Distributions. *R News* **2002**, *2*, 31–32.

30. Stephenson, A.; Ribatet, M. Evdbayes: Bayesian Analysis in Extreme Value Theory. R package version 1.1-0. 2012. Available online: http://CRAN.R-project.org/package=evdbayes (accessed on 17 May 2018).

31. Koutsoyiannis, D. Statistics of extremes and estimation of extreme rainfall: I. Theoretical investigation. *Hydrol. Sci. J.* **2004**, *49*, 575–590. [CrossRef]

32. Koutsoyiannis, D. Statistics of extremes and estimation of extreme rainfall: II. Empirical investigation of long rainfall records. *Hydrol. Sci. J.* **2004**, *49*, 591–610. [CrossRef]

33. Geweke, J. Evaluating the Accuracy of Sampling-Based Approaches to Calculating Posterior Moments. In *Bayesian Statistics*; Bernardo, J.M., Berger, J.O., Lawiv, A.P., Smith, A.F.M., Eds.; Clarendon Press: Oxford, UK, 1992; Volume 4.

34. Lee, C.-E.; Kim, S.U.; Lee, S. Time-dependent reliability analysis using Bayesian MCMC on the reduction of reservoir storage by sedimentation. *Stoch. Environ. Res. Risk Assess.* **2014**, *28*, 639–654. [CrossRef]

35. Plummer, M.; Best, N.; Cowles, K.; Vines, K. Coda: Output Analysis and Diagnostics for MCMC. R package version 0.12-1. 2017. Available online: http://CRAN.R-project.org/package=coda (accessed on 17 May 2018).

36. Knoesen, D.; Smithers, J. The development and assessment of a daily rainfall disaggregation model for South Africa. *Hydrol. Sci. J.* **2009**, *54*, 217–233. [CrossRef]

37. Güntner, A.; Olsson, J.; Calver, A.; Gannon, B. Cascade-based disaggregation of continuous rainfall time series: The influence of climate. *Hydrol. Earth Syst. Sci. Discuss.* **2001**, *5*, 145–164. [CrossRef]

38. Connolly, R.D.; Schirmer, J.; Dunn, P.K. A daily rainfall disaggregation model. *Agric. For. Meteorol.* **1998**, *92*, 105–117. [CrossRef]

 *hydrology*

Article

An Operational Method for Flood Directive Implementation in Ungauged Urban Areas

George Papaioannou [1,*], Andreas Efstratiadis [2], Lampros Vasiliades [1], Athanasios Loukas [1], Simon Michael Papalexiou [3], Antonios Koukouvinos [2], Ioannis Tsoukalas [2] and Panayiotis Kossieris [2]

[1] Laboratory of Hydrology and Aquatic Systems Analysis, Department of Civil Engineering, School of Engineering, University of Thessaly, 38334 Volos, Greece; lvassil@civ.uth.gr (L.V.); aloukas@civ.uth.gr (A.L.)

[2] Department of Water Resources and Environmental Engineering, School of Civil Engineering, National Technical University of Athens, 15780 Athens, Greece; andreas@itia.ntua.gr (A.E.); A.Koukouvinos@itia.ntua.gr (A.K.); itsoukal@mail.ntua.gr (I.T.); pkossier@itia.ntua.gr (P.K.)

[3] Department of Civil and Environmental Engineering, University of California, Irvine, CA 92697-2175, USA; simon@uci.edu

* Correspondence: gpapaioa@uth.gr; Tel.: +30-24210-74153

Received: 23 March 2018; Accepted: 19 April 2018; Published: 20 April 2018

Abstract: An operational framework for flood risk assessment in ungauged urban areas is developed within the implementation of the EU Floods Directive in Greece, and demonstrated for Volos metropolitan area, central Greece, which is frequently affected by intense storms causing fluvial flash floods. A scenario-based approach is applied, accounting for uncertainties of key modeling aspects. This comprises extreme rainfall analysis, resulting in spatially-distributed Intensity-Duration-Frequency (IDF) relationships and their confidence intervals, and flood simulations, through the SCS-CN method and the unit hydrograph theory, producing design hydrographs at the sub-watershed scale, for several soil moisture conditions. The propagation of flood hydrographs and the mapping of inundated areas are employed by the HEC-RAS 2D model, with flexible mesh size, by representing the resistance caused by buildings through the local elevation rise method. For all hydrographs, upper and lower estimates on water depths, flow velocities and inundation areas are estimated, for varying roughness coefficient values. The methodology is validated against the flood event of the 9th October 2006, using observed flood inundation data. Our analyses indicate that although typical engineering practices for ungauged basins are subject to major uncertainties, the hydrological experience may counterbalance the missing information, thus ensuring quite realistic outcomes.

Keywords: EU Floods Directive; flood risk management; extreme rainfall; SCS-CN; 2D hydraulic modelling; HEC-RAS; building representation; urban floods; ungauged streams; uncertainty

---

1. Introduction

Natural hazards have caused significant damages to natural and manmade environments during the last few decades. Floods are among the most destructive water-related hazards and are mainly responsible for the loss of human lives, infrastructure damages and economic losses [1]. Nowadays, there is a rising global awareness of flood damage mitigation due to the increase in frequency, magnitude, and intensity of flood events [2,3]. In the Mediterranean region, there are many small-medium sized watersheds, which have significant elevation changes (from mountainous areas to plains and even coastal areas). Many of these watersheds have urban and sub-urban areas in the lowland areas that are prone to floods. Furthermore, intense rainfall events typically generate flash floods events. Urban flooding is extremely difficult to manage due to different flood generation mechanisms and the diverse climatic, topographic, hydrologic and hydraulic conditions. In particular, Greece is a country with significantly varying geomorphological, physiographic and climatic conditions, all affecting the hydrological processes and the generation of floods [4]. According to the EM-DAT database, during the period 1900–2017 Greece experienced 26 major floods that caused 113 fatalities, affected about 23,000 people and cost $2.0 billion [5].

Rainfall (or flood) frequency estimation for the design of hydraulic structures is usually performed as a univariate analysis of extreme rainfall (flood) event magnitudes [6]. Evaluation of flood inundation areas and subsequently flood hazard is estimated by deterministic and/or probabilistic hydraulic approaches [7–11]. This is based on three modelling components: (i) synthetic storm generator; (ii) hydrological simulation model; and (iii) hydraulic simulation model.

In the context of the everyday engineering practice, the estimation of design flood hydrographs is typically employed by combining the Intensity-Duration-Frequency (IDF) approach with standard time profiles, for constructing synthetic rainfall events of a certain probability, the SCS-CN method for extracting the excess from the gross rainfall, and the unit hydrograph theory, for propagating the surface runoff to the basin outlet. In particular, the SCS-CN method, developed by the Soil Conservation Service [12] (currently referred to as Natural Resources Conservation Service, NRCS) is considered the prevailing modelling approach for ungauged basins. The overall scheme contains few parameters, which are generally extracted by regional formulas accounting for characteristic lumped properties of the study area [13]. In the event-based approach, the probabilistic measure of the return period, T, is set a priori to represent the acceptable risk for all relevant quantities (peak flow, flood volume, flow depths and velocities, inundated areas, etc.). Apparently, in ungauged areas the risk of the aforementioned flood-related quantities cannot be estimated statistically, i.e. on the basis of observed data. Therefore, the return period is assigned to the input, i.e. the rainfall, for which there are available records of observed rainfall maxima at several time scales. Nevertheless, this key assumption has been strongly criticized (i.e., [14]), since the total flood risk is in fact a joint probability of multiple, complex and interrelating mechanisms. Moreover, its estimation is strongly influenced by uncertainties that span over all facets of the simulation procedure, i.e. inputs, initial conditions (i.e., antecedent soil moisture), parameters and underlying modelling structures.

According to the EU Directive on floods (E.C. 2007/60), flood inundation modelling and mapping and associated flood risk assessment should be applied using suitable and efficient tools. These requirements have been mainly assessed using one-dimensional (1D) and two-dimensional (2D) hydraulic models (e.g., [7,10,15,16]). However, the selection of 1D-modelling approach can be misguided, leading to erroneous outcomes when applied in areas with composite river topography. Thus, under composite flow conditions, further investigation is needed in the selection of the modelling approach. In such cases, the use of a 2D-modelling approach is generally suggested due to the provision of more accurate or realistic results [10,11,17–26]. These models are able to simulate floodplain inundation and river hydraulics, as demonstrated in many studies (e.g., [9,15,27–30]). However, most of them have been carried out at gauged watersheds, taking advantage of hydrometric information, i.e., discharge data and stage/discharge relationships, which ensures accurate estimation of flood spatial extent. On the other hand, under limited data the applicability of these models becomes a difficult task [31,32], especially in urban and suburban areas. Similar to hydrological modelling, hydraulic modelling of floods is also affected by multiple sources of uncertainty (i.e., input data, model structure, model parameters) [33]. Furthermore, several factors in each source (type) of uncertainty affect the flood modelling process and the mapping results that increase/decrease the uncertainty of the outcome. Thus, the uncertainty of flood risk management implementations can be really high. The main input uncertainty factors that affect the flood inundation accuracy are: (a) river and riverine geometry determination-DEM accuracy; (b) roughness coefficient determination; (c) flood hydrograph estimation and accuracy.

An operational framework for flood inundation mapping in ungauged urban areas is proposed, developed and demonstrated in this study. The framework is developed in the context of the implementation of the EU Floods Directive in Greece and is demonstrated for Volos metropolitan area of Thessaly, Greece, where frequent flood episodes are observed due to intense storms. The methodology developed in this study will help us to better estimate and map flood inundation areas, evaluate the uncertainty in flood inundation mapping, to provide guidance for professionals involved in the flood management process and to apply design measures and policies for the protection of human life, property and economic activities.

2. Study Area

The study area consists of three watersheds, Xerias, Krafsidonas and Anavros, and is located in the region of Thessaly, Mangesia prefecture, Greece (Figure 1). The area of each watershed is approximately 117 km^2, 36 km^2 and 14 km^2 for Xerias, Krafsidonas and Anavros, respectively. The hydraulic-hydrodynamic modelling application involves three reaches that belongs to Xerias watershed, while Krafsidonas and Anavros streams consist of one reach per watershed (Figure 1). All selected streams drain through the city of Volos and contain multiple hydraulic structures and flood protection works. We point out that the city of Volos has experienced frequent flood events (e.g., 2003, 2006, 2009, 2012) due to heavy precipitation episodes that occurred in the last decades [4,10,25,26,34,35]. In particular, the extreme flash flood event that occurred in October of 2006 involved strong debris flow and mudslides that caused severe impacts on transportation networks, other technical infrastructures and agricultural areas. That specific event lasted from 06:00 UTC to 18:00 UTC, 9 October 2006, and generated a total rainfall amount of 232 mm [36].

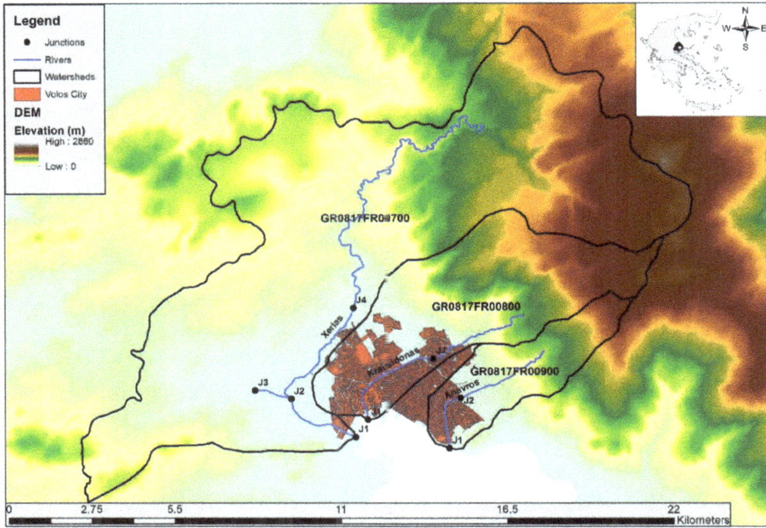

Figure 1. Study watersheds of Xerias, Krafsidonas and Anavros and the junction points and stream reaches that have been selected for flood inundation modelling and mapping.

Within hydrologic and hydraulic model simulations, we employed a semi-distributed schematization for each watershed of the study area, which allowed accounting for heterogeneities of modelling inputs and parameters. As an example, we have divided the river basin of Xerias into ten sub-basins, and configured a main hydrographic network comprising six reaches and seven junctions (Figure 2). The hydrological simulations spanned the full system, while the hydraulic simulations were employed across the downstream reaches, crossing the urbanized part of the basin.

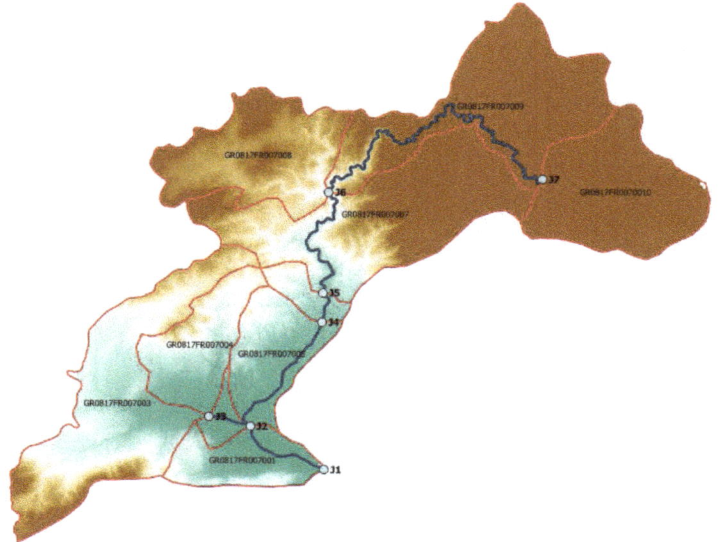

Figure 2. Map of Xerias watershed and modelling components (sub-basins, reaches, junctions).

3. Methodology

3.1. Overview of Flood Modelling Approach

In this study, an integrated flood hazard and risk modelling and mapping framework has been developed and implemented at ungauged urban and suburban streams/catchments. The main goal is to highlight the possible disastrous effect of fluvial floods on human health, economic activities, cultural heritage, and the environment for three typical design return periods (T = 50, 100, 1000 years). The single event-based deterministic approach is adopted, based on three modelling components: (i) synthetic storm generator; (ii) hydrological simulation model; and (iii) hydraulic simulation model. The major assumption of the framework is that the flood risk is connected to the determination of the input rainfall return period. Finally, the outcome of the framework is the flood risk maps (for T = 50, 100, 1000 years) corresponding to the "average" hydrological scenario as well as two "extreme" scenarios, which allow providing lower and upper uncertainty bounds of the estimated flood quantities for each return period of interest. The proposed framework, described in the next paragraphs, is expected to help water resource managers and decision makers to improve the design procedures for flood risk management of urban areas (ungauged streams/catchments) and flood mitigation strategies.

3.2. Design Rainfall

A key assumption of the event-based approach is that the flood risk is determined in terms of return period, T, of the design rainfall (hyetograph). The latter represents the temporal evolution of a hypothetical storm event of a certain duration D and time resolution Δt, which corresponds to the given return period. In this study, we have investigated a number of rainfall scenarios, setting D = 24 h (which is about five times larger than the time of concentration of the basin) and Δt = 15 min. Moreover, following the semi-distributed approach, we assigned spatially-varying rainfall inputs across sub-basins, thus accounting for the heterogeneity of the storm regime over the study basin, which is due to climatic reasons as well as relief and orography effects.

The computational procedure for extracting design hyetographs across sub-basins comprised three steps: (a) estimation of partial rainfall depths for all temporal scales and return periods of interest, on the basis of spatially-averaged IDF relationships; (b) derivation of a synthetic hyetograph, by placing the partial depths at specific time intervals across the given duration (i.e., 24 h); and (c) application of an empirical reduction formula, to transform point to areal estimations.

The IDF relationships (also referred to as ombrian curves) have been extracted within the implementation of the EU Flood Directive 2007/60 across the River Basin District of Thessaly [37]. The associated study comprised extensive collection, control and statistical analysis of observed extreme rainfall data from 71 meteorological stations over Thessaly. The raw data comprised annual series of maximum daily and two-day rainfall depths, as well as annual series of maximum intensities at 15 recording stations (pluviographs), that were available for time scales ranging from 5 min up to 48 h. The ombrian curves were expressed in terms of parameter values of a generalized statistical formula, providing estimations of point rainfall intensities for given time scale (duration) and return period. According to Flood Directive specifications (common for the whole of Greece), at each station we assigned the expression proposed by [38], representing the average rainfall intensity *i* over a certain time scale (also referred to as duration) *d*, and a given return period *T*, as the ratio of a probability function, $a(T)$, to a function of time scale, $b(d)$. i.e.,

$$i(d,\ T) = \frac{a(T)}{b(d)} = \frac{\lambda'\ (T^{\kappa} - \psi')}{(1 + d/\theta)^{\eta}} \tag{1}$$

In particular, the nominator $a(T)$ is the mathematical expression of a Generalized Extreme Value (GEV) distribution for rainfall intensity over some threshold at any time scale. The above formula contains five parameters (λ', ψ', κ, η, θ) that have been estimated through the following procedure:

Step 1: Global estimations of parameters η and θ were extracted on the basis of pluviographic data, by optimizing the fitting metric known as Kruskal-Wallis statistic [39] against the compound (unified) sample of extreme rainfall intensities for all available time scales.

Step 2: At each station, the shape parameter κ is initially obtained by fitting the GEV model to the maximum 24 h data and estimating its parameters by the *L*-moments method [40,41]. Next, we employ the correction technique developed by [42], in order to adjust the biased estimations of κ, thus prohibiting both the use of too high values and the generation of negative values, which are unfeasible, since the maximum rainfall cannot be bounded. We remark that such inconsistencies are mainly due to sample uncertainties, which are induced due to the small size of the observed rainfall maxima, the existence of outliers as well as measurement errors.

Step 3: Based on their point values of parameter κ, we employed a geographical classification of the stations to obtain regional values that are associated with climatic and topographic characteristics.

Step 4: For given parameters κ, η and θ, we employed the *L*-moments method to estimate the scale and location parameters, λ' and ψ', at each station.

In order to extract the confidence intervals of rainfall estimations, we employed a generalized Monte Carlo framework, since for the GEV distribution (as made for most of distributions) there are no analytical formulas [43]. Let *X* be a random variable following a distribution function F_X, θ the parameters of this distribution that have been estimated by a sample of *n* values of *X*, and *u* and γ are the desirable exceedance probability and confidence level, respectively. The computational procedure is the following:

Step 1: Using an appropriate generator of random numbers following the desirable distribution F_X, we produce m synthetic samples $x_i = \{x_{i1}, x_{i2}, \dots, x_{in}\}$, where n is the length of the historical data.

Step 2: From each synthetic sample x_i we estimate its statistical characteristics and the corresponding sample parameters θ_i of F_X, by applying the same procedure with the historical data (e.g., method of moments, L-moments, maximum likelihood, etc.).

Step 3: For the desirable probability u, we generate m synthetic values using the inverse cumulative distribution function, i.e.:

$$x_i(u) \; = \; F_X^{-1}(\theta_i, \, u) \tag{2}$$

Step 4: We estimate the confidence limits $x_U(u)$ and $x_L(u)$, by computing the larger $m\,(1-\gamma)/2$ and smaller $m\,(1+\gamma)/2$ values of the sorted sample of $x_i(u)$.

In the context of our analyses, at each station we initially extracted the compound sample of rainfall maxima, derived by multiplying each individual set of duration d by the duration function $b(d) = (1 + d/\theta)^\eta$. According to Equation (1), the derived data follows a Pareto distribution function, with parameters λ', ψ', κ, given by:

$$F_X \; = \; 1 - \left(x/\lambda' + \psi'\right)^{-1/\kappa} \tag{3}$$

In the context of Monte Carlo simulations, for each station we employed the inverse function of Equation (3) to generate 20,000 sets of synthetic rainfall data from the Pareto distribution, with length equal to the historical one. This function is given by:

$$x(u) \; = \; F_X^{-1}(u) \; = \; \lambda' \left[\frac{1}{(1-u)^\kappa} - \psi' \right] \tag{4}$$

Next, from each set we estimated the parameter values λ', ψ', κ, and associated confidence limits $x_U(u)$ and $x_L(u)$ for $T = 50$, 100 and 1000 years (or $u = 0.980$, 0.990 and 0.999, equivalently). By setting $\gamma = 80\%$, we obtained the 2000th smallest and largest value, respectively. Given that the confidence values refer to the compound sample, they are standardized against the duration. In order to obtain the desirable rainfall value for a specific duration, the derived values of $x_U(u)$ and $x_L(u)$ are multiplied by:

$$\xi(d) \; = \; \frac{d}{(1 + d/\theta)^\eta} \tag{5}$$

Using averaged IDF parameters and associated confidence limits per sub-basin we formulated the design storm hyetographs of 24 h duration, by employing two typical time profiles, i.e., the alternating block method, for $T = 50$ and 100 years, and the worst profile method, for $T = 1000$ years [44–46]. Both approaches require the estimation of partial rainfall depths for durations $\Delta t, 2\Delta t, \dots, N\,\Delta t = D$, which are appropriately allocated to formulate a hypothetical hyetograph that preserves the desirable return period at all temporal scales.

3.3. Hydrological Model Assumptions and Representation of Uncertainties

For each return period of interest ($T = 50, 100, 1000$ years) we have formulated three scenarios (herein referred to as low, average and high), in order to account for joint rainfall and hydrological uncertainties. In particular, we assumed that the design rainfall estimations provided by the IDF relationship correspond to the average scenario, while its 80% confidence limits, which are measure of rainfall uncertainty (more precisely, parameter uncertainty of the IDF expression), correspond to the two extreme scenarios. On the other hand, the hydrological uncertainty has been expressed in terms of three typical antecedent soil moisture conditions (dry, moderate, wet) that are employed within the rainfall-runoff modelling approach, as explained next. In this respect, we have formulated $3 \times 3 = 9$ scenarios, in total.

The transformation of the hyetograph to flood runoff was made by subtracting the hydrological deficits (i.e., the part of rainfall that is initially intercepted in the ground and by the canopy, and is then either infiltrated or evaporated), thus obtaining the so-called effective rainfall or rainfall excess. In this respect we employed the well-known SCS-CN approach, developed by the Soil Conservation Service [12]. This method uses two parameters, i.e. the maximum potential retention, S, and the initial abstraction, h_{a0}, to estimate the evolution of effective rainfall through the empirical formula:

$$h_e = \frac{(h - h_{a0})^2}{h - h_{a0} + S} \tag{6}$$

According to the standard SCS approach, we set $h_{a0} = 0.20\,S$, which is the threshold for surface runoff generation. Under this premise, the sole parameter of the method is the maximum potential retention of each sub-basin, estimated by the well-known formula:

$$S = 254\left(\frac{100}{CN} - 1\right) \tag{7}$$

where CN is the spatially-averaged curve number value of each specific sub-basin. The latter has been extracted on the basis of distributed soil and land cover information, following the typical classification by NRCS [47], by means of detailed lookup tables. In order to account for the soil moisture present in the soil profile before the start of an event, the SCS-CN method considers three antecedent soil moisture conditions types (AMC I, AMC II, and AMC III), referring to dry, average or wet conditions, respectively, which depend on the total 5-day antecedent rainfall and the season category (dormant or growing). The CN values given in lookup tables (hereafter symbolized CN II) refer to average conditions, while for the other two AMC types, SCS provides empirical conversion formulas to estimate the parameter value for dry (CN I) and wet conditions (CN III) In the proposed framework, the CN values for dry and wet conditions, combined with the low and upper confidence limits of rainfall, were used to generate design hyetographs that have been used to quantify the uncertainty around the "standard" (i.e., average) hydrological scenario.

For the transformation of the excess rainfall over the sub-basin to flood hydrograph at the outlet junction, we applied the unit hydrograph (UH) approach, using the dimensionless curvilinear unit hydrograph by NRCS (also referred to as Standard PRF 484), in which the time and discharge are expressed as ratios of time to peak and peak discharge, respectively. Key assumptions are that 37.5% of the total flood volume is produced within the rising limb, while the base time, t_b, is considered five times the time to peak, t_p. The UH is fully determined in terms of lag time, t_L, defined as the time distance from the centroid of the unit hydrograph of duration d, from the centroid of unit rainfall, which is by definition equal to $d/2$. Assuming also that $t_L = 0.6 t_c$, where t_c is the time of concentration, and under the assumption that the centroid of the unit hydrograph is approximately located in the peak, we approximate the time of peak by:

$$t_p = d/2 + 0.6 t_c \tag{8}$$

Moreover, given that $t_b = 5 t_p$, and by employing the mass balance equation for total volume equal to the effective rainfall volume, we can also estimate the peak discharge by:

$$q_p = 2.08\,A/t_c \tag{9}$$

where A is the basin area (in km^2).

According to widespread flood modelling practices for ungauged basins, t_c is characteristic time quantity of a river basin, typically defined as the longest travel time of the surface runoff from the hydraulically most remote point of a basin to its outlet, and computed by empirical formulas that estimate the basin response time as function of its geomorphological characteristics. A widely used empirical approach is the Giandotti formula, given by:

$$t_c = \frac{4\sqrt{A} + 1.5L}{0.8\Delta z} \tag{10}$$

where t_c is the time of concentration (h), A is the basin area (km^2), L is the length of the longest runoff distance across the basin (km), and Δz is the difference between the mean elevation of the basin and the outlet elevation (m). This formula has been proved quite suitable for reproducing observed peak flood flows in a number of small river basins in Cyprus; in particular, its predictive capacity was by far superior with respect to other widely-used empirical formulas of the literature [13]. In the study, we used the time of concentration, estimated by the Giandotti formula, as the reference response time of each area of interest, i.e., the entire catchment and its sub-basins. The computation of the associated geometric quantities A, L and Δz was estimated via typical processing tools in GIS environment. In the case of complex river networks, i.e., with confluences, we considered the longest flow path across each corresponding area.

Theoretical proof and empirical evidence imply that t_c is definitely not a constant property of the basin, but it varies significantly with the flow [13,48]. In fact, the variability of t_c is explained by the dependence of the kinematic wave celerity on the flow rate. Apparently, as surface runoff increases, the flow velocity across the river network and its tributaries also increases, which results in a faster response of the basin. For instance, [49] analyzed a large number of flood hydrographs and found that t_c varied by even one order of magnitude across flood events of different intensities. To account for the dependence of the response time of the basin against runoff, we employed the following semi-empirical formula, which arises from the kinematic wave theory, considering that t_c is inversely proportional to the design rainfall, i.e.,

$$t_c(T) \;=\; t_c\sqrt{\frac{i(5)}{i(T)}} \tag{11}$$

where $i(5)$ is the design rainfall intensity for return period $T = 5$ years, for which the time of concentration is estimated by the Giandotti formula, and $i(T)$ is the intensity of any higher return period, T. In this respect, the "reference" time of concentration by the Giandotti formula is assumed valid for high and medium frequency flood events, yet for low-frequency events, which interest hydrological design, this time has to be reduced. Evidently, as the time of concentration reduces, all associated time quantities are similarly reduced, namely, the lag time, the time to peak and the base time of the unit hydrograph. In fact, smaller lag times result in much narrower unit hydrographs, which in turn results in increased peak discharge, in order to preserve the unit flood volume. An example involving a hypothetical river basin is provided in Figure 3. This approach is definitely more consistent with reality, since it accounts for nonlinearities in runoff routing, which are ignored by the unit hydrograph theory.

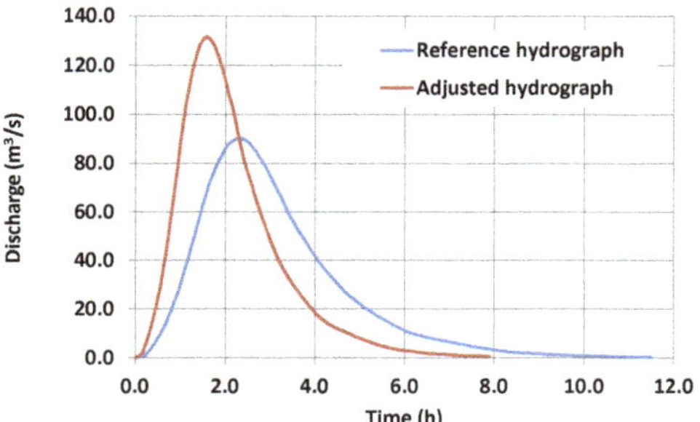

Figure 3. PRF484 unit hydrograph, corresponding to $T = 5$ years, and adjusted design hydrograph, corresponding to $T = 200$ years, in a hypothetical basin of 100 km^2, considering a reference time of concentration of 3.0 h and adjusted time of 1.8 h.

3.4. Hydraulic-Hydrodynamic Modelling

The two dimensional (2D) HEC-RAS model was developed by the Hydrologic Engineering Center (HEC) of United States Army Corps of Engineers [50] and has been applied in many studies for flood inundation modelling (e.g., [9,10,25–27,51–54]). Moreover, a benchmark analysis based on the two dimensional modelling capabilities, conducted by the U.S. Army Corps of Engineers, proved that HEC-RAS performed extremely well compared to the leading 2D models [55]. Therefore, the 2D HEC-RAS hydraulic-hydrodynamic model has been selected for flood inundation and mapping.

The HEC-RAS 5.0.3 computational engine is based on the full 2D Saint-Venant equations or the 2D diffusive wave equations [50]. Shallow water equations are simplifications of the Navier-Stokes equations. The basic assumptions that are followed in order to approximate the turbulent motion using eddy's viscosity are: (a) Reynolds averaged equation is used, (b) the flow is incompressible, (c) the density and the hydrostatic pressure are uniform and d) the horizontal length scale is bigger than the vertical length scale. Therefore, the application of these assumptions to the Navier-Stokes equations results to the differential form of Shallow Water equations [50]. The advection, viscous, and unsteady terms can be neglected in cases where the main terms of momentum equations are the bottom friction terms and the barotropic pressure gradient (gravity) term, resulting in the transformation of the momentum equation to a two-dimensional form of the Diffusion Wave Approximation. Thus, the Diffusive Wave Approximation of the Shallow Water (DSW) equations can be derived through the combination of mass conservation and the two-dimensional form of the Diffusion Wave Approximation. Finally, HEC-RAS 2D solver is using the sub-grid bathymetry approach [50].

The combination of the above assumptions results in an almost hydrostatic pressure and minor vertical velocity. In cases where strong wind forcing exists, the pressure is non-hydrostatic, and the baroclinic pressure gradients (variable density) are absent, a vertically-averaged version of the momentum equation is suitable. Given these conditions, the shallow water equations can be obtained by excluding the terms of vertical derivative and the vertical velocity. Accordingly, the momentum equations are expressed as follows [50]:

$$\frac{\partial u}{\partial t} + u\frac{\partial u}{\partial x} + v\frac{\partial u}{\partial y} = -g\frac{\partial H}{\partial x} + v_t\left(\frac{\partial^2 u}{\partial x^2} + \frac{\partial^2 u}{\partial y^2}\right) - c_f u + f_v \tag{12}$$

$$\frac{\partial v}{\partial t} + u\frac{\partial v}{\partial x} + v\frac{\partial v}{\partial y} = -g\frac{\partial H}{\partial x} + v_t\left(\frac{\partial^2 v}{\partial x^2} + \frac{\partial^2 v}{\partial y^2}\right) - c_f v + f_u \tag{13}$$

where u and v are the velocities in the Cartesian directions; g is the gravitational acceleration; v_t is the horizontal eddy viscosity coefficient; c_f is the bottom friction coefficient; and f is the Coriolis parameter.

One of the basic factors of input data uncertainty in flood inundation modeling and mapping, especially when 2D hydraulic hydrodynamic models are used, is the Digital Elevation Model (DEM) accuracy. The DEM estimation process involves several errors, especially in complex river and riverine areas, due to the topographical technique used [10,25,26,56]. The most common approaches followed for the river and riverine areas topography estimation and the DEM generation are ground surveying topographic approaches and photogrammetric techniques. However, the use of such techniques in flood inundation modelling, especially in complex river topographies involve some limitations such as the small spatial extent and the possibility to produce several errors and affect the accuracy of the produced DEM [57,58]. These limiting factors can be minimized with the use of high accuracy aerial photographs or orthophotos. The high accuracy aerial photographs or orthophotos can produce acceptable DEM resolutions for large scale flood inundation modelling and mapping at ungauged urban areas [59]. In this study, the DEM resolution used is 5 m and has been provided by National Cadastre and Mapping Agency S.A. (NCMA). The geometric resolution of the DEM is RMSEz ≤ 2.00 m with absolute accuracy ≤ 3.92 m for 95% confidence level. The raw data consist of the Digital Surface Model that includes canopy, manmade structures and other surface obstacles. First, the different DSMs derived from the 1:5000 aerial photos have been merged to a continue DSM. Then, the entire DSM has been processed to fill/sink the erroneous areas. Finally, the DSM has been re-corrected using typical elevation downgrading methods in order to create the DEM.

An important input data uncertainty factor in flood inundation modelling is the roughness coefficient and the parameterization process that follows. A typical approach for large scale applications that uses two-dimensional hydraulic models is the estimation of the roughness coefficient using CORINE land cover data and standard roughness coefficient tables (e.g., [30]). Therefore, in this study the average values of Manning's roughness coefficient (Table 1) have been estimated using CORINE land cover classification in combination with typical Manning's roughness coefficient tables [60]. Moreover, based on the EU Flood Directive guides the "upper" and "lower" boundaries of Manning's roughness coefficient were estimated. The "upper" and "lower" boundaries are estimated as −50% and +50% of the average Manning's roughness coefficient values (Table 1), respectively. Furthermore, a significant factor in hydraulic-hydrodynamic modelling applications for engineering purposes is the accurate representation of the river and riverine area hydraulic structures (bridges, culverts, weirs, flood protection works, etc.). Thus, all hydraulic structures of the study area were detected using aerial photographs, a GIS database of the technical works, field observations and information collected by several authorities. Then, the important hydraulic structures have been selected for accurate topographical survey based on the following guides:

- Hydraulic structures close to erroneous DEM area.
- Hydraulic structures close to historical flood points.
- Hydraulic structures inside the Potential High Flood Risk Areas [34].
- Hydraulic structures close to recently recorded flood episodes.
- Hydraulic structures that accurate topographical data are absent.
- Hydraulic structures within main water bodies.

Then, based on hydraulic structures geometry data, the entire DEM has been modified in order to include the flood protection works and the geometry of all hydraulic structures. Moreover, the two-dimensional HEC-RAS hydraulic-hydrodynamic model is capable of the representation of the bridges as a combination of culverts and weirs and to calculate water surface profiles for several system formulations [50].

Finally, flood inundation modelling and mapping at urban and suburban areas remains a big challenge due to the complexity of the entire system. Also, the hazardous effect of floods in urban and suburban areas has a significant social impact, big economic losses, and in some cases fatalities [1]. One of the most important factors in flood inundation modelling in built up areas is the building representation within the 2D hydraulic-hydrodynamic model. The most common building representation methods assume that each cell of the mesh/grid that is located inside a building block area is represented as [61–64]: (1) Solid object; (2) with local increase of the elevation; (3) with higher roughness coefficient values. In the solid representation method of the building blocks, the flow eliminates within the solid block. This can be achieved with several methods that depends on the numerical scheme used. In the local increase of building block representation method, each building block is modified in the DEM in order to have bigger elevation than the bare earth altitude. The third approach of building representation involves either the local rise of roughness coefficient (i.e., Manning, Chézy, Darcy-Weisbach) for the building blocks areas or by adding supplementary parameters of porosity and building drag coefficient to the numerical scheme of the model [63]. Recent studies concerning the building block methodologies [64] showed that all methods have advantages and disadvantages and none of them prevail among the others. Therefore, in this study that deals with large-scale applications, the second (local increase of the elevation) building block representation method has been selected.

Table 1. Average values of Manning's roughness coefficient based on CORINE land cover data.

LABEL1	LABEL2	LABEL3	Mannings n
1 Artificial surfaces	1.1 Urban fabric	1.1.1 Continuous urban fabric 1.1.2 Discontinuous urban fabric	0.013
	1.2 Industrial, commercial and transport units	1.2.1 Industrial or commercial units 1.2.2 Road and rail networks and associated land 1.2.3 Port areas 1.2.4 Airports	0.013
	1.3 Mine, dump and construction sites	1.3.1 Mineral extraction sites 1.3.2 Dump sites 1.3.3 Construction sites	0.013
	1.4 Artificial, non-agricultural vegetated areas	1.4.1 Green urban areas 1.4.2 Sport and leisure facilities	0.025
2 Agricultural areas	2.1 Arable land	2.1.1 Non-irrigated arable land 2.1.2 Permanently irrigated land 2.1.3 Rice fields	0.03
	2.2 Permanent crops	2.2.1 Vineyards 2.2.2 Fruit trees and berry plantations 2.2.3 Olive groves	0.08
	2.3 Pastures	2.3.1 Pastures	0.035
	2.4 Heterogeneous agricultural areas	2.4.1 Annual crops associated with permanent crops	0.04
		2.4.2 Complex cultivation patterns	0.04
		2.4.3 Land principally occupied by agriculture, with significant areas of natural vegetation	0.05
		2.4.4 Agro-forestry areas	0.06
3 Forest and semi natural areas	3.1 Forests	3.1.1 Broad-leaved forest 3.1.2 Coniferous forest 3.1.3 Mixed forest	0.1
	3.2 Scrub and/or herbaceous vegetation associations	3.2.1 Natural grasslands	0.04
		3.2.2 Moors and heathland	0.05
		3.2.3 Sclerophyllous vegetation	0.05
		3.2.4 Transitional woodland-shrub	0.06
	3.3 Open spaces with little or no vegetation	3.3.1 Beaches, dunes, sands	0.025
		3.3.2 Bare rocks	0.035
		3.3.3 Sparsely vegetated areas	0.027
		3.3.4 Burnt areas	0.025
		3.3.5 Glaciers and perpetual snow	0.01
4 Wetlands	4.1 Inland wetlands	4.1.1 Inland marshes 4.1.2 Peat bogs	0.04
	4.2 Maritime wetlands	4.2.1 Salt marshes 4.2.2 Salines 4.2.3 Intertidal flats	0.04
5 Water bodies	5.1 Inland waters	5.1.1 Water courses 5.1.2 Water bodies	0.05
	5.2 Marine waters	5.2.1 Coastal lagoons 5.2.2 Estuaries 5.2.3 Sea and ocean	0.07

3.5. Hydraulic Simulation of Lower Course of Volos City Streams and Evaluation Procedure

The methodology developed for large-scale fluvial flood inundation modelling applications for ungauged urban areas is applied in three streams (Xerias, Krafsidonas, Anavros) that cross Volos city, Greece (Figure 1). Typical techniques and methods for ungauged streams were used for the hydraulic-hydrodynamic modelling configuration. Because of severe lack of flood data, the proposed methodology is validated using a simulated historical flood event only for Xerias stream reaches. The application of the hydraulic simulation of the lower course of Volos city streams is described in the next paragraphs.

Xerias model domain (Figure 2) extends downstream of junction J4, and involves three reaches (J4-J2, J3-J2, J2-J1), while Krafsidonas and Anavros hydraulic model simulations extend downstream of their respective junction J2, and involve one reach per stream (J2-J1). The total length of the simulated stream reaches is approximately 8.5 km, 3.7 km, and 2.2 km for Xerias, Krafsidonas and Anavros, respectively. All examined reaches are crossing urban areas of Volos city. The HEC-RAS 2D model was used for flood inundation modelling and mapping with: (a) flexible mesh computation point spacing (Xerias = 14 m/Krafsidonas = 5 m/Anavros = 5 m), (b) 2D diffusion wave solution, and (c) computation interval 2 s. The input DEM spatial resolution used in this study is 5 m and in some cases where extensive flood mitigation works exist has been downgraded to 1 m. All building blocks of Volos city were represented with a 30 m local increase of the elevation (elevation rise method).

A significant problem of flood mitigation works and hydraulic structures (bridges, culverts, weir, etc.) representation in 2D flood inundation modelling is the pixel size and their false description within the DEM. A non-detailed DEM spatial resolution (big pixel size) in combination with the appearance of natural or artificial structures close to the flood mitigation works and hydraulic structures can lead to distortions of the elevation and by extension to their false representation within the DEM. In cases where the actual topography of the flood mitigation works and hydraulic structures exist, the limitation mentioned above can be eliminated with DEM editing. This study uses the XS interpolation surfaces module (HEC-RAS/RAS-Mapper) for DEM editing and by extension, the precise representation of flood mitigation works and hydraulic structures (bridges, culverts, weir, etc.) [65]. For better mesh resolution close to flood mitigation works and hydraulic structures, several Breaklines elements have been added. During the mesh generation process, the Breaklines elements force the orientation and the construction of cell faces along with them. Therefore, with the use of Breaklines elements the mesh is enforced near to flood mitigation works, hydraulic structures, and the building block areas [65].

Concerning the important hydraulic structures of the study areas, 10, 20 and 4 bridges have been taken into account in flood inundation modelling for Xerias, Krafsidonas, and Anavros streams, respectively. Inputs of hydraulic modeling were hydrographs provided by average hydrological simulation scenarios, using "average" roughness coefficients that were estimated according to CORINE 2000 land use classes. For all return periods, apart from the hydrographs provided by the lower and upper scenarios, we also perturbed the roughness values by -50% and $+50\%$, respectively, to obtain overall uncertainty bounds of inundated areas and associated hydraulic quantities, i.e., water depths and velocities.

A typical process followed in cases where severe lack of flood data exist is the use of qualitative criteria that are based on matching agreement of the 2×2 contingency table (Simulated vs. Observed) [7,10,17,25,26,66,67]. In this study the qualitative criterion Threat Score (TS) or Critical Success Index (CSI) [68] has been selected for the validation of the flood inundation modelling results. CSI is estimated by the 2×2 contingency table for all grids as following:

$$CSI = \frac{A}{(A + B + C)} \tag{14}$$

where A = Hit—event simulated to occur, and did occur, B = False alarm—event simulated to occur, but did not occur and C = Miss—event simulated not to occur, but did occur. Recent studies identified CSI as an efficient validation measure in flood inundation modelling and mapping when the focus is on the flood extent spatial distribution [7,10,25,26,66,67]. In this study, simulated flood inundation data that were derived from a historical flash flood event were used for validation and evaluation of the proposed methodology [10,25,26,35]. The CSI is used for the comparison of the inundated area between the constructed $T = 100$ year flood and the simulated historical flood event of approximately $T = 100$ year.

4. Volos City: Application and Results of the Modelling Framework

4.1. Semi-Distributed Hydrological Modelling of Volos City Watersheds

The parameters of the ombrian curves (Equation 1) for the design rainfall were estimated on the basis of extreme rainfall data at 57 meteorological stations across the River Basin District of Thessaly, comprising 224 samples of annual maxima at several time scales. Initially, we estimated the two parameters of the duration function, using finely-resolved rainfall maxima from the 15 recording stations. The optimal values of the two parameters were $\theta = 0.042$ και $\eta = 0.639$, which have been considered constant over the broader Thessaly. On the other hand, the estimation of the three parameters of the return period expression was based on 24-h data at 56 stations. In Figure 4, we contrast the initial and adjusted fitting of the GEV distribution to the daily rainfall maxima at the closest stations of the study area, i.e., Makrynitsa and Agchialos. In the first station, the initial estimation of parameter κ was -0.008, which has been corrected to 0.007, while in the second station the initial estimation was 0.513 (which is attributed to the very high value of the largest observed rainfall), which has been reduced to 0.190. The rainfall stations were next grouped into three climatic zones and assigned a representative value of κ. In particular, the three study basins are located at the median zone, with $\kappa = 0.09$. Accounting for the given values of parameters θ, η and κ, we finally provided point estimations of scale and location parameters λ' and ψ', respectively, and we also generated empirical 80% confidence intervals of point rainfall estimations for $T = 50$, 100 and 1000 years, by employing Monte Carlo simulations against the scale and location parameters λ' and ψ'.

In Table 2 we show the point estimations at Makrynitsa and Agchialos, and the 80% confidence limits for the 24 h rainfall, for the three return periods of interest. We assigned the same values for parameters κ, θ and η, i.e., 0.092, 0.042 and 0.639, respectively, while for the rest two parameters of the IDF relationship we used $\lambda' = 881.0$ and $\psi' = 0.788$ for Makrynitsa, and $\lambda' = 565.2$ and $\psi' = 0.840$ for Agchialos.

Table 2. Estimation of maximum 24-h rainfall at the two stations that are neighboring to Volos city, by employing the IDF relationship, and 80% confidence intervals (low and up values).

Station	T = 50 Years			T = 100 Years			T = 1000 Years		
	20% low	IDF	80% up	20% low	IDF	80% up	20% low	IDF	80% up
Makrynitsa	208.6	238.0	263.6	230.9	272.9	311.9	300.5	406.1	530.0
Agchialos	105.1	140.5	168.8	113.8	162.9	207.7	134.8	248.3	407.3

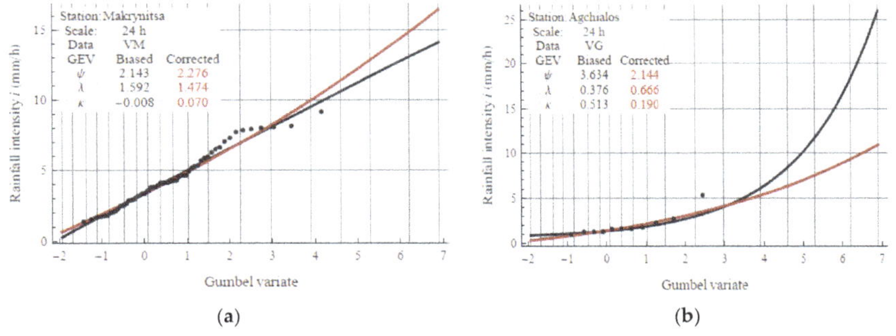

Figure 4. Fitting of GEV distribution to daily rainfall maxima and estimation of associated parameter values through the L-moments method (resulting to biased values) and the correction technique by Papalexiou and Koutsoyiannis [42], at the rainfall stations of (**a**) Makrynitsa; (**b**) Agchialos.

Using the point estimations of λ' and ψ', and the associated confidence limits of rainfall, we provided maps of distributed parameters over Thessaly (Figure 5), which allowed extracting spatially-distributed design rainfalls across the study catchments. In particular, we averaged the associated IDF parameters and standardized confidence limits over each sub-basin area, and next employed either the alternating block method (for $T = 50$ and 100 years) or the worst profile method (for $T = 1000$ years) in order to formulate the corresponding hyetographs.

Since the parameters of IDF curves, which are inputs for the estimation of design storms, have been estimated on the basis of the point data, they are valid at the point scale. However, it is well-known that for any given return period and duration, the spatially averaged rainfall over a given area is less than the maximum point rainfall depth. For this reason, we have reduced the point rainfall estimations, to provide areal estimations over the corresponding sub-basins, by applying a commonly used adjustment approach, based on the so-called areal reduction factor, φ. The latter is a dimensionless parameter, defined as the ratio of areal to point rainfall, which is decreasing function of the area and increasing function of duration. To facilitate calculations, we used the analytical formula by Koutsoyiannis and Xanthopoulos [69]:

$$\varphi = max\left(1 - \frac{0.048 A^{(0.36 - 0.01\,\ln A)}}{d^{0.35}}, 0.25\right) \tag{15}$$

where A is the area in km^2 and d is the rainfall duration in h. The above empirical relationship has been formulated on the basis of tabular data by UK-NERC [70], which captures a wide range of durations (in particular, from 1 min to 25 days) and catchment sizes (from 1 to 30,000 km^2). For instance, in the study basin of Xerias, with A = 116.8 km^2, we employed φ = 0.788 and 0.930, in order to reduce the design rainfall estimations for durations 1 and 24 h, respectively.

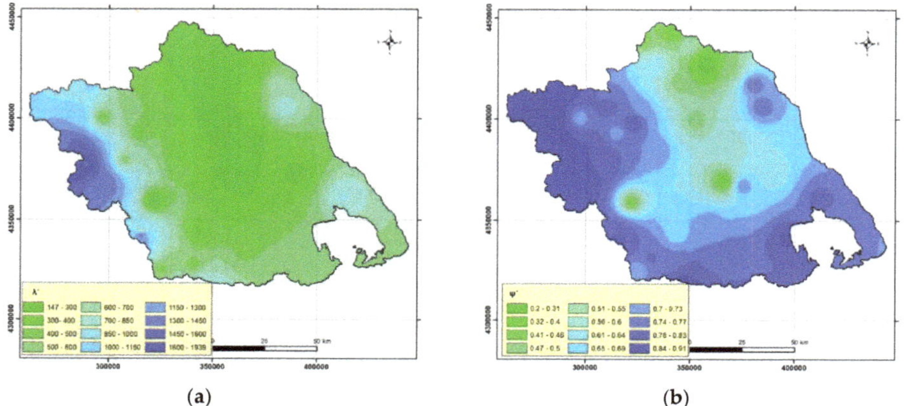

(a) (b)

Figure 5. Maps of distributed values of IDF parameters over the Water District of Thessaly: (**a**) scale parameter λ'; (**b**) location parameter ψ'.

In Table 3 we present the CN values that have been applied across the ten sub-basins of Xerias for the three AMC types, corresponding to the three hydrological scenarios of this study. It is interesting to highlight that the basin exhibits significant heterogeneity, since their CN II values range from 50 to 82, while the uncertainty induced to AMC is amplified in the areas with low CNs.

Table 3. Characteristic geomorphological properties and input parameters of Xerias sub-basins (A: sub-basin area; z_m: mean elevation; z_0: outlet elevation; L_{max}: maximum flow length; λ' and ψ': spatially-averaged scale and location parameters of IDF relationship; CN_I, CN_{II}, and CN_{III}: runoff curve number values for AMC type I, II and III, respectively).

id	A (km^2)	z_m (m)	z_0 (m)	L_{max} (km)	λ'	ψ'	t_c (h)	CN_I	CN_{II}	CN_{III}
1	6.1	66.0	0.0	5.6	698.1	0.757	2.81	49.3	69.8	84.2
2	1.4	26.4	8.7	1.7	695.4	0.754	2.17	62.5	79.9	90.1
3	20.4	199.8	21.3	8.9	613.6	0.738	2.94	48.8	69.4	83.9
4	8.0	140.0	21.3	5.1	686.9	0.749	2.17	46.6	67.5	82.7
5	7.5	73.3	8.7	5.2	754.0	0.763	2.91	65.8	82.1	91.3
6	2.2	130.2	51.9	3.1	789.5	0.768	1.50	60.5	78.5	89.4
7	22.3	447.7	58.7	10.6	808.6	0.771	2.20	29.2	49.6	69.4
8	13.6	338.4	170.7	7.7	697.7	0.743	2.54	31.3	52.0	71.4
9	20.0	722.7	170.7	15.1	788.1	0.766	2.15	32.4	53.3	72.4
10	15.3	1236.7	800.1	7.0	825.0	0.775	1.57	49.3	69.8	84.2

The application of the semi-distributed hydrological modelling procedure in the three watersheds led in the estimation of design hydrographs at the junctions used for the hydraulic simulation for all examined hydrologic conditions and return periods. For example, Figure 6 shows the design hydrographs at the outlet of Xerias watershed (Junction J1) for average soil moisture conditions (CN II) for the three study return periods. Similar design hydrographs were estimated for all examined junctions at the study area with similar shapes, but with different peak magnitudes.

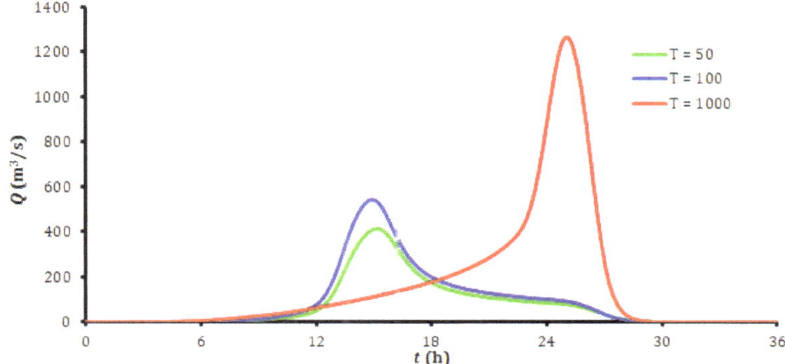

Figure 6. Design hydrographs at the outlet of Xerias river basin for average moisture conditions (CN II) and the study return periods.

4.2. Hydraulic Modelling of Lower Course of Volos City Streams

An operational method for Floods Directive implementation in ungauged urban areas is applied for different return periods (T = 50, 100, and 1000 years), three hydrologic conditions (AMC_I, AMC_{II}, AMC_{III}) that correspond to lower, average and upper estimations of the design rainfall, respectively, and three roughness values (-50%, average, $+50\%$). The examined scenarios aim to quantify the uncertainty induced to extreme rainfall analysis, antecedent soil moisture conditions and estimations of the roughness coefficient. Thus, this study investigates nine (9) different operational scenarios that incorporate important (yet not all) aspects of the total model uncertainty.

Due to lack of data, the methodology has been evaluated only for Xerias stream, based on the historical flood event that occurred October 9th, 2006. The observed hydrograph was estimated in previous works [10,25,26,35] and considered to correspond to an approximately 100 years flood event. The comparison of the simulated historical flood event and the simulated average design flood scenario (AMC_{II}, T = 100, average roughness value) is examined in the validation procedure with the skill score Critical Success Index. Table 4 presents the flooded areas (km^2) per river reach and the total flooded extent of Volos city for all examined hydrologic and hydraulic scenarios at the selected return periods. Flood extent variations that are presented in Table 4, and depicted in Figures 7 and 8, show that the hydrologic conditions scenarios, accounting for both rainfall and initial soil moisture uncertainty, have much stronger impact on the flood extent than the return period itself, which is only an indicator of the rainfall risk. Specifically, the flood extent ranges between 0.068 km^2 and 2.76 km^2 for dry (AMC_I) conditions, from 0.21 km^2 to 6.01 km^2 for average (AMC_{II}) conditions, and from 0.77 km^2 to 9.7 km^2 for wet (AMC_{III}) conditions (Table 4, Figure 7). This outcome is not surprising and confirms that the generation of a flood is strongly influenced by the soil moisture that is already stored at the beginning of rainfall.

Furthermore, based on the return period discretization it is observed that flood extent results of 50 and 100 years are very close to each other and the highest flood extent values with significant difference from the other return periods presented in 1000-year return period (Table 4, Figure 7). Specifically, the flood extent of all examined scenarios ranges between 0.068 km^2 and 5.3 km^2 for T = 50 years, from 0.081 km^2 to 6.34 km^2 for $T = 100$ years, and from 0.21 km^2 to 9.7 km^2 for $T = 1000$ years (Table 4, Figures 7 and 8). Figure 9 presents the simulated velocities only for average configurations of input rainfall, soil moisture conditions and roughness coefficients of return period: (a) $T = 50$ years, (b) $T = 100$ years, (c) $T = 1000$ years. Finally, the validation procedure that is based on the comparison between Xerias stream flood extent of the designed flood of $T = 100$ by employing the average input rainfall, soil moisture conditions and roughness coefficients and simulated flood extent of the 2006 historical flash flood event achieved a score in Critical Success Index of 0.77 and shown on Figure 10. This high score of CSI justifies the accuracy of the proposed operational methodology for flood directive implementation in urban and ungauged areas.

Table 4. Flooded areas (km^2) per river reach and total flooded extent of Volos city for all examined hydrologic and hydraulic scenarios at the selected return periods.

Code	River Name	Hydrologic Conditions/Roughness Coefficient Conditions	Return Period (Years)		
			50	100	1000
GR0817FR00700	Xerias	Dry (AMC$_I$)/ −50%	0.42	0.49	1.79
		Average (AMC$_{II}$)/ Average	2.15	2.63	4.84
		Wet (AMC$_{III}$)/ +50%	3.69	4.49	6.33
GR0817FR00800	Krafsidonas	Dry (AMC$_I$)/ −50%	0.085	0.087	0.75
		Average (AMC$_{II}$)/ Average	0.34	0.45	0.99
		Wet (AMC$_{III}$)/ +50%	0.93	1.34	2.91
GR0817FR00900	Anavros	Dry (AMC$_I$)/ −50%	0.068	0.081	0.21
		Average (AMC$_{II}$)/ Average	0.21	0.25	0.33
		Wet (AMC$_{III}$)/ +50%	0.77	0.82	1.2
Entire Volos city	Xerias & Krafsidonas & Anavros	Dry (AMC$_I$)/ −50%	0.57	0.66	2.76
		Average (AMC$_{II}$)/ Average	2.68	3.32	6.01
		Wet (AMC$_{III}$)/ +50%	5.3	6.34	9.7

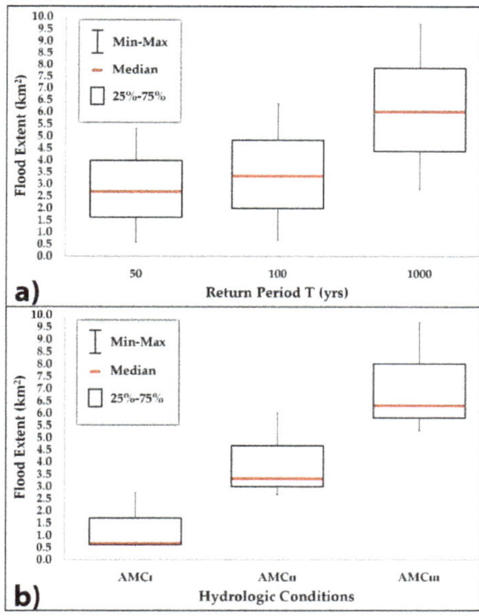

Figure 7. Box and Whisker plots of all examined scenarios according to flood extent (km^2): (**a**) for all return periods (50, 100, 1000 years) and, (**b**) for all hydrologic conditions (AMC$_I$, AMC$_{II}$, AMC$_{III}$).

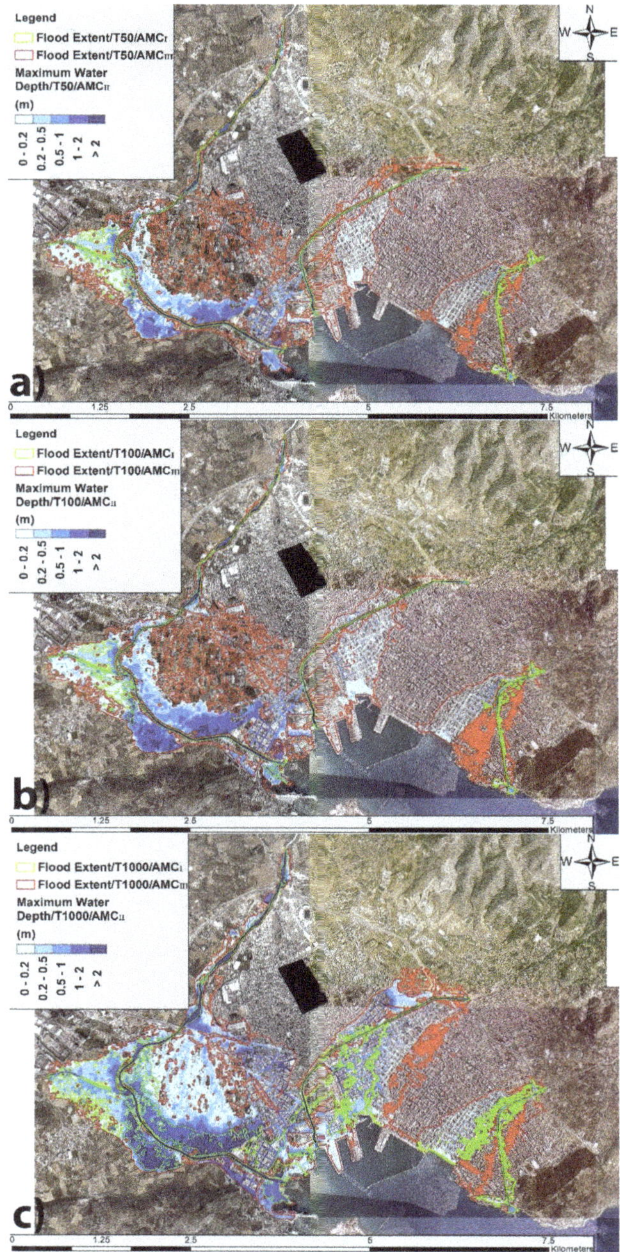

Figure 8. Flood extent and water depths for all configurations of input rainfall, soil moisture conditions and roughness coefficients of return periods: (**a**) $T = 50$, (**b**) $T = 100$, and (**c**) $T = 1000$ years.

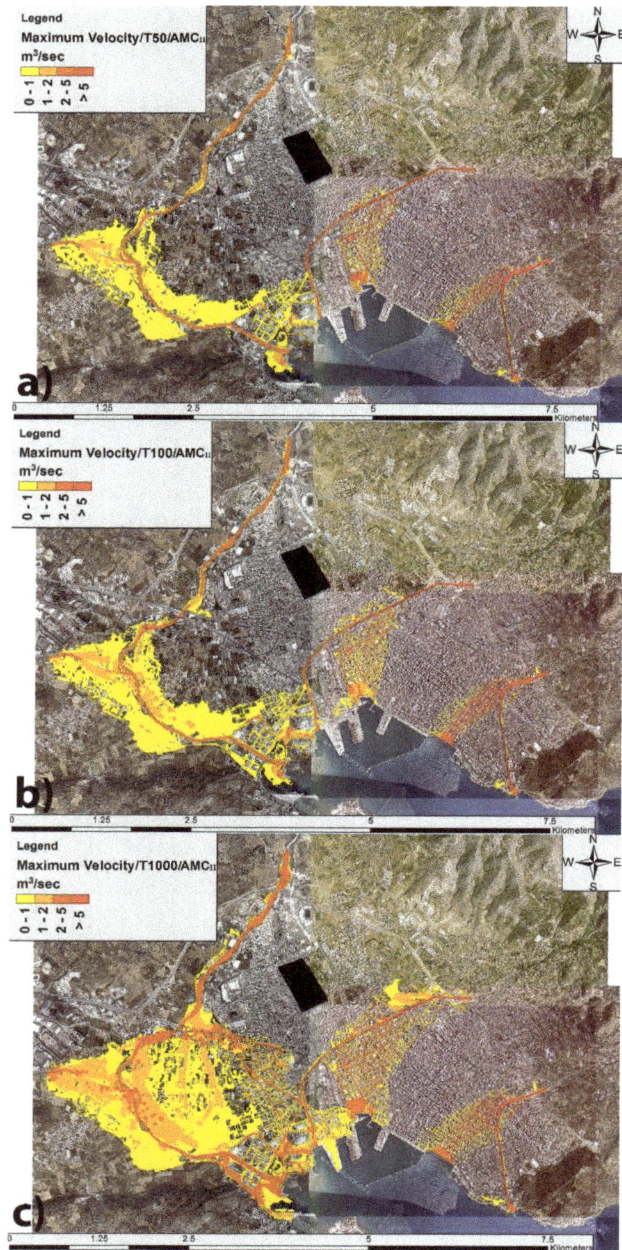

Figure 9. Simulated velocities only for average configurations of input rainfall, soil moisture conditions and roughness coefficients of return periods: (**a**) T = 50, (**b**) T = 100, and (**c**) T = 1000 years.

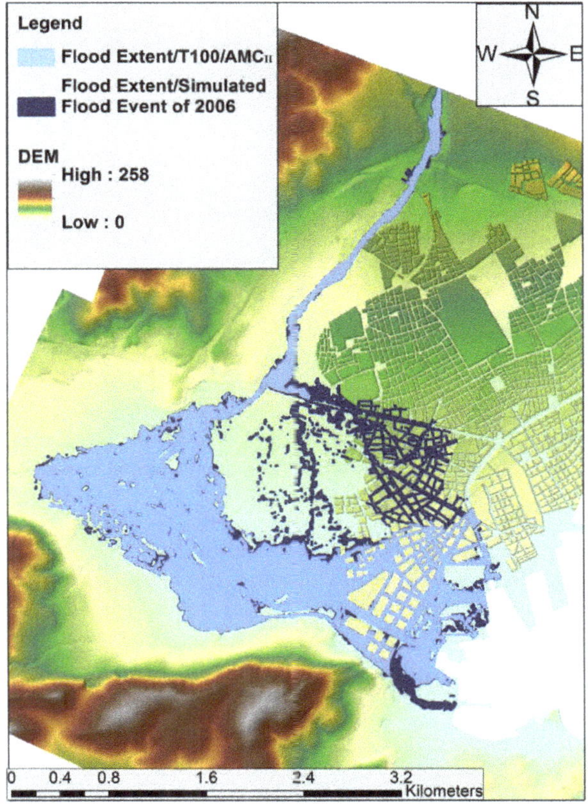

Figure 10. Xerias stream flood extent of the designed flood of $T = 100$ years, by employing the average input rainfall, soil moisture conditions and roughness coefficients and simulated flood extent of the 2006 historical flash flood event (CSI = 0.77).

Overall, in the context of flood inundation and mapping, this study tried to quantify the following major sources of uncertainty: (1) statistical uncertainty associated with the parameters of IDF relationships (i.e., scale and location parameters), originating from limited samples of observed extreme rainfall data; (2) hydrologic uncertainty associated with the initial soil moisture conditions of the hydrological model, resulting to a wide range of the key input parameter of the SCS-CN method, i.e., the potential maximum retention; and (3) parametric uncertainty associated with Manning's roughness coefficient, which is a typical input in all hydraulic simulation models. Results are quite diverse, since the uncertainty bounds of all key flood quantities (peak flows, flood volumes, inundated areas, etc.) strongly overlap the risk expressed in terms of return period of rainfall, while for large return periods, the lower and upper estimations may differ by one order of magnitude. Special attention should be given to the developed methodology and its application only for specific return periods and hydrologic-hydraulic conditions due to the great variability in the peak discharge estimation. A comparison of several methods and conditions prior to the decision-making phase could be proven useful flow flood management purposes. Hence, an ensemble of methods and scenarios should always be applied for engineering purposes, in order to choose the most appropriate technique in relation to the flood prone areas and proposed flood protection measures.

5. Concluding Remarks

In order to reduce the risk and adverse consequences of floods, the EU issued the so-called Flood Directive 2007/60/EC. Its implementation requires a proper estimation of flood hazards and representation of potential risks, which in turn makes essential the use of reliable hydrological methodologies for the estimation of associated flood quantities, as well as suitable and efficient hydraulic simulation tools for flood inundation modelling and mapping. In ungauged basins of relatively small extent that mainly affected by fluvial flash floods, which are the case in Greece, data-driven approaches in flood risk assessment, based on statistical analysis of observed floods, are not feasible. For this reason, flood estimations are exclusively estimated through classical engineering approaches, following the event-based deterministic paradigm.

In this study, a methodological approach for implementing the EU Floods Directive in Greece is developed, emphasized for flood risk management in urban areas, which is demonstrated for the Volos city case, where frequent flood episodes are observed. The methodology is based on typical hydrological and flood inundation modelling and mapping techniques for ungauged catchments. Spatially-distributed design hyetographs are applied for hydrologic and hydraulic 2D modelling of floods taking into account parametric and structural uncertainty. The average scenario against a simulated historical flood extent data was validated through the objective qualitative criterion of CSI that is counting the spatial distribution of the flooded area.

According to the flood extent values, it seems that the uncertainty induced in hydrological modeling, with respect to extreme rainfall estimations and antecedent soil moisture conditions, dominates against the return period. We remark that these two components are not the sole sources of uncertainty within rainfall-runoff transformations. This makes it essential to move to more rigorous methodological approaches (e.g., stochastic), instead of quantifying the flood risk on the basis of the return period of rainfall.

The overall results also proved that sensitivity analysis should be a mandatory process in flood modelling and mapping for urban areas due to the variation of the results because of the hydrologic conditions or the selected return period that is used. Finally, the results indicate the uncertainty introduced in flood risk management in urban areas using typical engineering practices.

Acknowledgments: This paper is part of the project "Management Plans of Flood Risks for River Basins in Thessaly, Western Sterea Hellas and Epirus Regions, Greece" co-funded by E.U. and the Greek Ministry of Energy and the Environment. This project constitutes the implementation of the EU Directive on floods (E.C. 2007/60) in the above regions of Greece. The research presented in the paper is partially supported by this project. The authors would like to thank the guest editors for their kind invitation and the three anonymous reviewers for their constructive and useful comments, which contributed to an improved presentation of the paper.

Author Contributions: G.P., A.E., L.V., and A.L. designed the study, developed the methodology, and wrote the manuscript. S.M.P., A.K., I.T. and P.K. performed the IDF analysis and contributed to the hydrologic analysis part. Furthermore, G.P. performed the hydraulic simulation and validation of the study area.

Conflicts of Interest: The authors declare no conflict of interest.

References

1. Tsakiris, G. Flood risk assessment: Concepts, modelling, applications. *Nat. Hazards Earth Syst. Sci.* **2014**, *14*, 1361–1369. [CrossRef]
2. Hall, J.; Arheimer, B.; Borga, M.; Brázdil, R.; Claps, P.; Kiss, A.; Kjeldsen, T.R.; Kriaučuniene, J.; Kundzewicz, Z.W.; Lang, M.; et al. Understanding flood regime changes in Europe: A state-of-the-art assessment. *Hydrol. Earth Syst. Sci.* **2014**, *18*, 2735–2772. [CrossRef]
3. Kreibich, H.; Di Baldassarre, G.; Vorogushyn, S.; Aerts, J.C.J.H.; Apel, H.; Aronica, G.T.; Arnbjerg-nielsen, K.; Bouwer, L.M.; Bubeck, P.; Caloiero, T.; et al. Earth's Future Special Section: Adaptation to flood risk: Results of international paired flood event studies. *Earth's Future* **2017**, *5*, 953–965. [CrossRef]
4. Diakakis, M.; Mavroulis, S.; Deligiannakis, G. Floods in Greece, a statistical and spatial approach. *Nat. Hazards* **2012**, *62*, 485–500. [CrossRef]

5. Centre for Research on the Epidemiology of Disasters (CRED). Summarized Table of Natural Disasters in Greece from 1900 to 2017, EM-DAT: The CRED/OFDA International Disaster Database–www.emdat.be–Université Catholique de Louvain–Brussels–Belgium. Available online: http://www.emdat.be (accessed on 12 January 2018).

6. Apel, H.; Thieken, A.H.; Merz, B.; Blöschl, G. Flood risk assessment and associated uncertainty. *Nat. Hazards Earth Syst. Sci.* **2004**, *4*, 295–308. [CrossRef]

7. Aronica, G.; Bates, P.D.; Horritt, M.S. Assessing the uncertainty in distributed model predictions using observed binary pattern information within GLUE. *Hydrol. Process.* **2002**, *16*, 2001–2016. [CrossRef]

8. Aronica, G.T.; Franza, F.; Bates, P.D.; Neal, J.C. Probabilistic evaluation of flood hazard in urban areas using Monte Carlo simulation. *Hydrol. Process.* **2012**, *26*, 3962–3972. [CrossRef]

9. Dottori, F.; Di Baldassarre, G.; Todini, E. Detailed data is welcome, but with a pinch of salt: Accuracy, precision, and uncertainty in flood inundation modeling. *Water Resour. Res.* **2013**, *49*, 6079–6085. [CrossRef]

10. Papaioannou, G.; Loukas, A.; Vasiliades, L.; Aronica, G.T. Flood inundation mapping sensitivity to riverine spatial resolution and modelling approach. *Nat. Hazards* **2016**, *83*, 117–132. [CrossRef]

11. Teng, J.; Jakeman, A.J.; Vaze, J.; Croke, B.F.W.; Dutta, D.; Kim, S. Flood inundation modelling: A review of methods, recent advances and uncertainty analysis *Environ. Model. Softw.* **2017**, *90*, 201–216. [CrossRef]

12. Soil Conservation Service (SCS). *National Engineering Handbook*; Section 4, Hydrology (NEH-4); U.S. Department of Agriculture: Washington, DC, USA, 1972.

13. Efstratiadis, A.; Koussis, A.D.; Koutsoyiannis, D.; Mamassis, N. Flood design recipes vs. reality: Can predictions for ungauged basins be trusted? *Nat. Hazards Earth Syst. Sci.* **2014**, *14*, 1417–1428. [CrossRef]

14. Gräler, B.; Van Den Berg, M.J.; Vandenberghe, S.; Petroselli, A.; Grimaldi, S.; De Baets, B.; Verhoest, N.E.C. Multivariate return periods in hydrology: A critical and practical review focusing on synthetic design hydrograph estimation. *Hydrol. Earth Syst. Sci.* **2013**, *17*, 1281–1296. [CrossRef]

15. Horritt, M.S.; Di Baldassarre, G.; Bates, P.D.; Brath, A. Comparing the performance of a 2-D finite element and a 2-D finite volume model of floodplain inundation using airborne SAR imagery. *Hydrol. Process.* **2007**, *21*, 2745–2759. [CrossRef]

16. Costabile, P.; Macchione, F. Enhancing river model set-up for 2-D dynamic flood modelling. *Environ. Model. Softw.* **2015**, *67*, 89–107. [CrossRef]

17. Horritt, M.S.; Bates, P.D. Evaluation of 1D and 2D numerical models for predicting river flood inundation. *J. Hydrol.* **2002**, *268*, 87–99. [CrossRef]

18. Merwade, V.; Cook, A.; Coonrod, J. GIS techniques for creating river terrain models for hydrodynamic modeling and flood inundation mapping. *Environ. Model. Softw.* **2008**, *23*, 1300–1311. [CrossRef]

19. Hunter, N.M.; Bates, P.D.; Horritt, M.S.; Wilson, M.D. Simple spatially-distributed models for predicting flood inundation: A review. *Geomorphology* **2007**, *90*, 208–225. [CrossRef]

20. Cook, A.; Merwade, V. Effect of topographic data, geometric configuration and modeling approach on flood inundation mapping. *J. Hydrol.* **2009**, *377*, 131–142. [CrossRef]

21. Mohammed, J.R.; Qasim, J.M. Comparison of One-Dimensional HEC-RAS with Two-Dimensional ADH for Flow over Trapezoidal Profile Weirs. *Casp. J. Appl. Sci. Res.* **2012**, *1*, 1–32. [CrossRef]

22. Néelz, S.; Pender, G.; Britain, G. *Desktop Review of 2D Hydraulic Modelling Packages*; DEFRA/Environment Agency: Bristol, UK, 2009; ISBN 9781849110792.

23. Néelz, S.; Pender, G.; Wright, N.G. *Benchmarking of 2D Hydraulic Modelling Packages*; DEFRA/Environment Agency: Bristol, UK, 2010; ISBN 9781849111904.

24. Néelz, S.; Pender, G. *Benchmarking the Latest Generation of 2D Hydraulic Modelling Packages*; DEFRA/Environment Agency: Bristol, UK, 2013; ISBN 9781849113069.

25. Papaioannou, G.; Vasiliades, L.; Loukas, A.; Aronica, G.T. Probabilistic flood inundation mapping at ungauged streams due to roughness coefficient uncertainty in hydraulic modelling. *Adv. Geosci.* **2017**, *44*, 23–34. [CrossRef]

26. Papaioannou, G.; Loukas, A.; Vasiliades, L.; Aronica, G.T. Sensitivity analysis of a probabilistic flood inundation mapping framework for ungauged catchments. *Eur. Water* **2017**, *60*, 9–16.

27. Di Baldassarre, G.; Schumann, G.; Bates, P.D.; Freer, J.E.; Beven, K.J. Cartographie de zone inondable: Un examen critique d'approches déterministe et probabiliste. *Hydrol. Sci. J.* **2010**, *55*, 364–376. [CrossRef]

28. Sarhadi, A.; Soltani, S.; Modarres, R. Probabilistic flood inundation mapping of ungauged rivers: Linking GIS techniques and frequency analysis. *J. Hydrol.* **2012**, *458*, 68–86. [CrossRef]

29. Domeneghetti, A.; Vorogushyn, S.; Castellarin, A.; Merz, B.; Brath, A. Probabilistic flood hazard mapping: Effects of uncertain boundary conditions. *Hydrol. Earth Syst. Sci.* **2013**, *17*, 3127–3140. [CrossRef]

30. Dimitriadis, P.; Tegos, A.; Oikonomou, A.; Pagana, V.; Koukouvinos, A.; Mamassis, N.; Koutsoyiannis, D.; Efstratiadis, A. Comparative evaluation of 1D and quasi-2D hydraulic models based on benchmark and real-world applications for uncertainty assessment in flood mapping. *J. Hydrol.* **2016**, *534*, 478–492. [CrossRef]

31. Bates, P.D.; Wilson, M.D.; Horritt, M.S.; Mason, D.C.; Holden, N.; Currie, A. Reach scale floodplain inundation dynamics observed using airborne synthetic aperture radar imagery: Data analysis and modelling. *J. Hydrol.* **2006**, *328*, 306–318. [CrossRef]

32. Aggett, G.R.; Wilson, J.P. Creating and coupling a high-resolution DTM with a 1-D hydraulic model in a GIS for scenario-based assessment of avulsion hazard in a gravel-bed river. *Geomorphology* **2009**, *113*, 21–34. [CrossRef]

33. Bates, P.D.; Horritt, M.S.; Aronica, G.; Beven, K. Bayesian updating of flood inundation likelihoods conditioned on flood extent data. *Hydrol. Process.* **2004**, *18*, 3347–3370. [CrossRef]

34. Special Secretariat for Water, Ministry of Environment and Energy (SSW-MEE). *Preliminary Assessment of the Flood Directive*; Athens: Ministry of Environment and Energy; Available online: http://www.ypeka.gr/Lin kClick.aspx?fileticket=T4DDG1hqQMY%3d&tabid=252&language=el-GR (accessed on 12 January 2018).

35. Papaioannou, G.; Vasiliades, L.; Loukas, A. Multi-Criteria Analysis Framework for Potential Flood Prone Areas Mapping. *Water Resour. Manag.* **2015**, *29*, 399–418. [CrossRef]

36. Harats, N.; Ziv, B.; Yair, Y.; Kotroni, V.; Dayan, U. Lightning and rain dynamic indices as predictors for flash floods events in the Mediterranean. *Adv. Geosci.* **2010**, *53*, 57–64. [CrossRef]

37. Efstratiadis, A.; Papalexiou, S.M.; Markonis, I.; Mamassis, N. Ombrian curves. In *Flood Risk Management Plan of River Basin District of Thessaly (GR08)–Phase A*; Special Secretariat for Water, Ministry of Environment and Energy (SSW-MEE): Athens, Greece, 2016.

38. Koutsoyiannis, D.; Kozonis, D.; Manetas, A. A mathematical framework for studying rainfall intensity-duration-frequency relationships. *J. Hydrol.* **1998**, *206*, 118–135. [CrossRef]

39. Hirsch, R.M.; Helsel, D.R.; Cohn, T.A.; Gilroy, E.J. Statistical analysis of hydrological data. In *Handbook of Hydrology*; Maidment, D.R., Ed.; McGraw-Hill: New York, NY, USA, 1993.

40. Hosking, J.R.M. L-Moments: Analysis and Estimation of Distributions Using Linear Combinations of Order Statistics. *J. R. Stat. Soc. B* **1990**, *52*, 105–124.

41. Vogel, R.M.; Fennessey, N.M. L moment diagrams should replace product moment diagrams. *Water Resour. Res.* **1993**, *29*, 1745–1752. [CrossRef]

42. Papalexiou, S.M.; Koutsoyiannis, D. Battle of extreme value distributions: A global survey on extreme daily rainfall. *Water Resour. Res.* **2013**, *49*, 187–201. [CrossRef]

43. Tyralis, H.; Koutsoyiannis, D.; Kozanis, S. An algorithm to construct Monte Carlo confidence intervals for an arbitrary function of probability distribution parameters. *Comput. Stat.* **2013**, *28*, 1501–1527. [CrossRef]

44. Sutcliffe, J.V. *Methods of Flood Estimation: A Guide to Flood Studies Report*; Institute of Hydrology: Wallingford, UK, 1978; Volume 49.

45. Chow, V.T.; Maidment, D.R.; Mays, L.W. *Applied Hydrology*; McGraw-Hill: Singapore, 1988.

46. Koutsoyiannis, D. A stochastic disaggregation method for design storm and flood synthesis. *J. Hydrol.* **1994**, *156*, 193–225. [CrossRef]

47. National Regulator for Compulsory Specifications (NRCS). *National Engineering Handbook: Part 630—Hydrology*; NRCS: Washington, DC, USA, 2004.

48. Michailidi, E.M.; Antoniadi, S.; Koukouvinos, A.; Bacchi, B.; Efstratiadis, A. Timing the time of concentration: Shedding light on a paradox. *Hydrol. Sci. J.* **2018**, *63*. [CrossRef]

49. Grimaldi, S.; Petroselli, A.; Tauro, F.; Porfiri, M. Time of concentration: A paradox in modern hydrology. *Hydrol. Sci. J.* **2012**, *57*, 217–228. [CrossRef]

50. Brunner, G. HEC-RAS River Analysis System: Hydraulic Reference Manual, Version 5.0. US Army Corps of Engineers–Hydrologic Engineering Center, 2016a; pp. 1–538. Available online: http://www.hec.usace.ar my.mil/software/hec-ras/documentation/HEC-RAS%205.0%20Reference%20Manual.pdf (accessed on 12 January 2018).

51. Alfonso, L.; Mukolwe, M.M.; Di Baldassarre, G. Probabilistic Flood Maps to support decision-making: Mapping the Value of Information. *Water Resour. Res.* **2016**, *52*, 1026–1043. [CrossRef]

52. Patel, D.P.; Ramirez, J.A.; Srivastava, P.K.; Bray, M.; Han, D. Assessment of flood inundation mapping of Surat city by coupled 1D/2D hydrodynamic modeling: A case application of the new HEC-RAS 5. *Nat. Hazards* **2017**, *89*, 93–130. [CrossRef]

53. Vozinaki, A.E.K.; Morianou, G.G.; Alexakis, D.D.; Tsanis, I.K. Comparing 1D and combined 1D/2D hydraulic simulations using high-resolution topographic data: A case study of the Koiliaris basin, Greece. *Hydrol. Sci. J.* **2017**, *62*, 642–656. [CrossRef]

54. Afshari, S.; Tavakoly, A.A.; Rajib, M.A.; Zheng, X.; Follum, M.L.; Omranian, E.; Fekete, B.M. Comparison of new generation low-complexity flood inundation mapping tools with a hydrodynamic model. *J. Hydrol.* **2018**, *556*, 539–556. [CrossRef]

55. Brunner, G. Benchmarking of the HEC-RAS Two-Dimensional Hydraulic Modeling Capabilities. US Army Corps of Engineers—Hydrologic Engineering Center, 2016b; pp. 1–116. Available online: http://www.hec.usace.army.mil/software/hec-ras/documentation/RD-51_Benchmarking_2D.pdf (accessed on 12 January 2018).

56. Tsubaki, R.; Fujita, I. Unstructured grid generation using LiDAR data for urban flood inundation modelling. *Hydrol. Process.* **2010**, *24*, 1404–1420. [CrossRef]

57. Md Ali, A.; Solomatine, D.P.; Di Baldassarre, G. Assessing the impact of different sources of topographic data on 1-D hydraulic modelling of floods. *Hydrol. Earth Syst. Sci.* **2015**, *19*, 631–643. [CrossRef]

58. Teng, J.; Vaze, J.; Dutta, D.; Marvanek, S. Rapid Inundation Modelling in Large Floodplains Using LiDAR DEM. *Water Resour. Manag.* **2015**, *29*, 2619–2636. [CrossRef]

59. Bates, P.D.; De Roo, A.P.J. A simple raster-based model for flood inundation simulation. *J. Hydrol.* **2000**, *236*, 54–77. [CrossRef]

60. Chow, W. *Open-Channel Hydraulics*, 1st ed.; McGraw-Hill: New York, NY, USA, 1959.

61. Hunter, N.M.; Bates, P.D.; Neelz, S.; Pender, G.; Villanueva, I.; Wright, N.G.; Liang, D.; Falconer, R.A.; Lin, B.; Waller, S.; et al. Benchmarking 2D hydraulic models for urban flooding. *Proc. Inst. Civ. Eng. Water Manag.* **2008**, *161*, 13–30. [CrossRef]

62. Bellos, V. Ways for flood hazard mapping in urbanised environments: A short literature review. *Water Util.* **2012**, 25–31.

63. Schubert, J.E.; Sanders, B.F. Building treatments for urban flood inundation models and implications for predictive skill and modeling efficiency. *Adv. Water Resour.* **2012**, *41*, 49–64. [CrossRef]

64. Bellos, V.; Tsakiris, G. Comparing Various Methods of Building Representation for 2D Flood Modelling In Built-Up Areas. *Water Resour. Manag.* **2015**, *29*, 379–397. [CrossRef]

65. Brunner, G. HEC-RAS River Analysis System 2D Modeling User's Manual. US Army Corps of Engineers—Hydrologic Engineering Center, 2016c; pp. 1–171. Available online: http://www.hec.usace.army.mil/software/hec-ras/documentation/HEC-RAS%205.0%202D%20Modeling%20Users%20Manual.pdf. (accessed on 12 January 2018).

66. Horritt, M.S.; Bates, P.D. Predicting floodplain inundation: Raster-based modelling versus the finite-element approach. *Hydrol. Process.* **2001**, *15*, 825–842. [CrossRef]

67. Alfieri, L.; Salamon, P.; Bianchi, A.; Neal, J.; Bates, P.; Feyen, L. Advances in pan-European flood hazard mapping. *Hydrol. Process.* **2014**, *28*, 4067–4077. [CrossRef]

68. Jolliffe, I.T.; Stephenson, D.B. *Forecast Verification*; Wiley-Blackwell: Chichester, UK, 2012; ISBN 9780470660713.

69. Koutsoyiannis, D.; Xanthopoulos, Th. *Engineering Hydrology*, 3rd ed.; National Technical University of Athens: Athens, Greece, 1999; p. 418. (In Greek)

70. U.K. National Environmental Research Council (UK-NERC). *Flood Studies Report*; Institute of Hydrology: Wallingford, CT, USA, 1975.

 *hydrology*

Article

Socioeconomic Impact Evaluation for Near Real-Time Flood Detection in the Lower Mekong River Basin

Perry C. Oddo [1,2,*], **Aakash Ahamed** [3] **and John D. Bolten** [2]

[1] Universities Space Research Association (USRA), Columbia, MD 21046, USA

[2] Hydrological Sciences Laboratory, NASA Goddard Space Flight Center, Greenbelt, MD 20771, USA; john.bolten@nasa.gov

[3] Department of Geophysics, Stanford University, Stanford, CA 94025, USA; aahamed@stanford.edu

* Correspondence: perry.oddo@nasa.gov; Tel.: +1-301-286-9088

Received: 16 March 2018; Accepted: 7 April 2018; Published: 10 April 2018

Abstract: Flood events pose a severe threat to communities in the Lower Mekong River Basin. The combination of population growth, urbanization, and economic development exacerbate the impacts of these events. Flood damage assessments, critical for understanding the effects of flooding on the local population and informing decision-makers about future risks, are frequently used to quantify the economic losses due to storms. Remote sensing systems provide a valuable tool for monitoring flood conditions and assessing their severity more rapidly than traditional post-event evaluations. The frequency and severity of extreme flood events are projected to increase, further highlighting the need for improved flood monitoring and impact analysis. In this study we integrate a socioeconomic damage assessment model with a near real-time flood remote sensing and decision support tool (NASA's Project Mekong). Direct damages to populations, infrastructure, and land cover are assessed using the 2011 Southeast Asian flood as a case study. Improved land use/land cover and flood depth assessments result in rapid loss estimates throughout the Mekong River Basin. Results suggest that rapid initial estimates of flood impacts can provide valuable information to governments, international agencies, and disaster responders in the wake of extreme flood events.

Keywords: near real-time; Mekong Basin; hydro-economic; socioeconomic; damage assessment; hydroinformatics

---

1. Introduction

Flood events are among the costliest natural disasters and pose significant threats to many low-lying and coastal communities [1–3]. The Mekong River Basin (MRB) is one of the most flood-prone regions in the world. Approximately 60 million people reside there, with many inhabiting areas along flood plains or within a few meters of sea level [4,5]. As populations and economic growth continue to increase, so do the expected damages from future events [6]. Nicholls et al. (2008) estimate that by the year 2070 Vietnam and Thailand alone will account for 6% of global assets and 12% of global population exposed to flooding [7].

The effects of climate change are also expected to exacerbate threats caused by flooding. Increased frequency, severity, and variability of storm surge, combined with rising sea levels, pose significant risks to coastal populations [6,8]. For inland areas, changes to precipitation patterns and monsoons can disturb the delicate balance between the necessary seasonal flood cycles and destructive extreme events [9]. In light of these potential risks, it is increasingly important for flood managers and decision makers to understand the socioeconomic effects of flooding on communities.

Impact assessments are commonly used to quantify the socioeconomic effects of flooding. Estimates of potential damages are critical to land-use planning and risk mapping, and serve as inputs for cost-benefit analyses of flood protection systems [3,10,11]. Post-event evaluations are commonly formulated by national governments or international agencies like the United Nations (UN), Red Cross, World Bank, or United States Agency for International Development (USAID) [12,13]. However, these evaluations often require a significant amount of time to produce, thereby delaying insights which could otherwise reduce vulnerability to future flood events. The ability to acquire rapid damage estimates can provide emergency responders, public administrators, and insurance companies valuable information in the wake of a hazardous event [14,15].

Geographic information system (GIS) and remote sensing technologies offer a way to synthesize geospatial and socioeconomic data more rapidly and efficiently. While several recent studies have demonstrated the efficacy of structured flood damage assessments in the MRB, these analyses were largely constrained to local or community-level evaluations (see: [16–18]). Impact evaluations across broader scales can be complicated by the need for trans-boundary coordination between countries, as is the case in the MRB [19]. Here, we propose a framework that couples the near real-time (NRT) flood detection capabilities of an existing web application (NASA's Project Mekong; http://projectmekongnasa.appspot.com) with a rapid damage assessment module to estimate flood impacts on a regional scale [2].

Project Mekong is a decision support tool that leverages the rapid revisit time and low latency afforded by NASA's Land, Atmosphere Near Real-Time Capability for Earth Observing Systems (LANCE) system. Imagery from the Moderate-resolution Imaging Spectroradiometer (MODIS) sensors on the Aqua and Terra satellites are obtained twice daily at 3-hr latency. Cloud filters are applied and a dynamic surface water classifier identifies flooding using the spectral Normalized Difference Vegetation Index (NDVI) signatures of permanent water bodies (MOD44W). For a more detailed description of the flood detection scheme used in the tool, see [2].

In addition to producing operational flood conditions in NRT, the Project Mekong system has previously been benchmarked against historical imagery. One notable example is the 2011 Southeast Asia flood, which was an event of historic magnitude and the effects of which are well documented in the literature [20–23]. Using the 2011 flood event as a case study, this analysis seeks to: (1) demonstrate the feasibility of assessing and visualizing socioeconomic impacts on a regional scale and (2) use readily-accessible data to establish a framework to produce damage estimates in NRT in conjunction with the Project Mekong tool.

2. Background and Motivation

2.1. Study Area

The Mekong is the world's 12th longest river and provides a critical source of water for one of the world's most densely populated landscapes. Originating in the Tibetan plateau, it extends over 4300 km and spans parts of China, Laos, Myanmar, Thailand, Cambodia, and Vietnam (Figure 1A) [24,25]. The basin drains an area of approximately 795,000 km^2 and can be divided into two primary catchments, known as the Upper and Lower Mekong Basins (LMB) [4]. The LMB can be further characterized by four physiographic regions: Khorat Plateau, Tonle Sap Basin, Northern Highlands, and Mekong Delta. The delta itself covers 39,000 km^2 and is home to over 18 million people [26]. The combination of low-lying terrain and high population density make communities in the LMB prone to river and coastal flooding [27].

The LMB is well adapted to seasonal inundation cycles, which typically occur between July and November. In the Mekong Delta, regular flooding affects up to 50% of the land surface [28]. The influx of nutrient-rich waters from these cycles replenishes fertile sediment for agriculture, recharges groundwater reservoirs, and provides habitat for aquaculture [21]. While many inhabitants of the LMB rely on seasonal inundation for their livelihoods, they remain highly susceptible to extreme events [29].

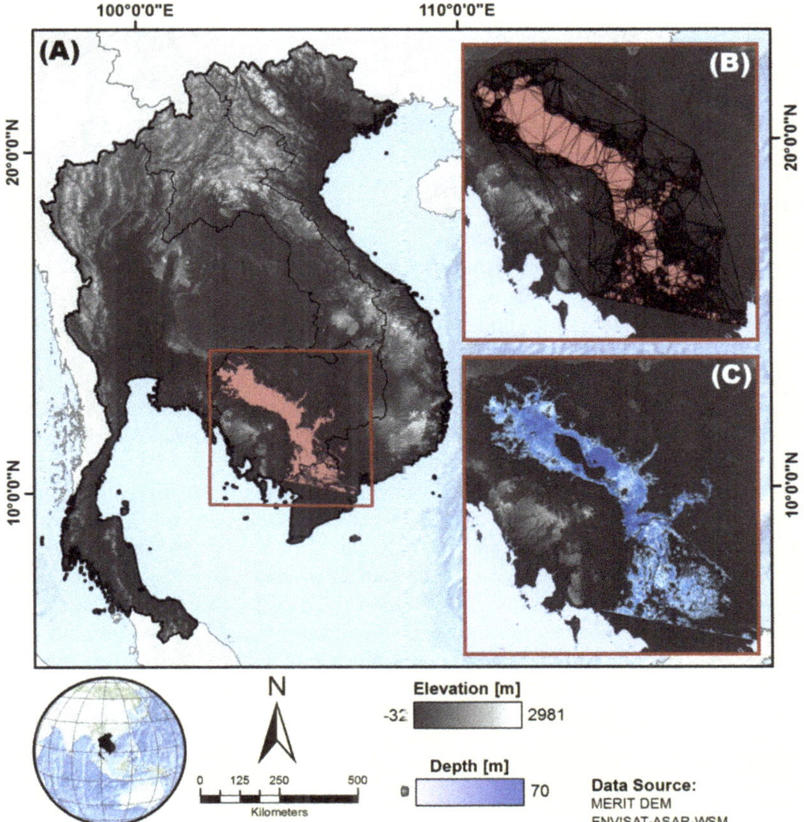

Figure 1. (**A**) Map of Mekong River Basin countries with flood extent from 2011 event. (**B**) Study extent showing results of the triangular interpolated network (TIN). (**C**) Depth raster produced by inundation depth analysis.

2.2. The 2011 Southeast Asia Floods

During the 2011 monsoon season, record flooding caused by a confluence of natural and human-made factors struck communities across Thailand, Myanmar, Cambodia, Laos, and Vietnam. La Ninã meteorological conditions resulted in a 143% increase in rainfall in Northern Thailand alone [30]. This extended period of above-average precipitation coincided with the onset of the southwest monsoon, which occurred between May and September over Thailand and the Andaman Sea [31].

Topography and land use played an important role in the distribution of flood damages. The gently sloping landscape resulted in an inundation of large geographic areas. Reservoir capacities were quickly overwhelmed and increased runoff channeled a high volume of water into the LMB [19]. Several densely-populated urban areas were significantly affected by these extreme hydrologic conditions, despite the existence of flood management infrastructure like levees and water gates. The area around Bangkok, Thailand, for instance, had subsided between 0.5–1.6 m during the preceding several decades [30]. The combination of increased vulnerability and the dense concentration of people and infrastructure is believed to have increased the resulting flood damages.

2.3. Flood Impact Assessments

Impact assessments are a critical component of flood management plans. Most modern flood management strategies rely on risk analyses that consider the probability of a given hazard (e.g., the 100-year flood) and its associated effects [10,32]. The expected damages from a flood event are a function of the vulnerability of a given population. Many factors can affect the vulnerability of a population, including socioeconomic variables like accumulated wealth, mobility, or the health status of households; robustness of infrastructure; flood characteristics like depth, duration, or flow velocity; and the existence of warning and response variables [16].

There are multiple ways to structure a flood impact assessment, all of which is subject to challenges and assumptions. Flood impacts are often described using a framework that classifies damages as 'direct vs. indirect' and 'tangible vs. intangible' [33]. Direct, tangible damages are those that occur as a result of direct contact with water and can be readily quantified by established metrics [3,10]. Other direct damages, such as the loss of human life, pose distinct ethical challenges when it comes to assigning damage values, leading them to be considered intangible [34,35]. Examples of indirect damages would be loss of income due to displacement, suspension of education, or issues of intergenerational justice [32,36]. Here, we focus primarily on the direct, tangible damages associated with flood inundation throughout the LMB.

Several countries have developed standardized flood impact frameworks which allow flood events to be compared across time [10]. In the U.S., the Federal Emergency Management Agency (FEMA) uses the Hazards U.S. (HAZUS) model for mitigation and recovery planning, as well as disaster preparedness and response [37]. The model is highly detailed, and is used to calculate exposure for a variety of different types of residential and commercial infrastructures. However, the most detailed version of the multi-hazard module requires "extensive additional economic and engineering studies by the user," [3] (p. 3741). Performing such analyses on a regional scale can, therefore, pose nontrivial practical and computational challenges. In the following section, we outline a streamlined workflow to rapidly produce flood damage estimates across the MRB.

3. Materials and Methods

Here, we adapt and employ a damage assessment framework originally developed by civil engineers in the Netherlands. This so-called "Standard Method," as outlined in Kok et al. (2004), calculates flood damages according to different land cover types and infrastructure categories [38]. Like the HAZUS model in the U.S., it relies on depth-dependent damage functions to determine severity of flood impacts. While other studies consider additional flood characteristics such as flow velocity or duration, this analysis considers only inundation depth as a simplifying assumption. This assumption is supported by Tang et al. (1992), who found that inundation depth was the primary driving variable for flood damages in the Bangkok area, particularly for commercial and agricultural areas [27] (p. 55). For an extended discussion of assumptions and limitations, see the Discussion and Conclusions section. The following sections describe how inundation depths are estimated and fed into the model to assess damages.

3.1. Inundation Depth Estimation

To illustrate the workflow for the near real-time assessment, we consider the 2011 Southeast Asia floods as a case study. Surface water extent estimates were obtained through The United Nations Institute for Training and Research (UNITAR), which supports satellite data collection of natural disasters through the Operational Satellite Applications Programme (UNOSAT). Imagery collected by the European Space Agency's ENVISAT Advanced Synthetic Aperture Radar Wide Swath Mode (ASAR-WSM) shows the extent of the surface inundation between 27 and 30 September, 2011 at a spatial resolution of 150-m [39] (Figure 1A). A geodatabase containing the vector data of the detected 2011 flood extent is available in [40].

QGIS software was used to process the ASAR flood extent vectors according to the method described in Cham et al. (2015) [41]. The flood extent polygon was converted to a polyline feature. Sample points were generated around the boundary of the flood extent at 250-m intervals. Twelve tiles from the "Multi-Error-Removed Improved-Terrain" digital elevation model (MERIT DEM) were stitched together to produce a regional elevation layer at 3 arc-second (90-m) resolution [42]. The MERIT DEM improves on many of the sources of error present in other global elevation datasets (e.g., speckling, striping, and vegetation biases). However, its vertical accuracy is not without uncertainty, especially when used in a flood modeling context. For a more detailed discussion of uncertainties see the Discussion and Conclusions section.

At each of the sampled points, the land surface elevation from the MERIT DEM was extracted. These elevation points were used to generate a triangular interpolated network (TIN) to serve as an estimate for the flood surface elevation (Figure 1B). Finally, flood depths were determined by subtracting the land surface elevation from the interpolated flood surface at each grid cell (Figure 1C). A cross-section schematic of this process is illustrated in Figure 2.

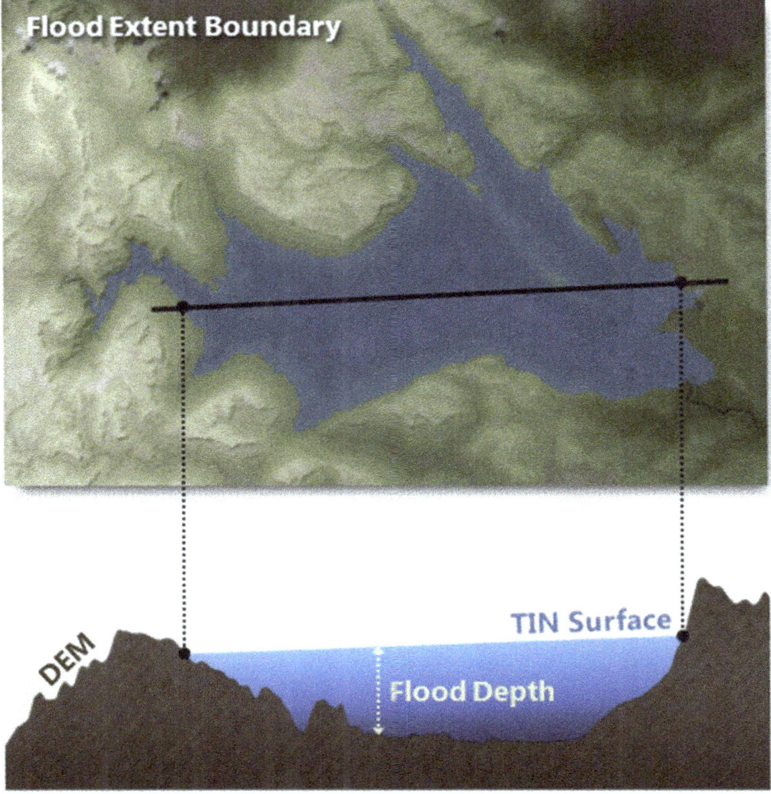

Figure 2. Plan view (**upper**) and cross-section schematic (**lower**) illustrating flood depth estimation using inundation extent. Figure adapted from Cham et al. (2015) [41]. DEM = digital elevation model.

3.2. Land Use/Land Cover Map

An updated land use/land cover (LULC) map produced by the Mekong River Commission (MRC) was used to determine damages according to different types of land cover. Land cover classifications were derived using imagery from Landsat 5 Thematic Mapper to produce a map with 30-m spatial resolution. Field surveys were conducted on 9357 points across the LMB to validate classifications [43]. In total, 19 unique land classifications were derived (Figure 3). The LULC map was resampled to match the DEM resolution using nearest neighbor method and was exported as an array. Arrays for land cover and inundation depths were imported in an R model to calculate damages on a per-pixel basis.

3.3. Infrastructure and Population Density

Population data produced by NASA's Socioeconomic Data and Applications Center (SEDAC) provide global, gridded estimates of population density at a resolution of 30 arc-second (~1 km) [44]. SEDAC also provides global datasets for roadways. Road data were clipped to the study area and classified as primary (highway), secondary (roadway), or residential/other according to the attribute 'fclass' designation (Figure 4) [45]. Road vectors were rasterized to match the resolution of the depth raster. Rasterized roads were exported as an array with indices corresponding to road type.

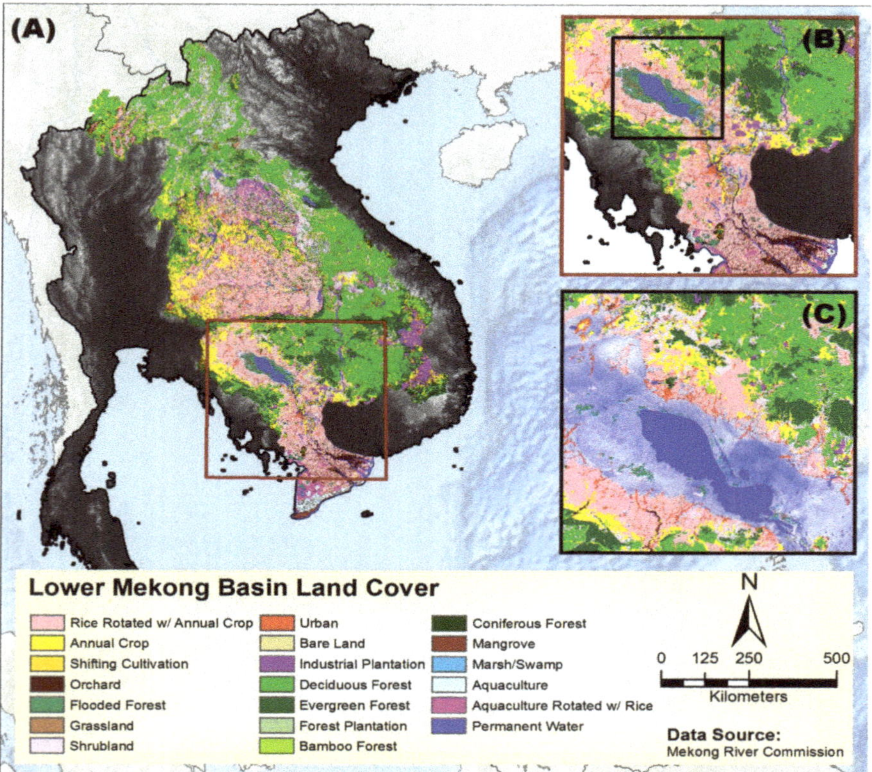

Figure 3. (**A**) Land use/land cover (LULC) map produced by Mekong River Commission (MRC, 2010). (**B**) Inset showing study extent and LULC details considered in this analysis. (**C**) Close view of the Tonle Sap Lake region, Cambodia.

Building infrastructure data were obtained from OpenStreetMap (OSM) [46,47]. Building locations and footprints were collected for the entirety of Thailand, Cambodia, Laos, and Vietnam and merged into a single dataset in QGIS. Building centroids were used to extract flood depths at point locations to estimate flood damages on a per-structure basis. Individual buildings were classified as either 'urban' or 'rural' structures according to a population density threshold of $1000/km^2$ (Figure 4) [48].

Leveraging open-source data presents a unique opportunity for understanding community-level impacts. However, since OSM data is user-generated, there are likely nontrivial data gaps and inconsistencies. Over 955,000 digitized structures across the Mekong region are included in this dataset, which would almost certainly underestimate the actual total number of residential and commercial buildings. For locations where OSM data does not exist but was classified as 'Urban' on the MRC LULC map, we supplement our understanding of infrastructure damages using the method described in Chen (2007). This approach calculates damages according to the area of urban land affected and uses a 40% correction factor to estimate the proportion of urban land occupied by infrastructure [49]. Damage estimates for both methods were combined to produce estimates for total building infrastructure damages.

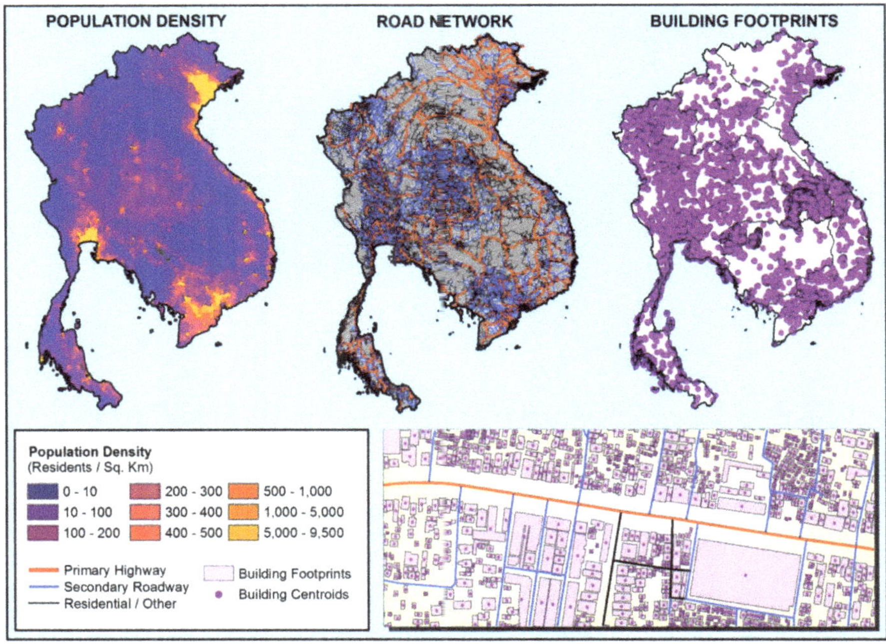

Figure 4. Population density (**left**), regional road networks (**center**), and building centroid and footprint data (**right, below**) considered in this study.

3.4. Damage Model

Flood damages are calculated as a function of three variables: damage factor category, a; maximum damage value, S; and the number of affected units, n

$$S = \sum_{i=1}^{n} a_i n_i S_i \tag{1}$$

   Damage factors for each category are determined by depth-damage functions (Figure 5). For each affected grid cell, the associated inundation depth is fed into the appropriate damage curve for the underlying land cover. The curves used in this study were derived primarily for use in the Huong River Basin, Vietnam, but we assume validity for other communities throughout the MRB [49]. For some of the land use classes considered in this study, no documented depth-damage relationship exists. In such cases, we adopt the closest analogue in order to approximate damage values (e.g., use 'forest' curve for 'orchard' class). This is a noted limitation of this approach and can also be seen as a motivation for the development of refined depth-damage curve datasets as well as more established land cover proxies.

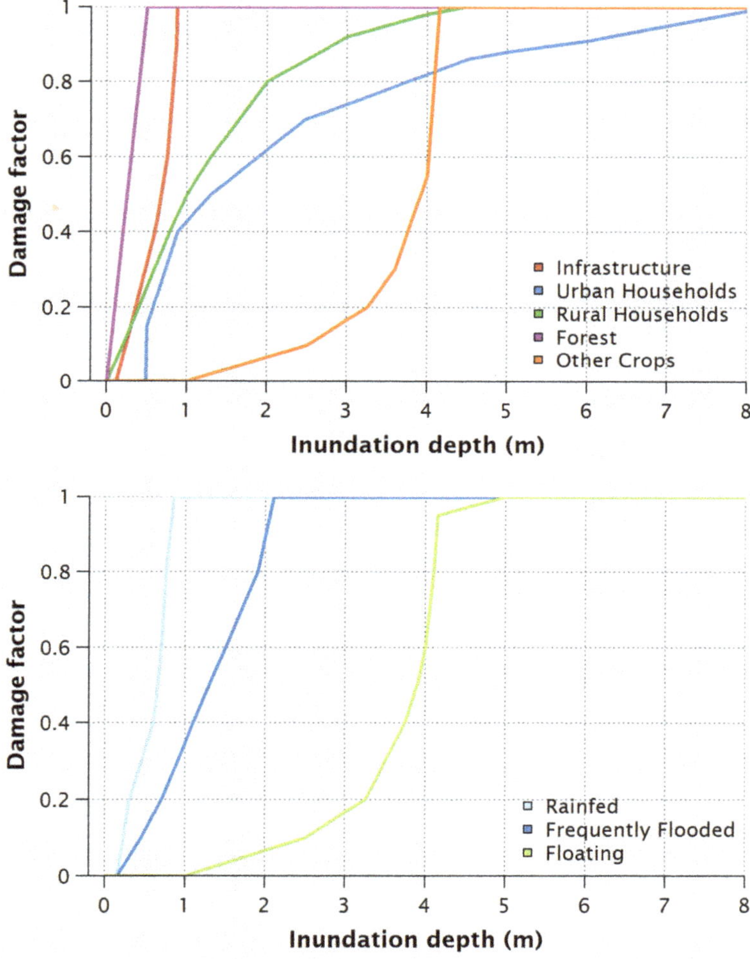

Figure 5. Damage factor curves for agriculture, forest, and infrastructure classes (**upper**) and rice varieties (**lower**) found in the Lower Mekong Basins (LMB). Curves digitized and adapted from Chen (2007) [49].

Maximum damage values, S_i, are also regionally-derived from studies performed in Vietnam and Thailand (Table 1) [16,17]. Maximum damage values indicate the total value assigned to a land cover type or infrastructure class per a unit area (e.g., crop destruction/m^2, roads impacted/m). Specific damage amounts are calculated based on either cost of replacement or cost of reconstruction [38,49]. For the purposes of this case study, we assume any crop that comes into contact with flood waters is considered 'totally destroyed.' However, the literature provides alternative damage values for partially destroyed crops. For a full table of the maximum damages values used in this study, see Table A1.

Table 1. Maximum damage values (S_i) used in this study.

Land Utility	USD/m^2	Source
Agriculture		
Rice, totally destroyed	0.078	
Crop, totally destroyed	0.109	
Other plants, totally destroyed	0.147	Leenders et al. (2009)
Rice, partially destroyed	0.027	
Crop, partially destroyed	0.030	
Other plants, partially destroyed	0.030	
Fishery		
Farm ponds and paddy fields	0.639	
Shrimp and shell fish	1.706	Leenders et al. (2009)
Freshwater fish	0.048	
Infrastructure		
Urban area	29	
Rural area	22	
Provincial road	80	
National road	400	Giang et al. (2009)
Railway	1000	
Other crops	0.02	
Forest	0.84	

4. Results

4.1. Land Cover Damages

The majority of inundation for the 2011 case study example extends from Tonle Sap Lake south toward the Mekong Delta (Figure 6). In total, approximately 23,000 km^2 was found to be inundated, with the majority (58%) classified as annual cropped rice. While rice paddies did comprise most of the inundated land area, they were only found to account for around 14% of the total damages (Table 2). The flooded forest belt surrounding the Tonle Sap Lake accounted for the majority of land cover damages (64%), due to its comparatively higher maximum damage value for Forests (0.84 vs. 0.078 USD/m^2), as well as the sensitivity of the 'Forest' damage curve to inundation depths between 0 and 0.5 m. The exact damage values for the forest class were derived from the Vietnamese Central Region Urban Environment Improvement Project (CRUEIP) as the unit cost to replace forest in the Thua Thien Hué Province [49,50]. As the expected damages for each land cover class are highly dependent on the associated maximum damage value, it is worth evaluating whether local values can be applied regionally. For a more detailed discussion of the limitations of this framework, see the Discussion and Conclusions section.

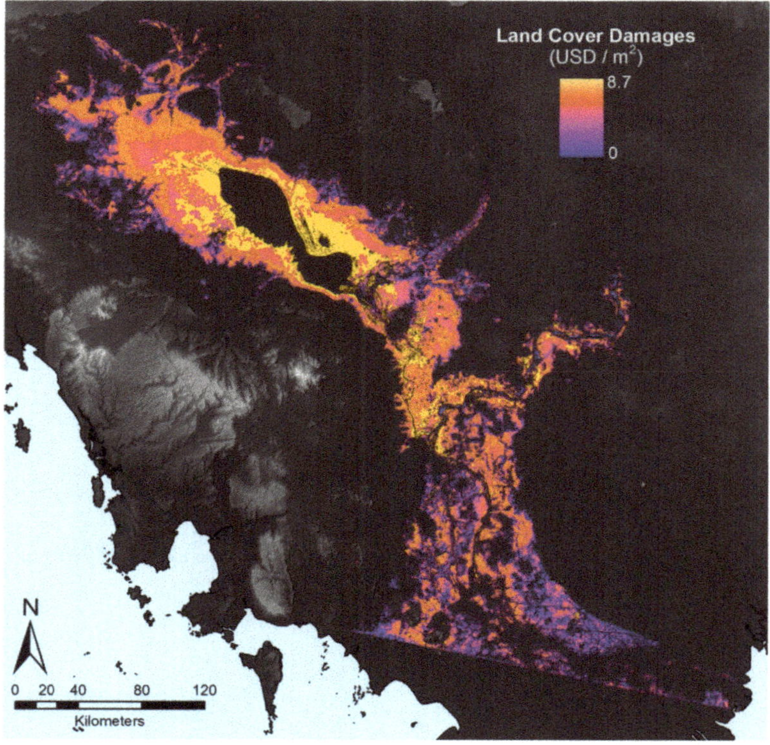

Figure 6. Results of damage assessment for land cover categories. Color gradient represents severity of damages in USD/m^2.

Table 2. Affected area and damage estimates for land utilities considered in this study.

Land Utility	Area (km^2)	Damages (USD)
Rice Rotated with Annual Crop	13,355.05	645,235,056
Annual Crop	1502.03	126,696,853
Shifting Cultivation	38.02	3,073,550
Orchard	332.35	6,572,509
Flooded Forest	3542.54	2,889,181,644
Grassland	1938.22	44,535,518
Shrub Land	1398.63	34,103,750
Urban	275.17	710,538,630
Bare Land	68.65	0
Industrial Plantation	1.42	24,608
Deciduous Forest	8.43	2,905,977
Evergreen Forest	2.28	1,530,465
Forest Plantation	-	-
Bamboo Forest	11.35	8,798,317
Coniferous Forest	-	-
Mangrove	1.71	842,254
Marsh/Swamp	482.85	12,703,670
Aquaculture	8.32	211,169
Aquaculture Rotated with Rice	26.39	27,770
Total	**22,993**	**4,486,981,740**

4.2. Infrastructure Damages

Over 275 km^2 of 'urban' land and 29,170 individual structures were exposed to inundation across the study area, according to the open source data. Total estimates of urban and residential damages amounted to $710 million. Nearly 5000 km of roads were flooded with nearly all being classified as either secondary roadways or residential streets. Roads were flooded at an average depth of 1.1 m, with an upper 95% quantile depth of 3.63 m.

4.3. Populations Affected

The ASAR-WSM flood extent encompassed an area of roughly 40,500 km^2. Based on population densities of the SEDAC dataset, it was estimated that approximately 4.1 million people resided within the inundated extent. Estimates from the United States Agency for International Development (USAID) place the total number of 'affected people' at over 4.73 million across Thailand, Cambodia, Laos, and Vietnam (USAID, 2011). Due to the satellite path, part of the inundated area along the Mekong Delta was not included in this analysis, which could potentially explain the discrepancy between these figures.

5. Discussion and Conclusions

The 2011 Southeast Asia flood provides a valuable case study for demonstrating the feasibility of near real-time damage assessments. Our results demonstrate that GIS-based approaches to such assessments can efficiently synthesize geospatial and economic data to produce damage estimates at time scales useful for first responders and decision makers. Furthermore, the impact assessment framework can be readily implemented at different spatial scales and locations, providing that associated depth-damage relationships and maximum damage values are known. While the method described here may have some advantages over traditional post-even evaluations, it is subject to uncertainties surrounding the estimated flood depths as well as the damage factors used.

5.1. Flood Depth Estimates

Flood depth estimates serve as the primary driving variable for the damage curves used in this study. While organizations like the MRC have an extensive network of hydrological monitoring stations (e.g., discharge, meteorological, or rain gauge stations), there is a limited distribution of real-time river gauges [51,52]. In the absence of widespread field-based observations of inundation depths from the 2011 flood, we compared the output of the TIN-derived depths with modeled estimates produced by the MRC. A simulation of a large flood event was generated using the MIKE11 hydrodynamic model, which produced estimates of inundation extent and depth at 100-m horizontal resolution. Geographic agreement between the ASAR-detected flood extent and the modeled extent was fair, with the modeled output failing to capture flooding north of the Tonle Sap Lake (Figure A1). Where inundation extents overlapped, the flood depths also showed good agreement (Figure A2). For this comparison, TIN-derived depths less than one meter were binned up to one meter, to match the MIKE11 output (the full distribution of the inundation depths produced by this study can be seen in Figure A3).

As previously mentioned, the MERIT DEM used to generate the flood depth estimates is also not without uncertainty. While this updated dataset achieves a nearly 20% improvement in land-area mapped with ±2 m or better vertical accuracy, higher-resolution elevation datasets would prove more useful for community-level flood assessment [42].

5.2. Damage Estimate Validation

Comparing modeled damages with government or agency estimates raises distinct methodological challenges. In the case of the 2011 floods, multiple storms occurred over the course of several anomalous months. It is, therefore, difficult to distinguish which damages were a direct result of specific flood events [30]. When damage figures are reported in post-event evaluations, statistics detailing affected population and infrastructure are reported by a wide array of sources (e.g., governments, agencies, and news outlets). Estimates can vary widely depending on the source and the timing [53]. Limited documentation can also make it unclear whether reported damages contain just direct tangible effects, or if estimates of indirect and intangible damages are included. Furthermore, methods by which agencies formulate estimates can often be ambiguous and can be based on little or no accurate information [49,50]. Therefore, we make no attempt to directly compare the estimates in this study to those produced by any government agencies. Instead, we emphasize that the lack of transparency surrounding many existing damage assessments highlights the benefits of using structured, standardized frameworks like HAZUS or the Standard Method.

While the inclusion of open-source data from OSM can potentially provide some community-level insight into flood impacts, the sparseness of user-generated data likely means that urban damage figures are underestimated. In particular, critical infrastructure like healthcare facilities, schools, transportation hubs, and energy infrastructure were all classified simply as 'urban.' The framework as designed, however, can readily be updated with newer data as they are generated, allowing for more granular valuations of high priority infrastructure.

As previously discussed, the lack of locally-specific depth-damage relationships can potentially obscure our findings. While several of the damage curves and maximum damage values used in this study are regionally-sourced, the diverse landscape in the LMB requires a more detailed understanding of how to value flood impacts. The high magnitude of the damages surrounding the Tonle Sap region, for example, illustrates the need for a more nuanced understanding of the unique depth-damage relationships for each specific land cover class. In this analysis, all forest classes were considered using the same damage curves, yet the specific morphology of the flooded forest ecosystem makes it highly adapted to seasonal inundation. It is, therefore, unlikely flood levels of under 1-m would result in the total losses assumed by the current 'Forest' damage curve.

Another potential limitation is the ethnography of the region. Mekong communities are well adapted to seasonal flooding and have been living in floodplains for thousands of years. Many residents have experienced periodic flooding in their lifetimes (e.g., in 2000), and are accustomed to living in flood conditions or relocating in times of flood [28]. Further, houses in the LMB can be semi-resilient to flooding by employing high stilts and sometimes being constructed as floating houses. These factors complicate the estimation of "affected" population, and on-the-ground efforts as well as higher-resolution remote sensing images should be employed where possible to fill these data gaps. One possibility is to employ a probabilistic approach in which affected populations are assessed in terms of possible ranges or likelihoods.

Ultimately, the damage values presented here require an explicit understanding of the limitations of the analysis. The previously discussed uncertainties in both model parameters and structure make it difficult to view the damages themselves as much more than rapid, initial estimates. However, these estimates can still provide valuable information by pinpointing areas of interest for more focused investigation. Several recent studies note that rapid assessments of economic losses can aid in the allocation of potentially scarce resources during the recovery and reconstruction phases of a flood event [54,55]. The case study application presented here illustrates how the damages could be assessed using historical imagery, but the same process could also be applied to flood forecast maps, further increasing its value for future risk planning.

6. Future Work

The aim of this analysis was to demonstrate the feasibility of a rapid damage assessment framework to assess and visualize flood impacts of the 2011 Southeast Asia floods. The analysis can be readily expanded to the entire Mekong region as it is integrated with the near real-time product, Project Mekong. The damage model used in this analysis has a number of simplifying assumptions that could merit further investigation. While inundation depth was the only driver of flood damages considered here, future analysis of time series data could provide insight into flood duration as a driver.

The updated LCLU maps provided by the LMB improve our understanding of damages to different land classes but further refinement is still possible, particularly with respect to crop production. The Mekong Delta is known as the "rice bowl" of Vietnam, and food security is an ongoing area of research in such a densely-populated part of the world [21,28]. Crop rotations and planting calendars play a large role in which varieties of rice are growing in any given month, so improved treatment of crop distributions would better constrain our damage estimates.

Many of the socioeconomic datasets used here are static and could benefit from improved spatial resolution. As new socioeconomic data (e.g., power lines, power plants, internet/cable lines, etc.) and satellite data (e.g., Sentinel 1B) become available, the system should be updated to include the most relevant and latest data.

Acknowledgments: The authors would like to thank Joseph Spruce and members of the Bolten research group for exploratory research and thoughtful discussion of this research. Support for this study provided by the NASA Applied Sciences Program with cooperation from the Asian Disaster Preparedness Center (ADPC). This work made use of the Open Science Data Cloud (OSDC) which is an Open Cloud Consortium (OCC)-sponsored project. OSDC is supported in part by grants from Gordon and Betty Moore Foundation and the National Science Foundation and major contributions from OCC members like the University of Chicago.

Author Contributions: P.C.O., A.A., and J.D.B. conceived and designed the research; A.A. designed the flood detection model while P.C.O. designed and integrated the impact assessment model; P.C.O., A.A., and J.D.B. analyzed the data; P.C.O. wrote the manuscript.

Conflicts of Interest: The authors declare no conflict of interest.

Appendix

Table A1. Land utilities with corresponding maximum damage values used in this study.

Land Utility	Maximum Damage Value (S_i)
Rice Rotated with Annual Crop	0.078
Annual Crop	0.109
Shifting Cultivation	0.109
Orchard	0.03
Flooded Forest	0.84
Grassland	0.03
Shrub Land	0.03
Urban	29
Bare Land	-
Industrial Plantation	0.3
Deciduous Forest	0.84
Evergreen Forest	0.84
Forest Plantation	0.84
Bamboo Forest	0.147
Coniferous Forest	0.84
Mangrove	0.639
Marsh/Swamp	0.03
Aquaculture	1.706
Aquaculture Rotated with Rice	0.639

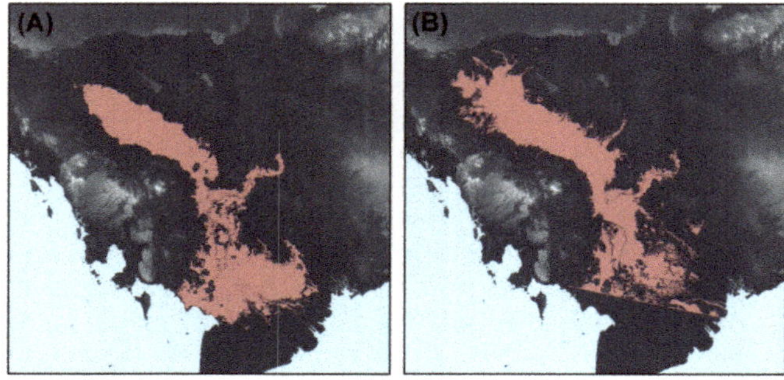

Figure A1. Comparison of inundation extent between the MIKE11 hydrodynamic simulation of a large flood event (**A**) and the Advanced Synthetic Aperture Radar (ASAR)-detected flooding (**B**).

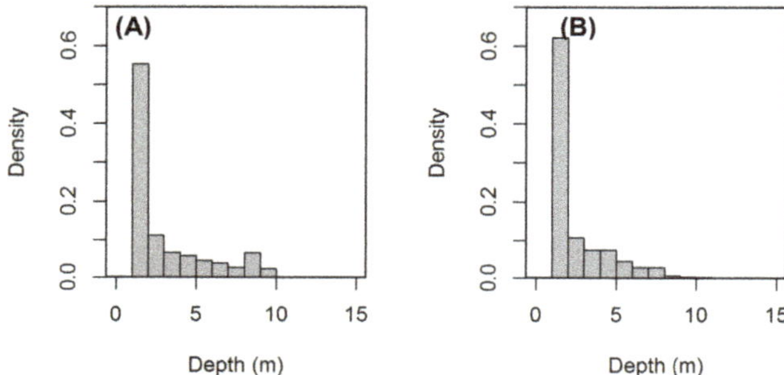

Figure A2. Histogram comparing MIKE11 hydrodynamic model simulation of a large-scale flood event (**A**) versus the TIN estimation from this study (**B**). For the purposes of this comparison, TIN estimations less than one meter were binned up to one meter to match the MIKE11 convention.

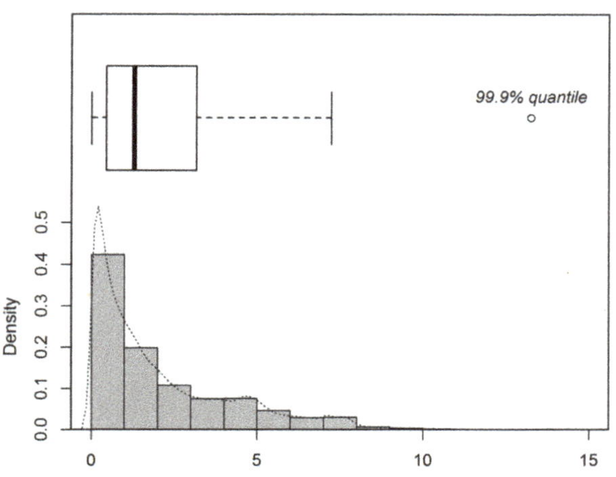

Figure A3. Full distribution of inundation depths produced by TIN estimates.

References

1. Guha-Sapir, D.; Vos, F.; Below, R.; Ponserre, S. *Annual Disaster Statistical Review 2011: The Numbers and Trends*; Centre for Research on the Epidemiology of Disasters (CRED): Brussels, Belgium, 2012.
2. Ahamed, A.; Bolten, J.D. A MODIS-based automated flood monitoring system for Southeast Asia. *Int. J. Appl. Earth Obs. Geoinf.* **2017**, *61*, 104–117. [CrossRef]
3. Jongman, B.; Kreibich, H.; Apel, H.; Barredo, J.I.; Bates, P.D.; Feyen, L.; Gericke, A.; Neal, J.; Aerts, J.C.J.H.; Ward, P.J. Comparative flood damage model assessment: Towards a European approach. *Nat. Hazards Earth Syst. Sci.* **2012**, *12*, 3733–3752. [CrossRef]
4. Mekong River Commission (MRC). *State of the Basin Report 2010*; Mekong River Commission: Vientiane, Laos, 2010.
5. Pech, S.; Sunada, K. Population growth and natural-resources pressures in the Mekong River Basin. *AMBIO J. Hum. Environ.* **2008**, *37*, 219–224. [CrossRef]
6. IPCC Managing the Risks of Extreme Events and Disasters to Advance Climate Change Adaptation. In *A Special Report of Working Groups I and II of the Intergovernmental Panel on Climate Change*; Field, C.B., Barros, V., Stocker, T.F., Dahe, Q., Dokken, D.J., Ebie, K.L., Mastrandrea, M.D., Mach, K.J., Plattner, G.-K., Allen, S.K., et al., Eds.; Cambridge University Press: Cambridge, UK, 2012; p. 582, ISBN 978-1-107-02506-6.
7. Nicholls, R.J.; Hanson, S.; Herweijer, C.; Patmore, N.; Hallegatte, S.; Corfee-Morlot, J.; Château, J.; Muir-Wood, R. *Ranking Port Cities with High Exposure and Vulnerability to Climate Extremes*; Organisation for Economic Co-operation and Development: Paris, France, 2008.
8. Perera, E.D.P.; Sayama, T.; Magome, J.; Hasegawa, A.; Iwami, Y. RCP8.5-Based Future Flood Hazard Analysis for the Lower Mekong River Basin. *Hydrology* **2017**, *4*, 55. [CrossRef]
9. Fayne, J.V.; Bolten, J.D.; Doyle, C.S.; Fuhrmann, S.; Rice, M.T.; Houser, P.R.; Lakshmi, V. Flood mapping in the lower Mekong River Basin using daily MODIS observations. *Int. J. Remote Sens.* **2017**, *38*, 1737–1757. [CrossRef]
10. Merz, B.; Kreibich, H.; Schwarze, R.; Thieken, A. Review article "Assessment of economic flood damage". *Nat. Hazards Earth Syst. Sci. Katlenburg-Lindau* **2010**, *10*, 1697. [CrossRef]
11. Oddo, P.C.; Lee, B.S.; Garner, G.G.; Srikrishnan, V.; Reed, P.M.; Forest, C.E.; Keller, K. Deep Uncertainties in Sea-Level Rise and Storm Surge Projections: Implications for Coastal Flood Risk Management. *Risk Anal.* **2017**. [CrossRef] [PubMed]
12. Gaume, E.; Borga, M. Post-flood field investigations in upland catchments after major flash floods: Proposal of a methodology and illustrations. *J. Flood Risk Manag.* **2008**, *1*, 175–189. [CrossRef]
13. World Bank. *Vietnam 2016: Rapid Flood Damage and Needs Assessment*; The World Bank Group: Washington, DC, USA, 2016; Available online: https://www.gfdrr.org/sites/default/files/publication/Vietnam%20Rapid%20Damage_FinalWebv3.pdf (accessed on 4 April 2018).
14. Kwak, Y.; Arifuzzanman, B.; Iwami, Y. Prompt Proxy Mapping of Flood Damaged Rice Fields Using MODIS-Derived Indices. *Remote Sens.* **2015**, *7*, 15969–15988. [CrossRef]
15. Poser, K.; Dransch, D. Volunteered geographic information for disaster management with application to rapid flood damage estimation. *Geomatica* **2010**, *64*, 89–98.
16. Leenders, J.K.; Wagemaker, J.; Roelevink, A.; Rientjes, T.H.M.; Parodi, G. Development of a damage and casualties tool for river floods in northern Thailand. In *Flood Risk Management: Research and Practice*; Taylor & Francis: London, UK, 2009; pp. 1707–1715, ISBN 978-0-415-48507-4.
17. Giang, N.T.; Chen, J.; Phuong, T.A. A method to construct flood damage map with an application to Huong River basin, in Central Vietnam. *J. Sci. Earth Environ. Sci.* **2009**, *25*, 10–19.
18. Wagemaker, J.; Leenders, J.; Huizinga, J. Economic valuation of flood damage for decision makers in the Netherlands and the Lower Mekong River Basin. In Proceedings of the 6th Annual Mekong Flood Forum, Phnom Penh, Cambodia, 27–28 May 2008; pp. 27–28.
19. Mekong River Commission (MRC). *Annual Mekong Flood Report 2011*; Mekong River Commission: Vientiane, Laos, 2011; p. 72.
20. Ahamed, A.; Bolten, J.; Doyle, C.; Fayne, J. Near Real-Time Flood Monitoring and Impact Assessment Systems. In *Remote Sensing of Hydrological Extremes*; Lakshmi, V., Ed.; Springer: Cham, Switzerland, 2017; pp. 105–118, ISBN 978-3-319-43743-9.

21. Chinh, D.T.; Bubeck, P.; Dung, N.V.; Kreibich, H. The 2011 flood event in the Mekong Delta: Preparedness, response, damage and recovery of private households and small businesses. *Disasters* **2016**, *40*, 753–778. [CrossRef] [PubMed]

22. Chinh, D.; Dung, N.; Gain, A.; Kreibich, H. Flood Loss Models and Risk Analysis for Private Households in Can Tho City, Vietnam. *Water* **2017**, *9*, 313. [CrossRef]

23. Kamoshita, A.; Ouk, M. Field level damage of deepwater rice by the 2011 Southeast Asian Flood in a flood plain of Tonle Sap Lake, Northwest Cambodia. *Paddy Water Environ.* **2015**, *13*, 455–463. [CrossRef]

24. Liu, S.; Lu, P.; Liu, D.; Jin, P.; Wang, W. Pinpointing the sources and measuring the lengths of the principal rivers of the world. *Int. J. Digit. Earth* **2009**, *2*, 80–87. [CrossRef]

25. Mekong River Commission (MRC). *Overview of the Hydrology of the Mekong Basin*; Mekong River Commission: Vientiane, Laos, 2005.

26. Kuenzer, C.; Guo, H.; Schlegel, I.; Tuan, V.Q.; Li, X.; Dech, S. Varying Scale and Capability of Envisat ASAR-WSM, TerraSAR-X Scansar and TerraSAR-X Stripmap Data to Assess Urban Flood Situations: A Case Study of the Mekong Delta in Can Tho Province. *Remote Sens.* **2013**, *5*, 5122–5142. [CrossRef]

27. Tang, J.C.; Vongvisessomjai, S.; Sahasakmontri, K. Estimation of flood damage cost for Bangkok. *Water Resour. Manag.* **1992**, *6*, 47–56. [CrossRef]

28. Dun, O. Migration and Displacement Triggered by Floods in the Mekong Delta. *Int. Migr.* **2011**, *49*, e200–e223. [CrossRef]

29. Guha-Sapir, D.; Below, R.; Hoyois, P. *EM-DAT: International Disaster Database*; Catholic University of Louvain: Brussels, Belgium, 2015.

30. Haraguchi, M.; Lall, U. Flood risks and impacts: A case study of Thailand's floods in 2011 and research questions for supply chain decision making. *Int. J. Disaster Risk Reduct.* **2015**, *14*, 256–272. [CrossRef]

31. Department of Disaster Prevention and Mitigation (DDPM). *National Disaster Risk Management Plan*; Department of Disaster Prevention and Mitigation, Ministry of Interior: Bankok, Thailand, 2015.

32. Sluimer, G.; Ogink, H.; Diermanse, F.; Keukelaar, F.; Jonkman, B.; Thanh, T.K.; Sopharith, T. Best Practice Guidelines for Flood Risk Assessment in the Lower Mekong River Basin. In Proceedings of the 7th Annual Mekong Flood Forum, Bankok, Thailand, 13–14 May 2009; p. 284.

33. Lekuthai, A.; Vongvisessomjai, S. Intangible flood damage quantification. *Water Resour. Manag.* **2001**, *15*, 343–362. [CrossRef]

34. Jonkman, S.N. Global perspectives on loss of human life caused by floods. *Nat. Hazards* **2005**, *34*, 151–175. [CrossRef]

35. Jonkman, S.N. Loss of Life Estimation in Flood Risk Assessment: Theory and Applications. Ph.D. Thesis, Delft University of Technology, Delft, The Netherlands, June 2007.

36. Bessette, D.L.; Mayer, L.A.; Cwik, B.; Vezér, M.; Keller, K.; Lempert, R.J.; Tuana, N. Building a Values-Informed Mental Model for New Orleans Climate Risk Management. *Risk Anal.* **2017**, *37*, 1993–2004. [CrossRef] [PubMed]

37. Federal Emergency Management Agency (FEMA). *HAZUS: Multi-Hazard Loss Estimation Model Methodology—Flood Model*; Federal Emergency Management Agency: Washington, DC, USA, 2003.

38. Kok, M.; Huizinga, H.J.; Vrouwenfelder, A.; Barendregt, A. *Standard Method 2004. Damage and Casualties Caused by Flooding*; Rijkswaterstaat: Delft, The Netherlands, 2004.

39. UN Operational Satellite Applications Programme (UNOSAT). *Flood Analysis for Cambodia 2014*; UNOSAT: Phnom Penh, Cambodia, 2014. Available online: http://floods.unosat.org/geoportal/catalog/search/resource/details.page?uuid=%7B85A44723-2428-4387-B509-70191C0F7B60%7D (accessed on 28 March 2018).

40. UNOSAT Flood Vectors-ASARWSM (27 September 2011) 2014. Available online: http://floods.unosat.org/geoportal/FP01/FL20111012KHM.gdb.zip (accessed on 28 March 2018).

41. Cham, T.C.; Mitani, Y.; Fujii, K.; Ikemi, H. Evaluation of flood volume and inundation depth by GIS midstream of Chao Phraya River Basin, Thailand. In *WIT Transactions on the Built Environment*; Brebbia, C.A., Ed.; WIT Press: Southampton, UK, 2016; Volume 2, pp. 1049–1960, ISBN 978-1-78466-157-1.

42. Yamazaki, D.; Ikeshima, D.; Tawatari, R.; Yamaguchi, T.; O'Loughlin, F.; Neal, J.C.; Sampson, C.C.; Kanae, S.; Bates, P.D. A high-accuracy map of global terrain elevations. *Geophys. Res. Lett.* **2017**, *44*, 5844–5853. [CrossRef]

43. Kityuttachai, K.; Heng, S.; Sou, V. *Land Cover Map of the Lower Mekong Basin*; Mekong River Commission: Phnom Penh, Cambodia, 2016; p. 82.

44. Center for International Earth Science Information Network (CIESIN). Gridded Population of the World, Version 4 (GPWv4): Population Density 2016. Available online: http://dx.doi.org/10.7927/H4NP22DQ (accessed on 7 November 2017).
45. CIESIN. ITOS Global Roads Open Access Data Set, Version 1 (gROADSv1) 2013. Available online: http://dx.doi.org/10.7927/H4VD6WCT (accessed on 7 November 2017).
46. OpenStreetMap Contributors. OpenStreetMap. Available online: https://www.openstreetmap.org/ (accessed on 7 November 2017).
47. Haklay, M.; Weber, P. Openstreetmap: User-generated street maps. *IEEE Pervasive Comput.* **2008**, *7*, 12–18. [CrossRef]
48. Organisation for Economic Co-operation and Development (OECD). *Definition of Functional Urban Areas (FUA) for the OECD Metropolitan Database*; Organisation for Economic Co-operation and Development: Paris, France, 2013.
49. Chen, J. *Flood Damage Map for the Huong River Basin*; University of Twente: Enschede, The Netherlands, 2007.
50. Central Region Urban Environmental Improvement Project (CRUEIP). *Supplementary Appendix [Resettlement plan]*; Report and Recommendation of the President to the Board of Directors on a Proposed Loan to the Socialist Republic of Viet Nam for the Central Region Urban Environmental Improvement Project; Asian Development Bank: Manila, Philippines, 2003.
51. Wang, W.; Lu, H.; Yang, D.; Sothea, K.; Jiao, Y.; Gao, B.; Peng, X.; Pang, Z.; Schumann, G.J.-P. Modelling Hydrologic Processes in the Mekong River Basin Using a Distributed Model Driven by Satellite Precipitation and Rain Gauge Observations. *PLoS ONE* **2016**, *11*, e0152229. [CrossRef] [PubMed]
52. MRC. Mekong River Real Time Water Level Monitoring. Available online: http://monitoring.mrcmekong.org/ (accessed on 12 February 2018).
53. Vinck, P. *World Disasters Report: Focus on Technology and the Future of Humanitarian Action*; International Federation of Red Cross and Red Crescent Societies: Geneva, Switzerland, 2013.
54. Raza, S.F.; Ahsan, M.S.; Ahmad, S.R. Rapid assessment of a flood-affected population through a spatial data model. *J. Flood Risk Manag.* **2017**, *10*, 219–225. [CrossRef]
55. Dutta, D.; Herath, S.; Musiake, K. A mathematical model for flood loss estimation. *J. Hydrol.* **2003**, *277*, 24–49. [CrossRef]

Article

Floods and Countermeasures Impact Assessment for the Metro Colombo Canal System, Sri Lanka

Mohamed Mashood Mohamed Moufar [1] and Edangodage Duminda Pradeep Perera [2,*

[1]   Sri Lanka Land Reclamation and Development Corporation, Colombo 10100, Sri Lanka; moufar5@yahoo.com
[2]   Institute for Water, Environment and Health, United Nations University, Hamilton, ON L8P 0A1, Canada
*   Correspondence: duminda.perera@unu.edu; Tel.: +1-905-667-5483

Received: 25 December 2017; Accepted: 20 January 2018; Published: 26 January 2018

Abstract: A 15th-century canal system in the Metro Colombo area of Sri Lanka was studied to identify its capacity in controlling floods. The canal system was modelled by MIKE FLOOD for 10, 25 and 50-year return periods of rainfalls to achieve respective floods. The impacts of the considered rainfalls were analyzed considering the flood levels, inundation distributions and affected people. Two simulation scenarios which were based on the river boundary conditions were carried out in the study and they were categorized as favourable and least favorable. It was identified that under the existing conditions, the canal system could handle only a 10-year rainfall flood event under the favourable condition. Therefore, the canal system's sustainability for future anticipated extreme events is suspicious. To mitigate such floods, four countermeasures were introduced and their impacts were analyzed. When the countermeasures were introduced one at a time, the flood water levels were lowered locally and they were not up to the flood safety levels of the surrounding area. When all four countermeasures were introduced together, the flood water levels were significantly lowered below the flood safety levels for a 50-year design rainfall under the favourable condition. The reduction of the inundated area was significant in the case of applying all four countermeasures together. In that case, a 46% inundation area reduction and a 49% reduction in the number of affected people were achieved.

Keywords: countermeasures; flood impacts; Metro Colombo canal system; Colombo city, Sri Lanka; urban floods

1. Introduction

The most common water-related hazards in urban areas, which causes substantial damages to human lives, health and infrastructures are urban floods. Urban flooding is a serious issue and a challenge to the development of cities and their residents. Common reasons for urban flooding are lack of drainage facilities, inadequate openings, inadequate water storage, intense rainfall, encroachment and blocking in the drainage system and backwater effect at outfalls. Urban floods have adverse impacts on the urban infrastructures such as transportation, electricity, water supply and drainage. Moreover, it adversely affects the lives of residents. These lead to extreme damages and disorder in the serviceability of urban infrastructures as well as transportation [1]. In recent years, more attention has been paid to consider the ability of urban drainage measures to reduce urban flood risks [2–4]. Effective adaptation measures can be made only after the nature of the impact is well understood. For urban planning and disaster preparedness, a quantitative assessment of the increase of flood hazards is important [5]. Urban cities are generally located in the flat areas in the middle or lower reaches of main rivers and much exposed to intensive floods. Urbanization in developing countries has been taken place in an unsustainable way, with a consequent degradation of the quality of life and the environment. It is essential to assess the impact of floods in urban areas and to prepare long-term plans for flood risk management through structural and non-structural countermeasures. This will assist policymakers to better understand the vulnerability of the urban regions under socio-economic and climatic changes because their contribution to the country's economy is enormous. In this study, we discuss a specific ancient canal system and related floods in the Metro Colombo area which is in the heart of the commercial capital of Sri Lanka.

The study was carried out based on hydrological modelling, flood impact assessment and estimating the efficiency of flood mitigation countermeasures. The main objective of this study was to assess the extent of inundations and flood levels for design rainfalls within the Metro Colombo area to identify the vulnerable areas and residents. Flood inundation maps were developed for the design rainfalls of having return periods of 10-year, 25-year and 50-year. It was essential to have flood inundation maps for the Metro Colombo canal basin to decide on new developments and flood mitigation/protection activities. The structural countermeasures to reduce the floods can be identified using the flood inundation maps. Also, the objective of this study extended to check the effectiveness of selected countermeasures which could be introduced to the canal system to mitigate floods. The feasibility of different countermeasures was assessed based on the reduction in inundation area and their impacts on the residents of seven Divisional Secretary's Divisions (DSDs).

To achieve these objectives, the entire canal system of the Metro Colombo area was modelled using MIKE11, MIKE21 and MIKE FLOOD models. The Digital Elevation Model (DEM) of the 30 m grid was introduced to MIKE21 as a two-dimensional domain. The DEM used in the study was a specially produced DEM, after a Lidar survey which was carried out covering the Metro Colombo area in 2011. Its horizontal accuracy was 1.0 m and the vertical accuracy was in the range of 0.15 m to 0.20 m. MIKE11 and MIKE21 models were coupled using MIKE FLOOD by defining the links between channels and floodplains. Sea levels at the outlets to the Ocean and the Kelani River discharges were used as boundary conditions for the model.

2. Study Area

The Metro Colombo canal system is in the Colombo city, in the wet zone of Sri Lanka. Colombo city is a coastal city located in the western part of Sri Lanka and it is the commercial capital of the country. The Metro Colombo canal system connects to the Kelani River which is the fourth largest river in Sri Lanka. The Kelani River drains into the Indian Ocean through Colombo and it is a highly influential river in Sri Lanka due to its economic and social importance. Figure 1 depicts the locations of the Colombo city, Kelani River basin and the Metro Colombo canal system. The Colombo district has a population of 2.3 million within 699 km^2 of area. It has a high population density of 3400 people/km^2 as of 2012 [6]. The Colombo city area is subjected to frequent floods during the south-west monsoon when it coincides with localized depressions. Figure 2 illustrates the annual rainfall in Colombo for 30 years from 1981 to 2010. An increment in the annual rainfall was identified for 30 years from 1981 to 2010. For each 10-year period, 1981–1990, 1991–2000 and 2001 to 2010 the yearly average rainfalls were 2113 mm, 2418 mm and 2377 mm respectively. Moreover, the maximum annual rainfalls were 2493 mm, 2888 mm and 3370 mm for each 10-year period. The increase in annual average rainfall and maximum annual rainfall indicate the vulnerability of the Colombo city to floods.

The history of the Metro Colombo canal system is mostly discussed from the time of King Veera Parakramabahu VII, who ruled Sri Lanka and established Kotte as the capital in the 15th-century. The Portuguese who ruled some parts of Sri Lanka from the 16th-century to the 17th-century improved the canal system for efficient transportation. The Dutch, who captured the island from the Portuguese, established Colombo as the capital and they enhanced the waterways systematically by adding structures and diversions in the 18th century. The British who ruled the island from the 18th-century to the 20th-century added several openings to the sea (for example, outfall at Wellawatta canal, Dehiwela canal) to control floods in the area [7].

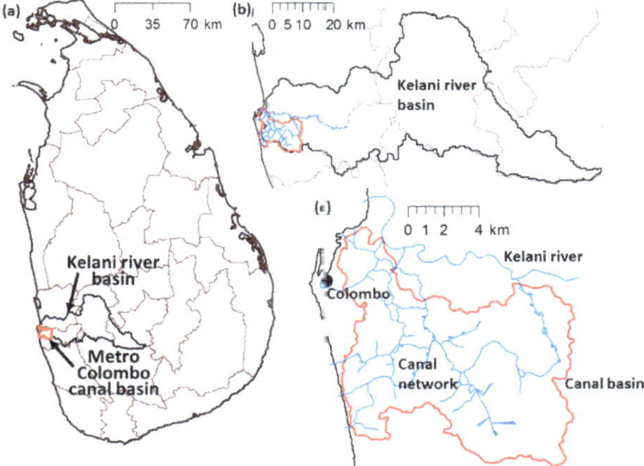

Figure 1. Study area: (**a**) Administrative districts of Sri Lanka; location of the Kelani River basin and the Metro Colombo canal basin; (**b**) Kelani River basin and the location of Metro Colombo canal basin and the canal system; (**c**) Metro Colombo canal system.

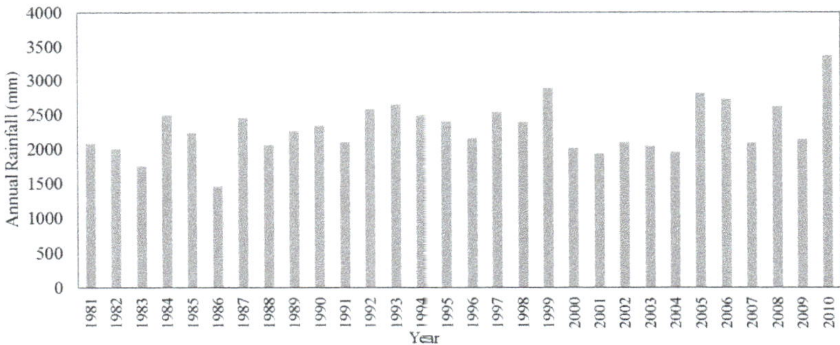

Figure 2. Annual average rainfall for Colombo for 30 years (1981–2010).

At present, the canal system acts as a passage to drain out flood water from the Metro Colombo area. The total length of it is 67 km. It has five main outfalls, among them three outlets—Dehiwela canal outlet, Wellawatta canal outlet and Mutwal tunnel outlet—drain the water into the Indian Ocean (Figure 3a). The other two, named as St-Sebastian north canal outlet and Madiwela east diversion canal outlet discharge flood water into the Kelani River. The water level measuring points are depicted in Figure 3b. The canal system has a zero-bed slope and the bed level of the canals is -1.0 m MSL (meters above Mean Sea Level). Figure 3c shows the schematic diagram of the canal system. There are three main marshy lands acting as retention areas for floods which are named as Kotte Ela marsh, Heen Ela marsh and Kolonnawa Ela marsh. Also, there are few detention ponds and lakes in the upper catchment area which can store flood water such as; Parliament Lake, Diyawanna Lake, Thalawathugoda Lake, Rampalawatta Lake, etc. The Metro Colombo basin, which is in the mostly urbanized area of Colombo, has an extent of 105 km^2. It is spread over seven DSDs named as Colombo, Thimbirigasyaya, Dehiwela, Kolonnawa, Sri Jayawardanapura Kotte, Maharagama and Kaduwela (Figure 3d).

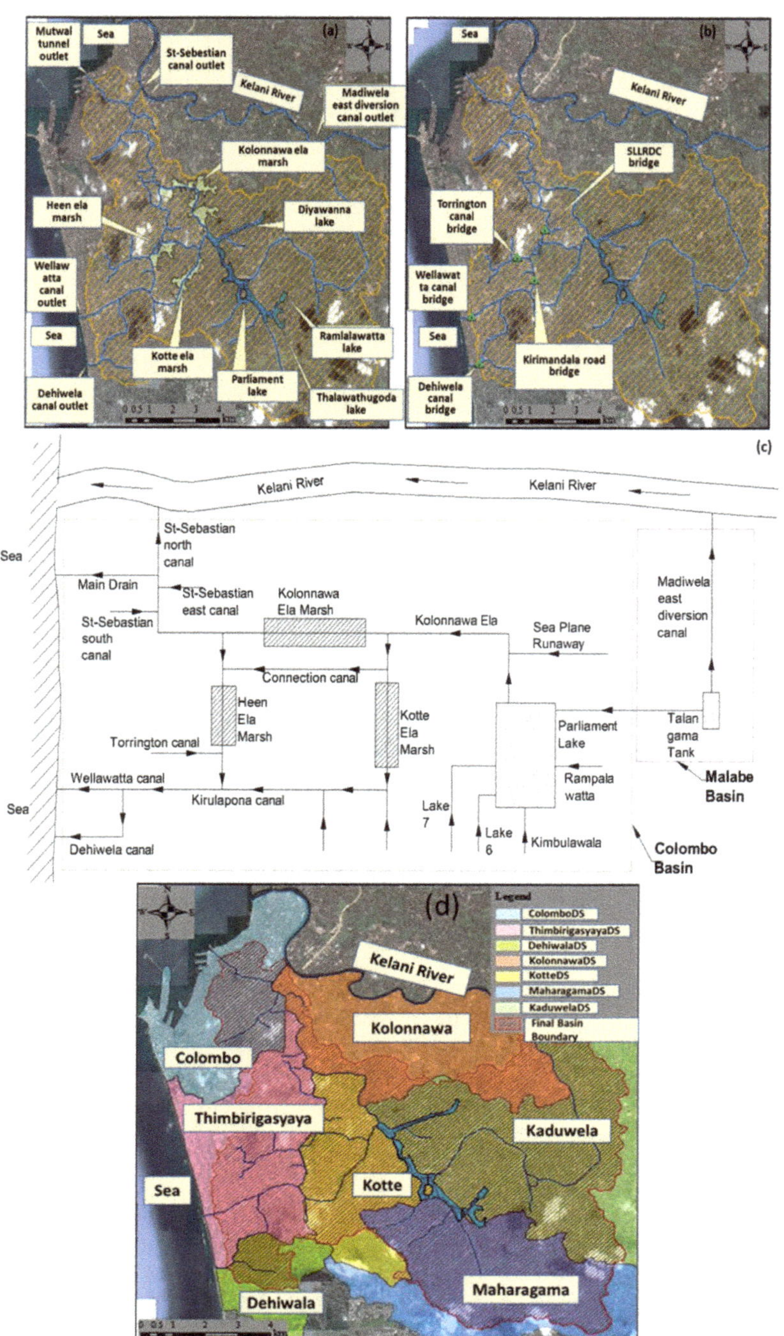

Figure 3. (**a**) Description of the Metro Colombo canal basin; (**b**) Water level gauging stations; (**c**) Schematic diagram of the Metro Colombo canal basin; (**d**) Divisional Secretary's Divisions (DSDs) in the basin.

The Metro Colombo area is subject to frequent floods during heavy rainfalls which causes economic losses and causalities. The situation is much worse when the water level of the Kelani River rises due to rainfall in the upper catchment. The geographical terrain of the Metro Colombo basin varies from 0 m MSL to 35 m MSL and a larger percentage of the basin sits below 3 m MSL. Therefore, it is difficult to have a proper gradient in the channel bed towards the outfalls to drain out the floods from the canal basin. Primary reasons for recent floods are increased surface runoff due to urbanization, diminishing of retention areas, growing trend for rainfall intensity and inadequate conveyance capacities of canals, structures and outfalls.

The Metro Colombo area has experienced severe floods in the recent past. On 14 May 2010 heavy monsoon rain caused floods in the area submerging roads and interrupting transportation. 94,000 people of about 15,000 families were affected due to the flood [8]. Again, between 10 November and 11 of 2010, about 440 mm of rainfall fell within 14 h. Almost all of the social and commercial activities were interrupted. The most severe consequences include the submergence of Sri Lankan Parliament premises, the closure of schools in the city, 123,000 people of 26,850 affected families, and the damage of approximately 257 households can be highlighted as the primary impacts [8]. Similar situations have occurred in 2014 and 2016 as well. Therefore, all the affected groups urged to study and understand the nature of floods in the Metro Colombo area.

3. Hydrological Modelling with MIKE FLOOD

3.1. Data Used for the Study

The catchment area of the Metro Colombo basin is 105 km^2 and there is only one rainfall gauging station located within the basin on the premises of the Metrological Department, Sri Lanka. The sea tide is measured at the gauging station located in the Colombo harbour. The flow discharge of Kelani River at Hanwella and the water levels at Nagalagam Street and Ambatale are measured by the Irrigation Department of Sri Lanka. Figure 4 shows the locations of gauging stations where rainfall, sea tide and river discharge are measured. The rainfall data of 15 min is available at the gauging station which is maintained by the Department of Metrology, Sri Lanka. Water level data from five gauging stations which had continuous data were used for model calibration and validation in this study. Figure 5 shows the locations of water level gauging stations. Design rainfalls for 5-year, 10-year, 25-year, 50-year and 100-year return periods were assessed using the records from the past 30 years (Table 1). In the analysis of design rainfalls, Generalized Extreme Value (GEV) distribution was employed by fitting the annual maximum of observed rainfall.

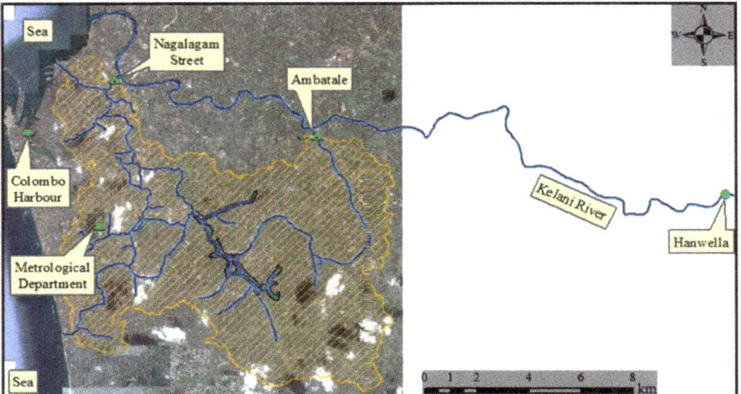

Figure 4. Locations of the rainfall gauge (Sri Lanka Metrological Department), sea tide gauge (Colombo harbour), water level and discharge gauging stations (Kelani River).

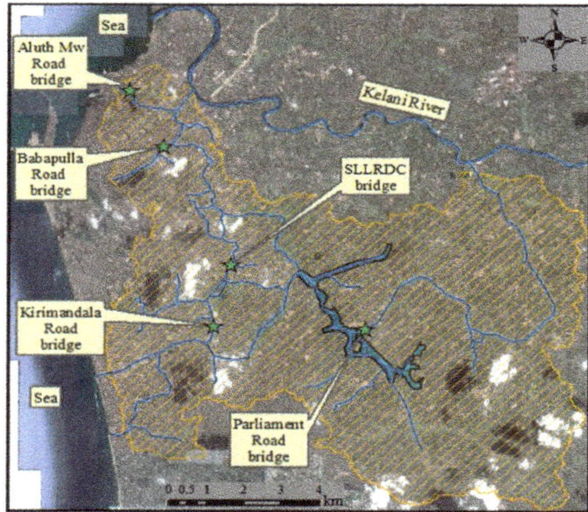

Figure 5. Water level gauging stations in the Metro Colombo canal basin.

Table 1. Daily design rainfalls in the study area.

Return Period (Year)	Design Rainfall (mm/Day)
5	197.3
10	256.4
25	363.4
50	476.5
100	626.1

3.2. MIKE Models

MIKE FLOOD is a hydrodynamic model, which consists of two sub-models: named as MIKE11and MIKE21 [9]. MIKE FLOOD is the coupling model that links MIKE11 and MIKE21. The MIKE FLOOD model combines best features and strengths of both MIKE11 and MIKE21 while minimizing the limitations of each model. The 1-D modelling tool, MIKE11 solves the one-dimensional form of the Saint-Venant equations along the channel [10]. The two-dimensional model MIKE21 is based on the two-dimensional solution of the Saint-Venant equations along the floodplain. The MIKE21 flow model is a fully dynamic general numerical modelling system, used for the simulation of hydraulic and environmental phenomenon in coastal areas, bays, lakes, etc. The mass conservation equation and momentum conservation equation describe the flow and water level variation in the MIKE21 model [11]. The MIKE11 model represents the conveyance along the channel while MIKE 21 exactly represents the 2D effects in out-of-bank flows in a floodplain. The model allows dynamic exchange internally in both directions between the 1D channel and 2D floodplain flow components [12]. Figure 6 illustrates the coupling of MIKE11 and MIKE21 models. The MIKE model applies the fully dynamic descriptions and solves vertically integrated equations such as the continuity equation (conservation of mass) and momentum equation (Saint-Venant equations) as shown in Equations (1) and (2) respectively [10].

$$\frac{\partial Q}{\partial x} + \frac{\partial A}{\partial t} = 0 \tag{1}$$

$$\frac{\partial Q}{\partial t} + \frac{\partial \left(\alpha \frac{Q^2}{A} \right)}{\partial x} + gA\,\frac{\partial h}{\partial x} + \frac{gQ|Q|}{C^2 AR} = 0 \tag{2}$$

where Q = Discharge in m^3/s, A = Wetted area in m^2, x = Longitudinal distance in m, t = Time in s, h = Water depth in m, g = Gravitational acceleration in m^2/s, R = Hydraulic radius in m, C = Chézy coefficient in m$^{\frac{1}{2}}$/s.

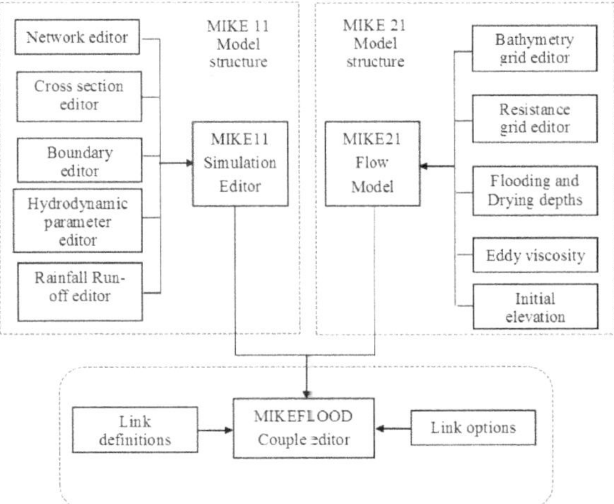

Figure 6. Schematic diagram of the model structure.

4. Methodology

The main goal of this study was to develop flood inundation maps showing flooded areas and flood depths for several rainfall events within the Metro Colombo area. Also, to check the effectiveness of selected countermeasures which were proposed to be introduced to the canal system. To achieve these goals several steps were adopted. Figure 7 systematically explains the steps followed in the study.

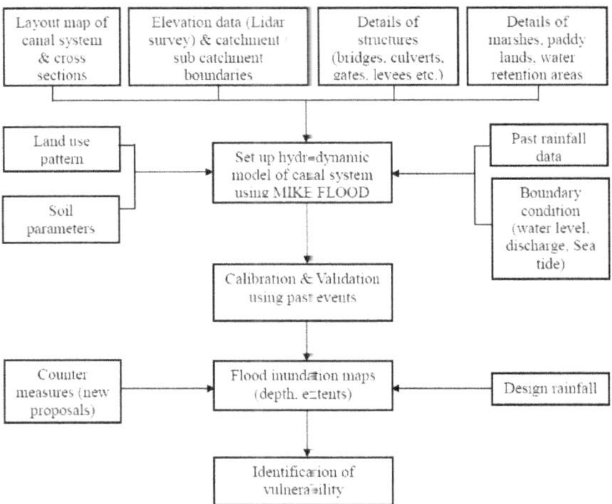

Figure 7. Methodology of the study.

5. Results and Discussion

5.1. Model Calibration for the Flood Event in November 2010

The set-up model for the Metro Colombo basin was calibrated for the flood event in November 2010 considering the measured water levels. The Metro Colombo area experienced heavy rainfall of 440 mm within 14 h between 10 November and 11 of 2010. The maximum hourly rainfall of the event was 116 mm. A grid file of Manning numbers, M (inversely proportional to the Manning coefficient, n) in the range of 15–25 $m^{1/3}/s$ were used for the resistance over the floodplain in the model. The model was calibrated considering the simulated water levels and observed water levels at five gauging stations along the canal system as shown in Figure 5.

Figure 8 shows the calibration results of simulated and observed water levels at the five gauging locations. Table 2 describes the estimated error indicators for simulated results at these locations. The error estimations indicate reasonably acceptable values for the simulation results. The model was validated for the flood event that occurred in May 2010 between the 14 and 18. A prolonged rainfall event was experienced in the Metro Colombo basin from the 13 of May 2010 up to the end of the month. The hourly maximum rainfall that occurred during this event was 76 mm. The validation results of simulated and observed water levels at the same five gauging locations are shown in Figure 9. The flood inundation maps for the selected events of calibration and validation are depicted in Figure 10. The non-availability of observed flood maps for the calibration and validation flood events was a limitation of the study since there was not a direct way to justify the inundation distributions. However, with the fact that the model showed accurate results for the water levels at five locations distributed across the basin, it was assumed that the calibrated MIKE FLOOD model produced agreeable inundations based on the given rainfalls and other boundary conditions.

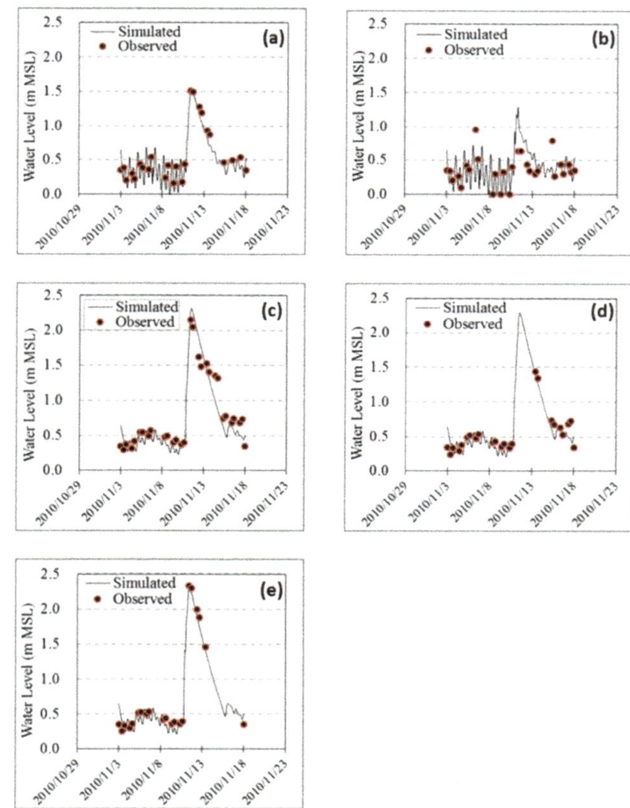

Figure 8. Water level comparison for the model calibration: (**a**) Galle Road Bridge in Wellawatta canal; (**b**) Galle Road Bridge in Dehiwela canal; (**c**) SLLRDC Bridge in Heen Ela canal; (**d**) Kirimandala Bridge in Heen Ela canal; (**e**) Railway Bridge in Torrington canal.

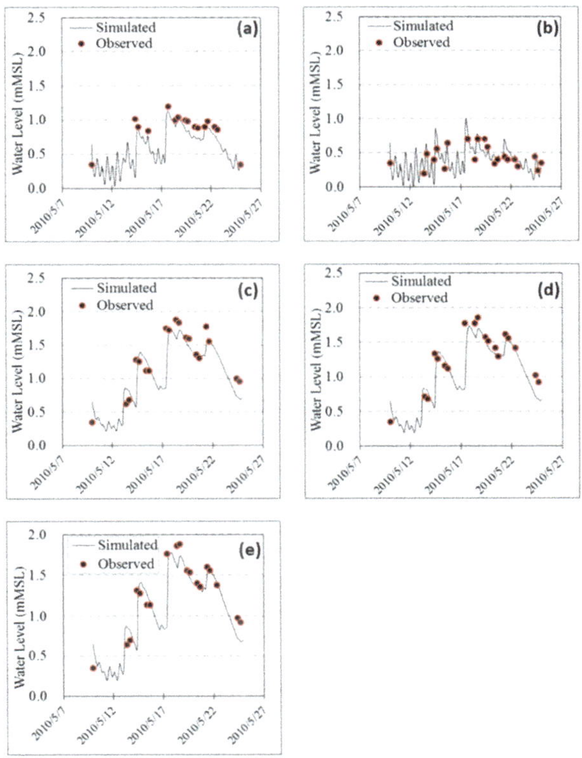

Figure 9. Water level comparison for the model validation: (**a**) Galle Road Bridge in Wellawatta canal; (**b**) Galle Road Bridge in Dehiwela canal; (**c**) SLLRDC Bridge in Heen Ela canal; (**d**) Kirimandala Bridge in Heen Ela canal; (**e**) Railway Bridge in Torrington canal.

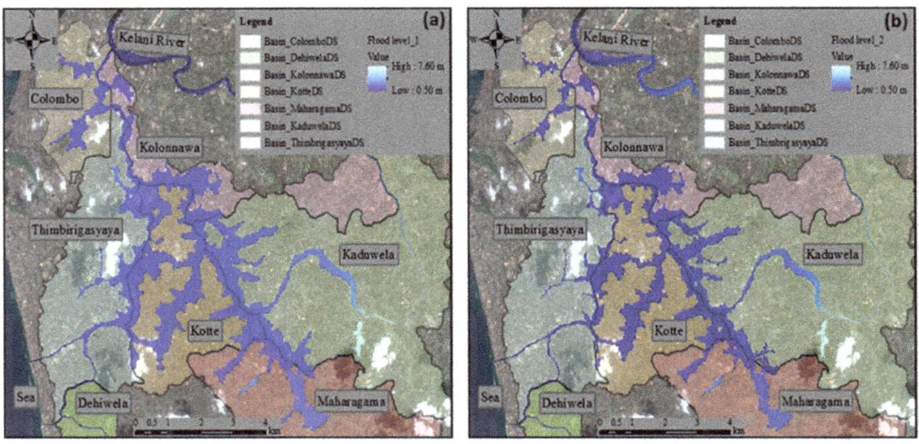

Figure 10. Inundation distributions for the model: (**a**) calibration; (**b**) validation events.

Table 2. Error estimators for the water level calibration.

Indices	Values from the Analysis				
	Galle Road Bridge in Wellawatta Canal	Galle Road Bridge in Dehiwela Canal	SLLRDC Bridge in Heen Ela Connection Canal	Kirimandala Road Bridge in Heen Ela Canal	Railway Bridge in Torrington Canal
Relative Root Mean Square Error (RRMSE)	0.175	0.640	0.211	0.189	0.088
N-S Coefficient Efficiency (EF)	0.938	−0.099	0.888	0.868	0.990
Coefficient of Determination (CD)	0.923	0.801	0.790	0.820	0.850

5.2. Simulation of Design Rainfall Floods

The calibrated model was employed to study the inundation distributions for different scenarios of rainfall and discharge boundary conditions. Frequency analysis was conducted, and design rainfalls were determined as shown in Table 1. In this study, we selected 10-, 25- and 50-year return period rainfall events for the scenario analysis. The rainfall patterns for each design rainfall were determined by the Alternating Block Method [13]. Two types of simulations were conducted with the design rainfall considering two discharge boundary conditions at the Hanwella gauging station of the Kelani River (Figure 4). Those two types were categorized as favourable and least favorable conditions. In both cases, considered rainfall was same, however the discharge boundary conditions at the Hanwella gauging station were different. In the favourable condition, a 30-year average river discharge (100 m^3/s) was used and for the least favorable case, a 30-year maximum discharge of 2000 m^3/s was used as the river boundary condition at Hanwella. The favourable condition represented the situation where the local rainfall in the Colombo city was dominant while the upstream of the Kelani basin was not experiencing considerable rain. The least favorable case represented the high rainfall in the upstream of Kelani basin as well as in the Metro Colombo area. The favourable condition gave a more relax boundary conditions at river outfalls of the canal system. There was a considerable backflow to the canal system from Kelani River in the least favorable condition due to high river discharge and therefore the outfalls to the river did not function in this situation. The model was simulated for several design scenarios considering the design rainfalls and boundary conditions as follows and the respective flood inundation maps were obtained.

1. Favourable/least favorable conditions with 10-year design rainfall.
2. Favourable/least favorable conditions with 25-year design rainfall.
3. Favourable/least favorable conditions with 50-year design rainfall.

Also, for each design scenario, the simulated water levels in the canal system at five important locations were obtained and a comparison was made between favourable and least favorable conditions. Those five locations are shown in Figure 5. Figure 11 compares the inundation distributions for the favourable and least favorable conditions for the 10, 25 and 50-year design rainfalls. In all cases, higher flood inundation areas were shown in the least favorable condition cases than the favourable conditions due to a high discharge boundary condition applied at the Hanwella gauging station. Table 3 compares the inundated areas for the considered cases. The Sri Jayawardenapura Kotte DSD showed the highest inundated area among all the cases according to the obtained results. The Sri Jayawardenapura Kotte city is the official capital of Sri Lanka and various governmental institutions are in this DSD including the Sri Lanka Parliament. Maharagama, Kaduwela, Kolonnawa and Thimbirigasyaya DSDs showed moderate inundations in all cases while the Colombo and Dehiwela DSDs showed a mild impact according to the inundation area comparisons. Inundation area increment ratios for the 25 and 50-year rainfalls for the favourable condition relative to the 10-year rainfall were 1.88 and 2.63. In the case of the least favorable case, the same increment ratios relative to the 10-year rainfall were 1.22 and 1.41.

Table 4 compares the water levels at the five locations under different cases. According to the results shown in Table 4, the Parliament Road Bridge showed the highest water levels. In the case of the 50-year rainfall, above 2 m water levels resulted in both favourable and least favorable cases. In the 50-year least favorable case scenario, the water levels in all five locations exceeded 2.5 m, increasing the risk of high floods in the canal basin.

Table 3. Comparison of inundation areas for design rainfalls under favourable and worse conditions.

Divisional Secretary's Division (DSD)	Inundation Area (km^2) 10 Years		Inundation Area (km^2) 25 Years		Inundation Area (km^2) 50 Years	
	Favourable Condition	Least Favorable Condition	Favourable Condition	Least Favorable Condition	Favourable Condition	Least Favorable Condition
Colombo	0.2	1.8	0.6	2.0	1.0	2.1
Thimbirigasyaya	0.4	1.7	1.0	2.1	1.7	2.5
Dehiwela	(0.072)	(0.073)	(0.073)	(0.078)	(0.078)	(0.082)
Kolonnawa	0.5	1.5	0.8	1.6	1.0	1.7
Sri Jayawardanapura Kotte	1.4	2.7	2.1	3.1	2.7	3.6
Maharagama	0.5	0.7	0.8	1.0	1.0	1.2
Kaduwela	0.5	1.2	1.3	1.9	1.8	2.4
Total	3.5	9.6	6.6	11.7	9.2	13.5

Table 4. Comparison of water levels for design rainfalls under favourable and least favorable conditions.

Locations Where Flood Levels Compared	Flood Level (m MSL) 10 Years		Flood Level (m MSL) 25 Years		Flood Level (m MSL) 50 Years	
	Favorable Condition (m)	Least Favorable Condition (m)	Favorable Condition (m)	Least Favorable Condition (m)	Favorable Condition (m)	Least Favorable Condition (m)
Parliament Road bridge	1.82	2.10	2.17	2.43	2.48	2.70
Kirimandala Road bridge	1.58	2.01	1.97	2.33	2.31	2.60
SLLRDC bridge	1.61	2.04	1.99	2.35	2.33	2.62
Babapulla Road bridge	1.50	2.26	1.87	2.42	2.17	2.61
Aluth Mw Road bridge	1.41	2.32	1.75	2.43	2.04	2.57

The affected residents in the seven DSDs under the considered cases are summarized in Table 5. Calculation of the affected people was carried out assuming an equal distribution of the population in each DSD area. The Dehiwela DSD showed a minimum impact due to its location in the Metro Colombo basin and it experienced minor floods according to the simulation results. All other DSDs showed high vulnerability for floods. The residents of the Sri Jayawardenapura Kotte, Kolonnawa and Colombo DSDs had high vulnerabilities to the floods in all cases compared to other DSDs. Under the favourable conditions, Sri Jayawardenapura Kotte DSD had the highest potential to be in the highly vulnerable area since the number of affected people in that DSD showed high in number. However, in the least favorable case, the Colombo DSD showed the highest number of affected people. As far as the total number of affected people was concerned, increments relative to the 10-year rainfall under favourable conditions, for 25-year and 50-year cases, were 1.65 and 2.35. Similar ratios for the least favorable cases for 25-year and 50-year were 1.17 and 1.32.

Table 5. Comparison of affected people for design rainfalls under favourable and worse conditions.

Divisional Secretary's Division (DSD)	Affected People 10 Years		Affected People 25 Years		Affected People 50 Years	
	Favourable Condition	Least Favorable Condition	Favourable Condition	Least Favorable Condition	Favourable Condition	Least Favorable Condition
Colombo	3749	29,084	10,141	31,337	15,559	33,499
Thimbirigasyaya	4219	16,907	10,626	21,409	16,957	25,625
Dehiwela	29	33	42	50	88	129
Kolonnawa	3640	10,602	5608	11,220	7241	11,881
Sri Jayawardanapura Kotte	8854	17,009	13,367	19,647	16,983	22,358
Maharagama	2380	3449	3942	4921	5304	6069
Kaduwela	1504	3319	3537	5311	5003	6552
Total	28,594	80,403	47,263	93,895	67,135	106,113

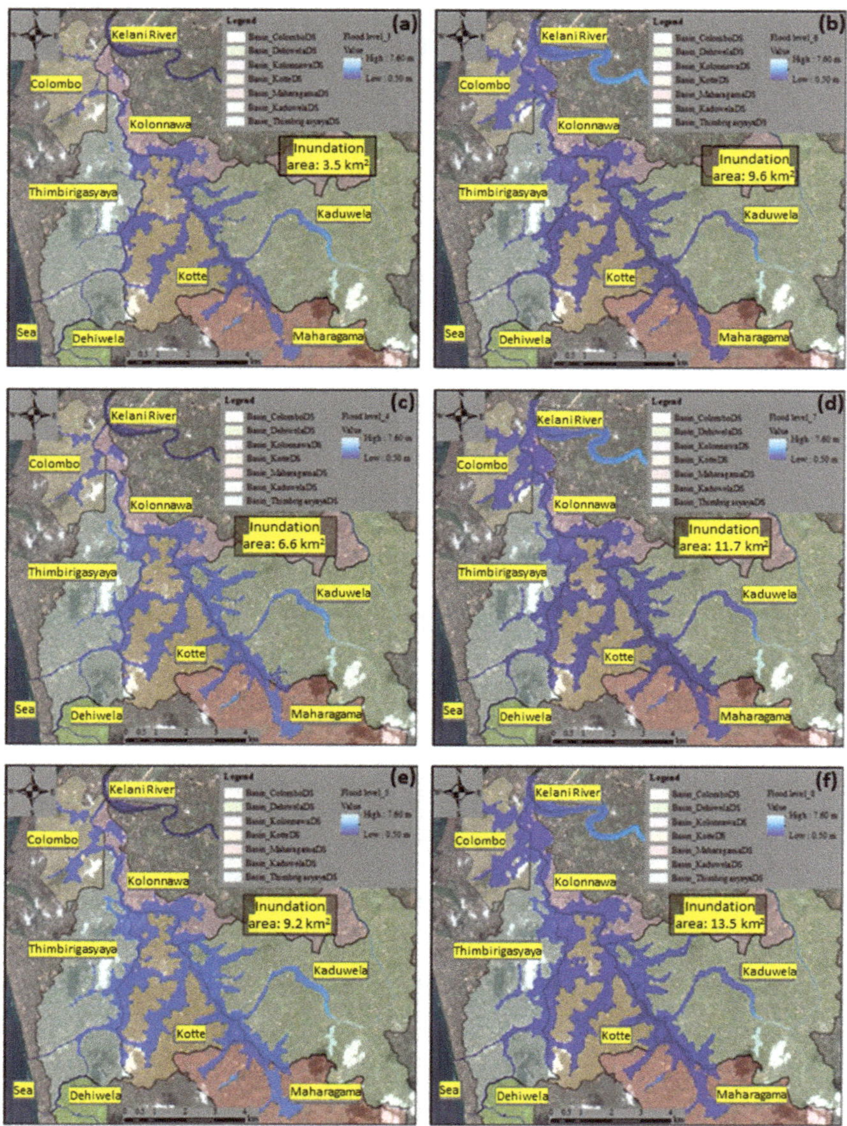

Figure 11. Inundation maps for different scenarios: (**a**) 10-year rainfall with favourable condition; (**b**) 10-year rainfall with least favorable condition; (**c**) 25-year rainfall with favourable condition; (**d**) 25-year rainfall with least favorable condition; (**e**) 50-year rainfall with favourable condition; (**f**) 50-year rainfall with least favorable condition.

5.3. Flood Inundation Analysis with Countermeasures

The floods can be minimized by introducing feasible countermeasures to the canal system such as diversions, lakes, tunnels, gates and pumps. Four feasible countermeasures as listed below were proposed to the Metro Colombo canal system in this study to understand their effectiveness in reducing the floods (Figure 12).

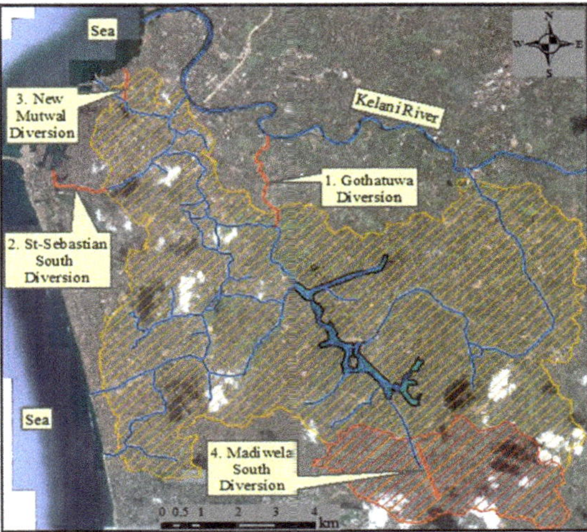

Figure 12. Locations of the proposed countermeasures.

1. The Gothatuwa diversion canal was introduced to the model as an open rectangular canal of 50 m wide to the centre part of the canal system. This diversion canal was connected to the main canal at Kolonnawa at its upstream and discharged water to the Kelani River. The bed level of the Gothatuwa diversion canal was set up to −1.0 m MSL as same as the entire canal system and the total length of the canal was 2.9 km approximately.
2. The St-Sebastian south diversion canal was introduced to the model as covered box drain of 8 m wide which connected the upstream end of the St-Sebastian South canal to the sea through Colombo harbour. The total length of the diversion canal was around 1.5 km and the bed level was set to −1.0 m MSL. The St-Sebastian South canal would flow through many commercial areas and residential areas as well.
3. The New Mutwal diversion was modelled with 3.0 m diameter underground tunnel and open rectangular canal of 6.0 m wide at both ends. The total length of this diversion canal was 900 m and carried water from the main canal to the sea. The length of the tunnel was 700 m and the invert levels were −1.5 m MSL and −2.0 m MSL at the upstream and downstream ends respectively. The invert level at the sea outfall of the downstream rectangular canal was −1.0 m MSL and therefore the water would flow from the underground tunnel through the rectangular canal with syphon action.
4. The Madiwela south diversion scheme: The sub-catchment in the upstream of Parliament Lake was divided from the Metro Colombo basin as shown in Figure 12 and the flood water of that particular sub-catchment was diverted to the southern part. The area of this sub-catchment was 14 km² and the model was reconstructed by removing this sub-catchment. The point where this sub-catchment was divided from the main basin was much closer to the Parliament and therefore by eliminating this sub-catchment, it was possible to reduce a certain amount of flow towards Parliament Lake.

The above countermeasures were introduced to the model separately and the model was simulated for the 50-year design rainfall under the favourable condition (when the Kelani River water level was low) since the favourable condition was the most probable scenario. The flood inundation area which was obtained using model results after the introduction of each countermeasure were compared with the model results of existing canal system. Then the model was set up by introducing all countermeasures together and the flooded area and flood levels were compared with the results of the existing condition. Table 6 summarizes the reduction of water levels at the five gauging stations with the introduction of countermeasures to the model while Table 7 explains the efficiency of the countermeasures in reducing the inundation area. The impact to the people in the basin with the introduction of countermeasures is shown in Table 8. The inundations maps for the peak flood distribution for each scenario are illustrated in Figure 13.

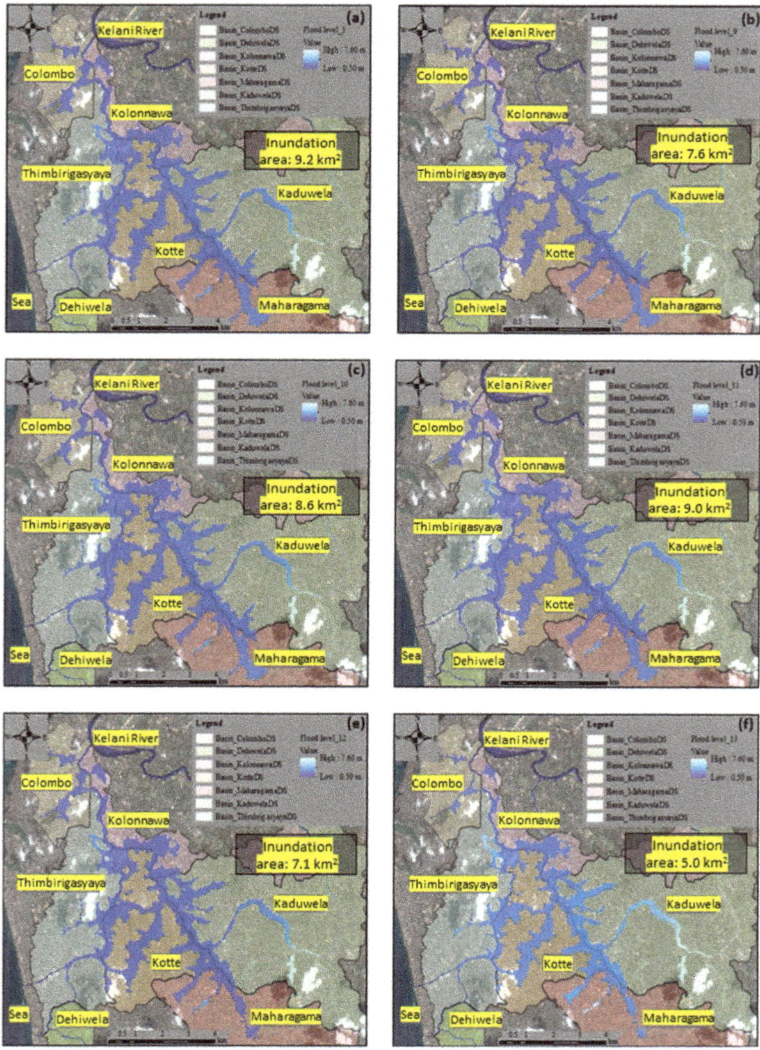

Figure 13. Comparison of the impact of the introduced countermeasures, inundation maps for (**a**) existing condition; (**b**) with Gothatuwa diversion canal; (**c**) with St-Sebastian south canal; (**d**) with Mutwal diversion; (**e**) with Madiwela south diversion; (**f**) with all four countermeasures.

Table 6. Water level comparison of different countermeasures.

Water Level Gauging Stations	50-Year Favorable, Existing Condition (m)	Gothatuwa Diversion (m)	St-Sebastian South Diversion (m)	New Mutwal Diversion (m)	Madiwela South Diversion (m)	All Countermeasures (m)
Parliament Road Bridge	2.48	2.37 (4.4%)	2.46 (0.8%)	2.47 (0.4%)	2.26 (8.9%)	2.07 (16.5%)
Kirimandala Road Bridge	2.31	2.06 (10.8%)	2.28 (1.3%)	2.30 (0.4%)	2.13 (7.8%)	1.85 (19.9%)
SLLRDC Bridge	2.33	2.07 (11.2%)	2.30 (1.3%)	2.32 (0.4%)	2.15 (7.7%)	1.86 (20.2%)
Babapulla Road Bridge	2.17	1.93 (11.1%)	1.98 (8.8%)	2.12 (2.3%)	2.05 (5.5%)	1.56 (28.1%)
Aluth Mw Road bridge	2.04	1.91 (6.4%)	1.93 (5.4%)	1.48 (27.4%)	1.97 (3.4%)	1.13 (44.6%)

Table 7. Inundation area comparison for different countermeasures.

Divisional Secretary's Division (DSD)	50-Year Favorable Existing Condition (km²)	Gothatuwa Diversion (km²)	St-Sebastian South Diversion (km²)	New Mutwal Diversion (km²)	Madiwela South Diversion (km²)	All Countermeasures (km²)
Colombo	1.0	0.7	0.7	0.9	0.8	0.3
Thimbirigasyaya	1.7	1.2	1.6	1.6	1.3	0.9
Dehiwela	0.0	0.0	0.0	0.0	0.0	0.0
Kolonnawa	1.0	0.8	0.9	1.0	0.9	0.6
Sri Jayawardanapura Kotte	2.7	2.2	2.6	2.7	2.4	1.8
Maharagama	1.0	1.0	1.0	1.0	0.5	0.4
Kaduwela	1.8	1.7	1.8	1.8	1.2	1.0
Total	9.2	7.6 (17%)	8.6 (7%)	9.0 (2%)	7.1 (23%)	5.0 (46%)

Table 8. Impact residents in DSDs comparison for different countermeasures.

Divisional Secretary's Division (DSD)	50-Year Favorable (km²)	Gothatuwa Diversion (km²)	St-Sebastian South Diversion (km²)	New Mutwal Diversion (km²)	Madiwela South Diversion (km²)	All Countermeasures (km²)
Colombo	15,559	11,193	12,001	14,114	13,121	4793
Thimbirigasyaya	16,957	12,115	15,107	16,496	13,589	9437
Dehiwela	88	88	88	88	88	88
Kolonnawa	7241	5792	5496	6832	6259	4363
Sri Jayawardanapura Kotte	16,983	13,991	15,657	16,831	14,959	11,057
Maharagama	5304	5174	5287	5297	2435	1900
Kaduwela	5003	4631	4923	4967	3348	2805
Total	67,136	52,984 (21%)	61,559 (8%)	64,625 (4%)	53,799 (20%)	34,443 (49%)

5.3.1. Introduction of the Gothatuwa Diversion Canal

The model was first simulated without the Gothatuwa diversion canal (with existing canal system) for 50-year design rainfall with the favourable condition and the inundated area was obtained as 9.2 km² (Figure 13). After introducing the Gothatuwa Diversion canal, the inundated area reduced to 7.6 km² and there was a 17% reduction in inundated area due to the introduction of this diversion canal. Inundation area and the affected people in each DSD for a 50-year rainfall without and with the Gothatuwa diversion are shown in Table 8. There was a reduction of 21.1% in affected people within the entire basin with the introduction of Gothatuwa diversion. There were about 17.6%, 28% and 28.5% reductions in the numbers of affected people in Sri Jayawardenapura Kotte, Colombo and Thimbirigasyaya DSDs which were highly urbanized DSDs in the Metro Colombo area. The flood water level in the canal system was lowered by a considerable level due to the introduction of the Gothatuwa diversion canal. Approximately 11% of the reduction in water levels were achieved at the Kirimandala Road Bridge, SLLRDC Bridge and the Babapulle Road Bridge water level gauging stations after introducing the Gothatuwa diversion canal. These three locations are much closer to several administrative, state and commercial institutes and flood safety level of these areas are around

2.0 m MSL. Introduction of the Gothatuwa diversion would reduce the water level very close to the flood safety level and thereby the area would be safer against floods.

5.3.2. Introduction of the St-Sebastian South Diversion Canal

The St-Sebastian South diversion canal was connected to the main canal system at the most western part of the Metro Colombo basin and therefore the effectiveness of this diversion canal was higher in the surrounding area of the diversion canal than other areas. Therefore, the total reduction in the inundated area, considering the entire basin, was only 8%. However, a 30% local reduction of the inundation area was achieved in the Colombo DSD with the introduction of the St-Sebastian South diversion canal. The decrease in the number of affected people in the Colombo DSD showed a high percentage (23%) while other DSDs did not show a significant decrease. As far as the water level was concerned, the Babapulla Road bridge showed an 8.8% reduction while others did not show significant reductions. The Babapulla Road Bridge is located very close to the connection point of the St-Sebastian diversion canal. The surrounding area of the Babapulla Road Bridge is highly developed with several commercial activities and there are some settlements of low-income people in the vicinity of this location. The flood safety level of commercial places is around 2.0 m MSL while the flood safety level of few settlements is 1.8 m MSL.

5.3.3. Introduction of the New Mutwal Diversion

According to the results shown in Tables 7 and 8, there was no significant reduction in inundation area and affected people locally or in the entire basin due to the introduction of the New Mutwal diversion. The New Mutwal diversion tunnel was connected to the Main drain canal of the Metro Colombo canal system and the connection point was very near to the Aluth Mawatha Road Bridge. Some low-level areas in the sides of the main drain canal are highly packed with houses and shanties and these settlements are more vulnerable to floods. The flood safety level of some of the houses and shanties is 1.5 m MSL. The flood level near the Aluth Mawatha Road Bridge was 2.04 m MSL for the design event of 50-year return period with the favourable condition and was reduced to 1.48 m MSL with the introduction of the New Mutwal Tunnel. The reduction in water level, therefore, was 27% in this area and it can be said that the introduction of the New Mutwal diversion would reduce the flood risk of this area enormously even though its influence on the other areas was less.

5.3.4. Implementation of the Madiwela South Diversion Scheme

The model was simulated by eliminating the sub-catchment area of 14 km^2 from the Metro Colombo basin in the upstream of Parliament Lake, where sub-catchment was entitled to the Madiwela south diversion scheme. By eliminating this sub-catchment, the flood inundation area was reduced to 7.1 km^2 from its original area of 9.2 km^2. The percentage of the reduction in flood inundation area was 23% with the implementation of the Madiwela south diversion scheme. There was a 20% reduction of affected people within the entire Colombo basin with the implementation of the Madiwela south diversion scheme. The flood water level in the Parliament Lake was reduced by around 8.9% while the reduction of water levels near the Kirimandala Road Bridge and SLLRDC Bridge were around 7.8%. Flood safety levels of Parliament and the area near the Kirimandala Road Bridge and SLLRDC Bridge is 2.0 m MSL. However, this countermeasure did not reduce the water level up to the flood safety level. The flood level could be further reduced up to the flood safety level by adding few more countermeasures with the implementation of the Madiwela south diversion scheme.

5.3.5. Introduction of All Four Countermeasures

The model was simulated by introducing all four countermeasures together (Gothatuwa diversion, St-Sebastian south diversion, New Mutwal diversion and Madiwela south diversion) for the design rainfall of 50-year under the favourable condition. The inundation area reduced to 5.0 km^2 from 9.2 km^2 (existing condition) and the percentage of reduction was around 46%. There was a reduction of 49% of affected people within the entire Colombo basin with the introduction of all four countermeasures. Also, it showed considerable reductions in inundation areas and affected people in each DSD with the introduction of all four countermeasures.

Table 6 explains the flood levels for the existing condition, with the introduction of all four countermeasures at selected locations and the percentage reduction in flood levels. The flood water level in Parliament Lake was reduced to 2.07 m MSL from 2.48 m MSL (16.5% reduction) with the introduction of all countermeasures. However, the flood safety level near the Parliament area is 2.0 m MSL and therefore, it is still at risk from flooding with a 50-year rainfall event. There was a 20% reduction in flood water levels near the Kirimandala Road Bridge (2.31 m MSL was reduced to 1.85 m MSL) and the SLLRDC Bridge (2.33 m MSL was reduced to 1.86 m MSL) with the introduction of all countermeasures. The surroundings of the Kirimandala Road Bridge and the SLLRDC Bridge are more commercialized areas and would be much safer for floods with a 50-year design rainfall since the flood levels were lower than the flood safety level of 2.0 m MSL. The area near to Babapulla Road Bridge is highly commercialized and compacted with some residential settlements. The flood water level near this area was reduced to 1.56 m MSL from 2.17 m MSL (28% reduction) with the introduction of all countermeasures. The flood safety level of some housing settlements in this area is 1.80 m MSL and therefore the area would be safer against the floods of a 50-year rainfall event with the favourable condition.

With the introduction of all countermeasures, the flood water level near the Aluth Mw Road Bridge was reduced to 1.13 m MSL from 2.04 m MSL (45% of the reduction in water level) for a 50-year rainfall event. The surroundings of the Aluth Mw Road bridge are a highly residential area with some settlements in low lying areas and therefore the flood safety level in this area is 1.50 m MSL. Since the flood level was much reduced than the flood safety level with the introduction of all countermeasures, the area would be safer against floods where rainfall would have a 50-year return period and favourable condition in the Kelani River.

5.4. Introduction of All Countermeasures for the 2010 November High Rainfall Event

The Metro Colombo basin experienced heavy rainfall in a short period of time between the 10 and 11 of November 2010. The total rainfall amounted 440 mm and poured into the basin within 14 h. The total rainfall amount can be compared with a design rainfall with a 35-year return period. The maximum hourly rainfall of the event was 116 mm. The model simulated the rainfall event of November 2010 for the existing conditions and the flood inundation area was estimated as 9.0 km^2. The flood inundation area reduced to 5.5 km^2 after introducing all four countermeasures to the canal system for the event of November 2010. Figure 14 compares the inundations for the November 2010 rainfall under existing conditions and with countermeasures.

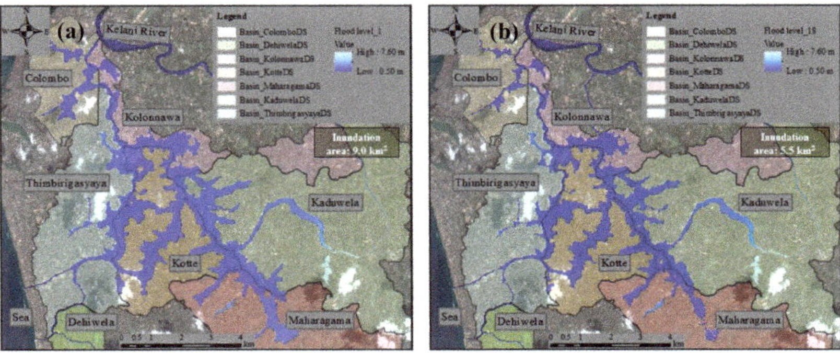

Figure 14. Inundation maps for 2010 rainfall: (**a**) with existing condition; (**b**) with all four countermeasures.

Table 9 shows the inundation area and the number of affected people in each DSD division under the existing conditions and with all four countermeasures for the past rainfall event of November 2010. There was a reduction of 41.2% of affected people within the entire Colombo basin with the introduction of all four countermeasures. Also, it shows considerable reductions in inundation areas and affected people in each DSDs with the introduction of all four countermeasures. With the introduction of all four countermeasures to the canal system, the flood level at selected locations were significantly lowered as shown in Table 10. Therefore, it can be expressed that, if all four countermeasures had been implemented before November 2010, a reasonable part of the basin would not have flooded to the extend that it did.

Table 9. Inundation area and affected people in DSDs for 2010 November rainfall.

Divisional Secretary's Division (DSD)	Inundation Area (km^2)		Number of Affected People		Percentage Reduction (%)
	Existing Condition	With All Four Countermeasures	Existing Condition	With All Four Countermeasures	
Colombo	1.0	0.4	15,383	6754	56.1
Thimbirigasyaya	1.6	1.0	16,153	9771	39.5
Dehiwela	0.0	0.0	44	42	4.5
Kolonnawa	1.0	0.7	7084	4854	31.5
Sri Jayawardanapura Kotte	2.7	1.9	16,804	12,027	28.4
Maharagama	1.0	0.4	5256	2155	59.0
Kaduwela	1.8	1.1	4861	2989	38.5
Total	9.0	5.5	65,586	38,593	41.2

Table 10. Comparison of flood levels for 2010 November rainfall.

Locations Where Flood Levels Compared	Flood Level (m MSL)		Percentage Reduction (%)
	Existing Condition	With All Four Countermeasures	
Parliament Road bridge	2.50	2.17	13.2
Kirimandala Road bridge	2.30	1.92	16.2
SLLRDC bridge	2.31	1.93	16.7
Babapulla Road bridge	2.15	1.65	23.4
Aluth Mw Road bridge	2.06	1.22	40.7

6. Conclusions

The Metro Colombo canal system was modelled using the one-dimensional MIKE 11 and the terrain of the entire basin was modelled using MIKE 21, the two-dimensional hydrodynamic model. Then both the 1-D and 2-D models were coupled using MIKE FLOOD by defining the links between them. The model was calibrated and validated against observed water levels at five gauging stations located in the Metro Colombo canal basin for the rainfall events in 2010. The coupled model was simulated for several scenarios considering rainfalls of 10-year, 25-year and 50-year return periods. Sea level and river discharge boundary conditions were employed for the aforementioned scenarios.

The safety flood level in the basin is 2.0 m MSL and was satisfied only with the 10-year rainfall event under the favourable condition according to the obtained results. In all the least favorable-case simulations, the water levels at the observation locations exceeded the 2.0 m MSL. The most vulnerable DSDs in the basin were Sri Jayawardenapura Kotte, Colombo and Thimbirigasyaya since the number of affected people in those DSDs were significantly high according to the obtained results. Sri Jayawardenapura Kotte could be ranked as the most vulnerable DSD in the context of the inundated area. Dehiwela DSD was the least affected DSD in the basin according to the obtained results. In all the considered cases, least favorable cases showed high impacts due to extremeness of the considered boundary condition. Since the water level of Kelani River was high in the least favorable condition, the river outfalls at Nagalagam Street and Ambatale did not discharge water at their full capacity due to the backwater effect. Therefore, the extent of the inundation areas were greater in least favorable condition than in the favourable condition in all the scenarios. Flood inundation maps expressed that more areas were inundated closer to the river outfall in least favorable condition than the favourable condition. Also, the flood water levels were higher in least favorable condition than the favourable condition in all five locations considered, and it was higher especially at the locations closer to river outfall. The flood water levels for several scenarios at different locations were compared and it was expressed that the existing canal system was only capable of carrying the flood water of a 10-year rainfall event (favourable condition) without any inundation in the area.

To investigate the feasible countermeasures to mitigate the floods in the Metro Colombo basin, four measures were tested in the study. The flood water level at different selected locations was reduced when the countermeasures were introduced to the canal system. Each countermeasure gave a clear reduction in water level at the location where that particular countermeasure had been introduced and a limited reduction in other places. There was an estimated 11% reduction in flood water level at the Kirimandala Road Bridge, the SLLRDC Bridge and the Babapulla Road Bridge with the introduction of the Gothatuwa diversion, however in the other two locations (such as Parliament Road Bridge and Aluth Mawatha Road Bridge) the reduction was relatively small. The introduction of the St-Sebastian South diversion was a better proposal to lower the flood water level in the surrounding area of the Babapulla Road Bridge while the introduction of the New Mutwal diversion reduces the flood water level in the surroundings of the Aluth Mawatha Road Bridge. Implementation of the Madiwela South diversion scheme was a feasible proposal to lower the flood water level in the Parliament area, surroundings of the Kirimandala Road Bridge and the SLLRDC Bridge and there was a reduction in floodwater levels of around 9% due to this countermeasure. However, countermeasures mentioned above were not strong enough to reduce the water level lower than their respective flood safety levels at five different locations when they were introduced individually. When all four countermeasures were introduced together in the model, the outcomes were preferable, and the flood water levels were lower than their respective flood safety levels. At the Aluth Mw Bridge, the reduction of water level was about 44.6% while the other four locations also reductions were greater than 16%.

The reduction of inundated area was significant in the case of applying all four countermeasures together. In that case, a 46% area reduction was achieved. In the case of individual effectiveness, the Gothatuwa diversion showed a 17% reduction of inundation and the Madiwela south diversion showed a 23% reduction while the other two countermeasures were able to achieve less than a 10% area reduction. When the countermeasures were applied for the 2010 flood event, there was a 5.5% reduction of inundated area and a 41.2% reduction of affected people was achieved.

In the present simulations, we used the available 30 m DEM and tested a limited number of scenarios. Since the Metro Colombo basin is an urbanised area, with a fine resolution DEM with detailed land use data and hydrological data, we would have achieved more accurate results than the obtained results. Applying finer DEM, several patterns of rainfall and boundary conditions at the Kelani River and seaside, the uncertainties would have been minimized. Another limitation of the study was the lack of opportunity to verify the inundation areas with the simulated inundations. In future, flood distribution observations should be conducted and utilised in these kind of studies to validate flood models. However, despite the uncertainties, we were able to produce essential information for the Metro Colombo canal basin to identify the vulnerable areas and communities. The results of the study will provide a baseline for future studies.

Acknowledgments: The authors acknowledge the academic and material support given by the International Centre for Water Hazards and Risk Management (ICHARM) at Public Works Research Institute (PWRI) of Japan, National Graduate Institute for Policy Studies (GRIPS) of Japan and Japan International Cooperation Agency (JICA) who provided the financial support for the first author to study for his postgraduate degree in Japan (2013–2014). We acknowledge the kindness of DHI for providing technical and software assistance for us to the completion of the study. Also, we are pleased to thank DHI-Japan for the valuable advice given to us to achieve our study goals. The authors are very grateful to the anonymous reviewers for their valuable comments and suggestions to improve the manuscript to the present level.

Author Contributions: E.D.P.P. and M.M.M.M. conceived and designed the study; M.M.M.M. performed the hydrological simulations; E.D.P.P. and M.M.M.M. analyzed the data and simulation results; E.D.P.P. wrote the paper.

Conflicts of Interest: The authors declare no conflict of interest.

References

1. Birgani, Y.T.; Yazdandoost, F. A framework for evaluating the persistence of urban drainage risk management systems. *J. Hydro-Environ. Res.* **2014**, *8*, 330–342. [CrossRef]
2. Ashley, R.M.; Balmforth, D.J.; Saul, A.J.; Blanskby, J.D. Flooding in the future e predicting climate change, risks and responses in urban areas. *Water Sci. Technol.* **2005**, *52*, 265–273. [PubMed]
3. Arnbjerg-Nielsen, K.; Fleischer, H.S. Feasible adaptation strategies for increased risk of flooding in cities due to climate change. *Water Sci. Technol.* **2009**, *60*, 273–281. [CrossRef] [PubMed]
4. Peck, A.M.; Bowering, E.A.; Simonovic, S.P. A flood risk assessment to municipal infrastructure due to changing climate part II: A case study. *Urban Water J.* **2014**, *11*, 519–531. [CrossRef]
5. Huong, H.T.L.; Pathirana, A. Urbanization and climate change impacts on future urban flooding in Can Tho city, Vietnam. *Hydrol. Earth Syst. Sci.* **2013**, *17*, 379–394. [CrossRef]
6. Ministry of Finance and Planning. *Population Atlas of Sri Lanka—2012*; Department of Census and Statistics, Ministry of Finance and Planning: Colombo, Sri Lanka, 2012.
7. Lanka Library Sri Lanka, Dutch Waterway in Sri Lanka. Available online: http://www.lankalibrary.com/geo/dutch/dutch4.htm (accessed on 19 January 2014).
8. Disaster Management Centre, Ministry of Disaster Management—Sri Lanka. 2010. Available online: http://www.dmc.gov.lk/index_english.htm (accessed on 12 February 2014).
9. DHI Water & Environment. *MIKE FLOOD, Modeling of River Flooding—Step by Step Training Guide*; DHI Water & Environment: Hørsholm, Denmark, 2011.
10. DHI Water & Environment. *MIKE 11—A Modelling System for Rivers and Channels, Reference Manual*; DHI Water & Environment: Hørsholm, Denmark, 2009.
11. DHI Water & Environment. *MIKE 21 Flow Model—Hydrodynamic Model User Guide*; DHI Water & Environment: Hørsholm, Denmark, 2011.
12. Tuteja, N.K.; Shaikh, M. Hydraulic modelling of the spatio-temporal flood inundation patterns of the Koondrook Perricoota forest wetlands—The living Murray. In Proceedings of the 18th World IMACS Congress and MODSIM09 International Congress on Modelling and Simulation, Cairns, Australia, 13–17 July 2009; Available online: http://www.mssanz.org.au/modsim09/J1/tuteja.pdf (accessed on 20 January 2014).
13. Chow, V.T.; Maidment, D.R.; Mays, L.W. *Applied Hydrology*; McGraw-Hill: New York, NY, USA, 1988.

 *hydrology*

Article

RCP8.5-Based Future Flood Hazard Analysis for the Lower Mekong River Basin

Edangodage Duminda Pradeep Perera [1,*], Takahiro Sayama [2], Jun Magome [3], Akira Hasegawa [4] and Yoichi Iwami [5]

[1]   United Nations University–Institute for Water, Environment and Health, Hamilton, ON L8P 0A1, Canada
[2]   Disaster Prevention Research Institute, Kyoto University, Kyoto 611-0011, Japan; sayama.takahiro.3u@kyoto-u.ac.jp
[3]   International Research Centre for River Basin Environment, University of Yamanashi, Kofu 400-8511, Japan; magome@yamanashi.ac.jp
[4]   International Centre for Water Hazard and Risk Management, Public Works Research Institute, Tsukuba, Ibaraki 302-8516, Japan; hase55@pwri.go.jp
[5]   Nagasaki Prefectural Civil Engineering Department, Nagasaki 850-8570, Japan; youichi-iwami@pref.nagasaki.lg.jp
*   Correspondence: duminda.perera@unu.edu; Tel.: −1-905-667-5483

Received: 13 September 2017; Accepted: 21 November 2017; Published: 23 November 2017

Abstract: Climatic variations caused by the excessive emission of greenhouse gases are likely to change the patterns of precipitation, runoff processes, and water storage of river basins. Various studies have been conducted based on precipitation outputs of the global scale climatic models under different emission scenarios. However, there is a limitation in regional- and local-scale hydrological analysis on extreme floods with the combined application of high-resolution atmospheric general circulation models' (AGCM) outputs and physically-based hydrological models (PBHM). This study has taken an effort to overcome that limitation in hydrological analysis. The present and future precipitation, river runoff, and inundation distributions for the Lower Mekong Basin (LMB) were analyzed to understand hydrological changes in the LMB under the RCP8.5 scenario. The downstream area beyond the Kratie gauging station, located in the Cambodia and Vietnam flood plains was considered as the LMB in this study. The bias-corrected precipitation outputs of the Japan Meteorological Research Institute atmospheric general circulation model (MRI-AGCM3.2S) with 20 km horizontal resolution were utilized as the precipitation inputs for basin-scale hydrological simulations. The present climate (1979–2003) was represented by the AMIP-type simulations while the future (2075–2099) climatic conditions were obtained based on the RCP8.5 greenhouse gas scenario. The entire hydrological system of the Mekong basin was modelled by the block-wise TOPMODEL (BTOP) hydrological model with 20 km resolution, while the LMB area was modelled by the rainfall-runoff-inundation (RRI) model with 2 km resolution, specifically to analyze floods under the aforementioned climatic conditions. The comparison of present and future river runoffs, inundation distributions and inundation volume changes were the outcomes of the study, which can be supportive information for the LMB flood management, water policy, and water resources development.

Keywords: climate change; flood hazards; high-resolution AGCM; inundation analysis; Lower Mekong river basin

1. Introduction

Reports published by the Intergovernmental Panel on Climate Change (IPCC) [1,2] have mentioned that the intensity and frequency of heavy precipitation will increase in the future due to climate change. The projections given by the IPCC reports were based on simulation results from several general circulation models (GCMs), and many studies have assessed changes in hydrological characteristics due to climate change by using those GCM outputs. Such changes may be catastrophic by the end of the 21st century due to climate change impacts unless sustainable mitigation actions are not taken in a prompt manner. Researchers suggest that there is a high possibility that the frequency and magnitude of flood disasters will increase globally [2–4]. Therefore, quantitative and qualitative assessments of changes in flood characteristics under climate change in river basins are critically important, and are strongly requested by policy-makers, river engineers, and flood fighters for practical river basin management against increasing flood risk. In this context, a study of future extreme flood events is essential. Overwhelming emissions of greenhouse gases are likely to alter the global climate and, consequently, the hydrological characteristics of river basins would be affected significantly, causing severe droughts and floods. Changes in patterns, intensity, and frequency of precipitation should be assessed to understand future extreme floods more deeply because such climate variations are threats to the existence of all ecosystems. Understanding of future rainfall and river discharge variations, trends, and volumes is also essential, considering the impacts on river basins from extreme floods, long-lasting droughts, water storage, and other events due to climate change. Flood risk management procedures that are carefully designed will benefit communities vulnerable to extreme floods in the future. Scientific understanding of climatic conditions and technical advances in climatic and hydrological modelling can make climate change studies more sophisticated and rational.

GCMs are used as the main tool to produce climatic variables under various greenhouse gas emission scenarios. Past, present, and future conditions of climatic variables can be produced by GCMs under given various driving forces. Still, their future predictions remain uncertain to some degree due to uncertainties in the emission scenarios [1]. Moreover, factors, such as limited spatial resolutions, simplified physics and thermodynamic processes, numerical schemes, or incomplete knowledge of climate system processes can increase the uncertainty of the GCM outputs [5]. Uncertainties arising from GCMs and emission scenarios have been investigated by several studies [6–9]. Despite uncertainties in GCM predictions, their approximations for future climatic conditions are still useful to understand and prepare for possible future climatic hazards.

Studies have shown that possible impact of climate change varies significantly, depending on the selection of emission scenarios [10,11]. The IPCC 5th Assessment Report (AR5), published in 2013, introduced "Representative Concentration Pathways (RCPs)" as a new set of greenhouse gas emission scenarios. There are four RCPs, namely, RCP2.6, RCP4.5, RCP6.0, and RCP8.5. Those RCPs were determined considering the possible range of radiative forcing values (W/m^2) in the year 2100, with their numbers referring to low, intermediate, moderate, and excessive levels of greenhouse gas emissions. RCP8.5 corresponds to the highest greenhouse gas emissions pathway among the four. Its main assumptions are relatively slow income growth, high population, modest rates of technological advances, and high energy consumption which leads to long-term high emissions of greenhouse gases in the absence of climate change policy [12]. It is termed as the 'baseline' scenario since no other specific climate mitigation targets are included in RCP8.5. Therefore, the RCP8.5 scenario should be used in a study where the most extreme disasters under climate change are analyzed.

GCMs are generally recognized as capable of producing global climatic parameters reasonably well. However, they are not advanced enough to produce accurate climatic parameters on a local or regional scale due to their coarse resolutions. Hence, the application of different downscaling techniques is inevitable to use coarse-resolution GCM outputs. Basically, two categories of downscaling techniques, i.e., statistical downscaling and dynamical downscaling, are commonly used. Dynamical downscaling extracts regional-scale information from coarse-resolution GCMs and produces regional climatic dynamics [13]. Statistical downscaling develops empirical relationships between local climate variables (e.g., surface air temperature and precipitation) and large-scale predictors (e.g., pressure fields), and applies those relationships to GCM outputs [14]. The comparison of those two techniques and their merits and demerits are explained in several studies [15–17].

Uncertainties caused by coarse-resolution GCMs and limitations of downscaling techniques can be minimized by utilizing super high-resolution GCM outputs. To this end, the 20 km high-resolution atmospheric general circulation model (AGCM), a state-of-art AGCM developed by the Meteorological Research Institute of Japan ard commonly called MRI-AGCM3.2S [18,19], can be utilized to produce future climatic conditions without conducting downscaling. It covers the entire globe with 20 km spatial resolution and its model output is highly expected to serve for improving flood impact assessment.

Even though various studies have been conducted using climatic models, more research on regional- and local-scale hydrology should be conducted through the combined application of high-resolution AGCMs and physically-based hydrological models under extreme climatic conditions. To this end, the application of high-resolution MRI-AGCM3.2S precipitation outputs for RCP8.5 with a physically-based river runoff model and a flood simulation model is ideal to study future extreme flood events. This study was planned to analyze future possible extreme floods by applying this ideal combination to the Lower Mekong Basin (LMB). In the recent past, several studies discussed the climate change impact on the Mekong basin [20–24]. However, those studies aimed to elaborate climate change by only considering rainfall or river runoff variations under different climate scenarios, and less effort has been devoted to simulating floods in the Mekong basin considering the RCP8.5 scenario, which projects extreme flood events. Moreover, previous studies have used climatic models with coarse resolutions, which require downscaling. This study was conducted to overcome such limitations in previous studies.

This study analyzed the climate change impact on the LMB in a cross-boundary region of Cambodia and Vietnam, which lies downstream of the Kratie gauging station (Figure 1), utilizing bias-corrected 20-km high-resolution MRI-AGCM3.2S precipitation outputs with two physically-based hydrological models named the block-wise TOPMODEL (BTOP) [25] and the rainfall-runoff-inundation (RRI) model [26]. We produced inundation maps for different climatic conditions, and analyzed river runoffs, rainfalls, and inundations in the LBM for time durations defined by MRI-AGCM3.2S, i.e., present (1979–2003) and future (2075–2099). Flooding in the LMB causes intolerable difficulties to people living in vulnerable areas, causing human casualties in some cases, and imposes damage to agriculture, fisheries and physical properties. Therefore, it is essential to identify flood events in the future and assess the possible hydrological impacts they may cause. Assessment results will provide insights to update existing plans and policies for agriculture, fisheries, and future developers to build a more resilient basin community.

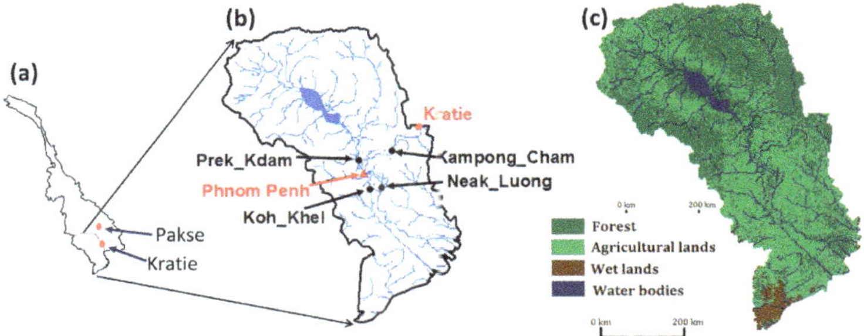

Figure 1. Study area and river gauging stations (**a**) Mekong River Basin (BTOP model domain) , (**b**)LMB area (RRI model domain), and (**c**) LMB land use map used in RRI model.

2. Study Area

The Mekong River is the 12th largest river in the world with the largest river basin in Southeast Asia, which expands across six riparian countries of China, Myanmar, Laos, Thailand,

Cambodia, and Vietnam. It is also the seventh longest river in Asia with a length of about 4620 km. The hydrological nature of the basin is governed by the monsoon climate, generating a mono-peak flood pulse during the monsoon season from July to September. The Mekong River extends from the Tibetan Plateau in China to the Mekong Delta in Vietnam. The river basin is located between the latitudes of 8° N to 34° N and the longitudes of 94° E to 110° E. The alpine climate prevails in the northern part of the basin, and a large tropical floodplain lies in the downstream part of the basin. The basin area is approximately 795,000 km^2 and its annual average discharge to the South China Sea is 475 km^3. Geographically, the basin can be divided into upper and lower parts. The upper part has a steep slope from the headwaters, while the river bed is more or less flat along the slope from Kratie to the ocean through the Mekong Delta shared by Cambodia and Vietnam. The lower part of the basin belongs mostly to a tropical monsoon climate zone, where the year is divided into dry and wet seasons. The wet season lasts from approximately early May to October, and the dry season from November to April. The wet season climate is dominated by the summer monsoon, arriving partly from the southwest and partly from the southeast. The uppermost part of the basin is located on the Tibetan plateau, where the precipitation pattern is similar to that in the lower part of the basin with most of the precipitation occurring during the summer. Due to lower temperatures at higher elevations, the precipitation during winter falls mainly as snow. Due to the monsoonal climate and the steepness of the river bed in the upper basin, the hydrograph of the Mekong River is single-peaked with large differences between high and low flow values. Average annual precipitation from 1964–2005 in the basin ranged from 850 mm to 2500 mm [27]. The flood season in the Mekong River Basin lasts from June to November and accounts for 80–90% of the total annual flow [28]. The annual flood season is especially important for the LMB because it shapes the environment and inhabitants in the basin. The LMB, the focus area of the present study, is shown in Figure 1a,b, which also illustrates respective model domains for the river runoff model of BTOP and the inundation simulation model of RRI.

3. Methodology and Models

3.1. Methodology

The most widely used approach in simulating the hydrological impacts of climate change is to combine GCM outputs with hydrological models. The proposed methodology in this study employs models with different scales to simulate climate change impact on the LBM. The global-scale AGCM, the regional-scale BTOP model, and the local-scale RRI model were employed to analyze plausible future extreme flood situations. Cascade linking of global-, regional-, and local-scale models has been proven productive in carrying out inundation simulations and hydrological assessments for the LMB, which is systematically explained in Figure 2.

The MRI-AGCM3.2S precipitation outputs based on the RCP8.5 scenario were utilized for the future climate simulation in this study while the present climatic simulation was conducted using MRI-AGCM3.2S under the AMIP-type scenario [29]. The horizontal grid size is about 20 km in MRI-AGCM3.2S [19,30]. While the resolution is satisfactory for regional-scale hydrological modelling, the MRI-AGCM3.2S precipitation still has bias. However, MRI-AGCM3.2S precipitation can be bias-corrected without modification of its horizontal resolution, which is the main advantage of the MRI-AGCM3.2S dataset. According to Chen [31], in the process of studying climate change impact, the main uncertainties are from GCMs and downscaling. GCMs provide information at a resolution that is too coarse to give results that can be used directly in hydrological modelling [32]. In this study, however, that was not an issue due to the utilization of 20-km high-resolution AGCM outputs. BTOP, a physically-based hydrological model, was used to simulate river runoff for the present and future periods based on bias-corrected 20-km MRI-AGCM3.2S precipitations for the entire Mekong Basin. BTOP-simulated daily river discharges for present and future at Kratie station were used as the boundary condition for the RRI model, which simulated LMB floods with relevant rainfall. Considering the computational time, available resources, and the physically-based nature of the BTOP model, it was selected to perform the simulation for the entire Mekong Basin at a 20-km regional scale. The RRI model, which is more sophisticated in simulating river runoff and inundations under the diffusive wave model framework, however, needs a longer computational time and, thus, was employed only for the LMB area with a 2-km gridded model at a local scale. Figure 1a,b illustrates the respective model domains for the BTOP and RRI models. The RRI model simulated the river flow system downstream of the Kratie station. The flow system consisted of flow reversal in the Tonle Sap River, bank overflows, flood occurrences due to river runoff and rainfall, and varying inflows from Kratie station (as the boundary condition).

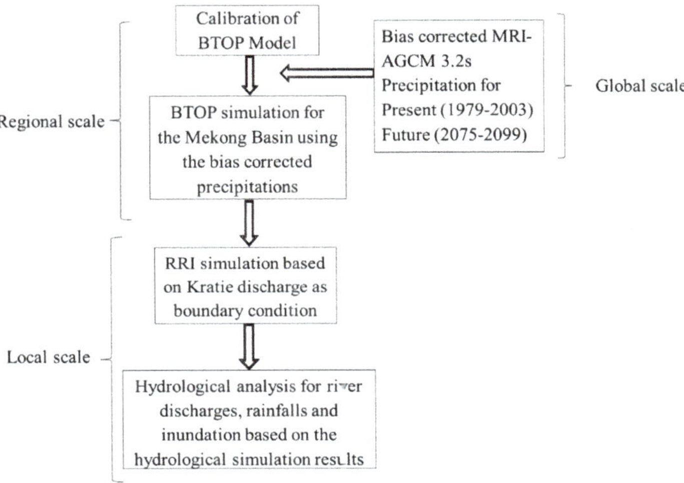

Figure 2. Flowchart of the modelling process.

The study utilized observed and calculated datasets for precipitation and river runoffs. The observed precipitation data from the time period of 1951 to 2007 was obtained from the Asian Precipitation—Highly-Resolved Observational Data Integration towards Evaluation of Water Resources—APHRODITE [33], which is a gridded daily precipitation dataset based on rain gauges in Asia. The data are in the resolution of 20 km. APHRODITE was developed by the Japan Meteorological Agency and the Meteorological Research Institute of Japan, which succeeded the APHRODITE project [33–35], an observational dataset of daily precipitation and the prototype of gridded daily rain gauge datasets on the global land surface. The hydrological models used in this study were calibrated to the observed daily discharge data received from the Mekong River Commission (MRC).

3.2. MRI-AGCM3.2S

MRI-AGCM3.2S has the finest AGCM resolution so far available and its outputs have been used in several studies to project future climatic conditions [36–38]. The model is based on a hydrostatic primitive equation system using a spectral transform method of spherical harmonics [19]. MRI-AGCM3.2S data were made available for the present and future, each of which was a 25-year period of 1979–2003 and 2075–2099, respectively. An Atmospheric Model Inter-comparison Project (AMIP)-type [29] simulation using the observed boundary conditions from 1979 to 2003 was treated as the present climate experiment outputs, labelled as "SPA_m01" in this study for the present. The sea surface temperature (SST) is used in the MRI-AGCM3.2S as the temperature boundary condition. For the future climate experiments under the RCP8.5 greenhouse gas emission scenario, four different SST distributions from 28 coupled models in Couple Model Inter-comparison Project Phase 5

(CMIP5) [39] were considered. As explained in Mizuta [18], it is expected that an AGCM generates different precipitation distributions as a response to different SST distribution. Considering different SSTs, the SST uncertainty under the RCP8.5 emission scenario can be assessed.

The RCP8.5 future climate projections from 2075 to 2099 with the SST distributions were grouped as Cluster 1, 2, 3, and Total (28-model average), which were respectively labelled as "SFA_rcp85_c1", "SFA_rcp85_c2", "SFA_rcp85_c3", and "SFA_rcp85" [18] in this study. These clusters represent different sets of SSTs. Cluster 1 (eight-model average) is characterized by a nearly uniform warming in the northern and southern hemispheres; Cluster 2 (14-model average) shows an El Nino-like pattern with a larger warming belt in the central equatorial Pacific; Cluster 3 (six-model average) is dominated by a larger warming in the northern hemisphere than in the southern hemisphere; and Total is the intermediate between the clusters [18]. These characteristics of the four clusters are explained by Mizuta [18].

MRI-AGCM3.2S precipitation outputs were corrected for their bias using a statistical method developed by Inomata [40]. The statistical method appropriately corrects biases in both monthly and extreme daily precipitations. The concept of the bias correction method is to adjust the probability distribution of GCM daily precipitation to that of its observed counterpart. The method was tested for the Yoshino River basin of Japan, and the results showed appropriate corrections of GCM biases in both monthly and extreme daily precipitations. Datasets of daily precipitation bias-corrected using this statistical method have been widely used for hydrological model simulations and risk assessments to assess the climate impacts of floods and droughts in Asia [41–43].

3.3. BTOP Mode

The BTOP model is a distributed hydrological model, which was developed based on the TOPMODEL [25,44,45]. This model uses a topographic index with a block-wise concept and simulates watershed-scale rainfall runoff processes, including snowmelt, overland flow, soil moisture in the root zone and unsaturated zones, subsurface flow, river flow routing, and dam operation. For river flow routing, a modified Muskingum-Cunge (MC) routing method is integrated to conserve water at each river segment [46]. The detailed description of the BTOP model is provided by Takeuchi [25,45]. Until now, the BTOP model has been employed in various hydrological applications, such as large-basin long-term simulations [47], poorly or ungauged basins [25,45], flood hazard assessment [48], drought analysis with standardized indices [49], dam operation for flood and drought reduction [49–51], now-casting, basin-scale scenario analyses of future projection on hydro-meteorological conditions [52], and nutrient loading [53]. The BTOP model, employed in the present study, covered the entire Mekong Basin under 20 km resolution, and it was calibrated and validated for the river runoff at the Pakse gauging station (Figure 1).

3.4. RRI Model

The rainfall-runoff-inundation (RRI) model is a two-dimensional model which is capable of simulating rainfall-runoff and flood inundation simultaneously. The model deals with slopes and river channels separately [26,54]. At a grid cell, in which a river channel is located, the model assumes that both the slope and river are positioned within the same grid cell. The channel is discretized as a single line along its center line of the overlying slope grid cell. Figure 3 depicts a schematic diagram of the RRI model's concept of the river channel and the slope. The flow of the slope grid cells is calculated with a two-dimensional diffusive wave model, while the channel flow is calculated with a one-dimensional diffusive wave model. For better representation of rainfall-runoff-inundation processes, the RRI model simulates lateral subsurface flow, vertical infiltration flow, and surface flow. The lateral subsurface flow, which is typically more important in mountainous regions, is treated in terms of the discharge-hydraulic gradient relationship, which considers both saturated subsurface and surface flows. On the other hand, vertical infiltration flow is estimated by using the Green-Ampt model [55,56]. The flow interaction between the river channel and the slope is estimated based on different overflowing formulae, depending on water-level and levee-height conditions. A storage cell-based inundation model [57] was used to calculate lateral flows on slope grid cells. The model equations were derived based on the following mass balance equation, momentum equations, and gradually-varied unsteady flow. The RRI model setup and its equations are explained in Sayama [26,54]. The various applications of the RRI model in hydrological studies can be found in several studies [26,41,54,58,59].

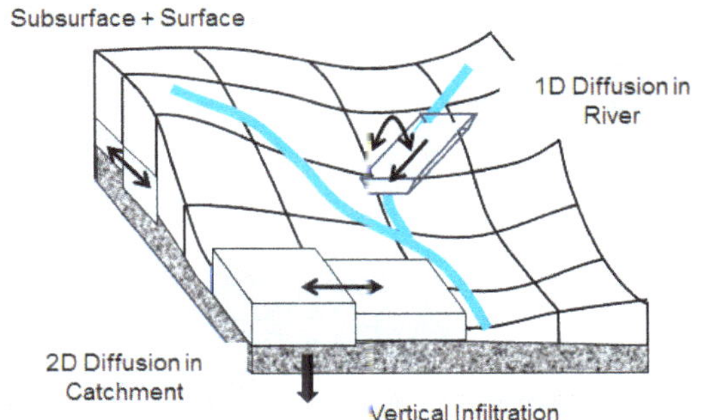

Figure 3. Schematic diagram of rainfall-runoff-inundation (RRI) model.

The model equations are derived based on the following mass balance Equation (1) and momentum Equations (2) and (3) for gradually varied unsteady flow:

$$\frac{\partial h}{\partial t} + \frac{\partial q_x}{\partial x} + \frac{\partial q_y}{\partial y} = r - f \tag{1}$$

$$\frac{\partial q_x}{\partial t} + \frac{\partial u q_x}{\partial x} + \frac{\partial v q_x}{\partial y} = -gh\frac{\partial H}{\partial x} - \frac{\tau_x}{\rho_w} \tag{2}$$

$$\frac{\partial q_y}{\partial t} + \frac{\partial u q_y}{\partial x} + \frac{\partial v q_y}{\partial y} = -gh\frac{\partial H}{\partial y} - \frac{\tau_y}{\rho_w} \tag{3}$$

The second terms of the right side of Equations (2) and (3) are calculated with Manning's Equations (4) and (5).

$$\frac{\tau_x}{\rho_w} = \frac{gn^2 u \sqrt{u^2 + v^2}}{h^{1/3}} \tag{4}$$

$$\frac{\tau_y}{\rho_w} = \frac{gn^2 v \sqrt{u^2 + v^2}}{h^{1/3}} \tag{5}$$

h = height of the water from the local surface

q_x, q_y = unit width discharges in x and y directions

u, v = flow velocity in x and y directions

r = rainfall intensity

f = infiltration rate

H = height of the water from the datum

ρ_w = density of water

g = gravitational acceleration

τ_x, τ_y = shear stress in x and y directions

n = Manning's toughness parameter

4. BTOP and RRI Model Setup for the Study Area

4.1. BTOP Model Application

Magome [42] calibrated and validated the BTOP model for the entire Mekong basin considering the observed and simulated river runoff at the Pakse gauging station (Figure 1). The river network dataset, used in the BTOP modelling for the whole Mekong basin, was up-scaled from 3-arcsec (~90 m) HydroSHEDS [60] to 10 arcmin (~20 km), while preserving river network features, the up-stream catchment area, and river length and slope calculated from the original HydroSHEDS, which employed the river-network upscaling algorithm developed by Masutani and Magome [46]. Land cover data from the USGS International Geosphere-Biosphere Program (IGBP) and soil type data from the Food and Agriculture Organization soil map [61] were used for the root zone depth and soil properties. Precipitation data were used from APHRODITE [34] for calibration and validation. The CRU TS3.1 climate forcing data [62], a fourteen-day global Normalized Difference Vegetation Index (NDVI) dataset from the Global Inventory Modelling and Mapping Studies (GIMMS) [63], were used to simulate long-term potential evapotranspiration using the Shuttleworth-Wallace model. Model parameters, dischargeability D, decay factor m, drying function parameter α, Manning's coefficient n, and the groundwater parameter b, were tuned during the calibration for reasonable performance using observed discharge time series of Pakse station, provided by the Mekong River Commission. The BTOP model was calibrated for the period of 11 years from 1980 to 1990 and validated for a 10-year period starting from 1991. The Nash-Sutcliffe Coefficient (NSC) was estimated to be 86.9% and 90.3% for calibration and validation, respectively [43].

4.2. RRI Model Application

The resolution corresponded approximately to 2.0 km $\times$ 2.0 km in the RRI model developed for the LMB. The model domain for the LMB was about 187,000 km^2. As the model was being set up, the digital elevation model (DEM), flow direction, and flow accumulation were delineated from HydroSHED's 30 s resolution [55] and up-scaled to a 60 s (~2 km) resolution [64]. Other model inputs were precipitation, potential evapotranspiration, and river channel dimensions. Four land-use types, i.e., forests, agricultural lands, wetlands, and water bodies, were considered in the LMB inundation simulation as shown in Figure 1c.

The model parameters adjusted during the calibration process include: hydraulic conductivity k, Manning's roughness for the river bed n_r, and Manning's roughness for the slope n_s. The Manning's roughness coefficients used in this study were based on the different land use types, as illustrated in Chow [65]. RRI model parameters were manually tuned until the simulation discharges reasonably matched the observed values. The RRI model was calibrated for river runoff and inundation distribution. Figure 4 depicts simulated and observed daily discharges for the selected stations named Kampong Cham, Prak Kdam, Chroy Changver, Neak Luoung, and Koh Khol (Figure 1). The model performance for the calibration and validation processes was estimated using the relative root mean squire error (RRMSE), Nash-Sutcliffe Coefficient (NSC), and coefficient of determination (R^2), as shown in Table 1. The estimated model performance indicators for the model calibration and validation showed acceptable values relative to the optimal value of each index. The RRI model was able to achieve the flood peak in all the simulated flood events reasonably. Prek Kdam is at a special location, where the river flows towards the Tonle Sap Lake during the flood season, and in the reverse direction during the dry season. Tonle Sap is the largest freshwater lake in Southeast Asia which covers an area of 8800 km^2. At Phnom Phen, Cambodia, it connects to the mainstream of the Mekong River. According to the Cochrane [66], the volume of water flowing into Tonle Sap from the Mekong mainstream during the rainy season (June to October) is nearly six times greater than the volume of water during the dry season (November to May). The Prek Kdam's discharge direction varies according to the season and it was correctly simulated by the RRI model. The discharge towards the Tonle Sap during the rainy season is indicated by negative values, while the discharge in the mainstream during the dry season is considered as positive, as shown in Figure 4. The RRI model was able to simulate the water flow from the Tonle Sap Lake correctly according to the obtained results for the Prek Kdan station. Since the river flow division of the Bassak River was not exactly known, combined discharges of Koh Khel and Neak Luong were considered in the calibration and validation. Kampong Cham, Chroy Changver, Neak Luong, and Koh Khol stations show very good error estimations, while Prek Kdam shows relatively poor error estimations due to the dynamic nature of its location. The justification of the inundation distribution simulated by RRI was carried out by comparing the inundation maps published by MRC for 1998 (dry year) and 2000 (flood year). Comparison of the inundations in these two years showed reasonable matches with the MRC published figures.

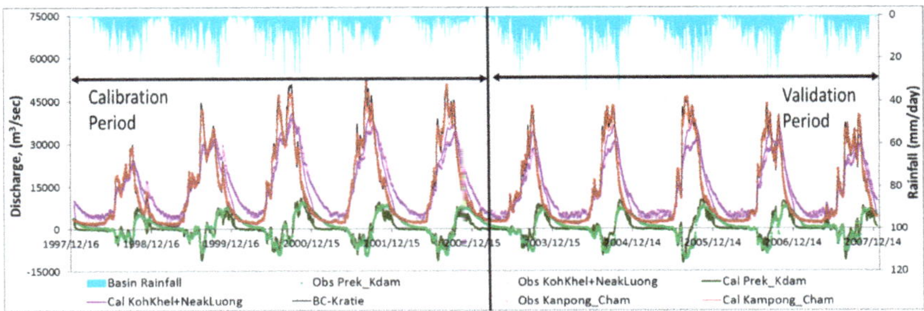

Figure 4. Comparison of observed and simulated discharges at the river gauging stations located in the LMB.

Table 1. Performance indicators for RRI simulated river runoff of selected gauging stations in the LMB.

Performance Indicators	Kampong Cham		Prak Kdam		Chroy Changver		Neak Luoung + Koh Khol	
	Calibration	Validation	Calibration	Validation	Calibration	Validation	Calibration	Validation
RRMSE	0.16	0.14	2.14	3.47	0.12	0.16	0.25	0.21
NSC	0.96	0.98	0.67	0.65	0.97	0.93	0.82	0.92
R2	0.96	0.99	0.72	0.72	0.97	0.97	0.93	0.95

5. Discussion

Variability, trends, and shifts in precipitation are crucial in assessing climate change impacts on water availability, floods, droughts, and agricultural productivity. Several basic statistical and physical properties of rainfall should be considered in characterizing the variability of rainfall in the context of climate change [67]. As far as the inter-annual variability of the Mekong River Basin's precipitation is concerned, 25-year average and $\pm\sigma$ (standard deviation) were selected to illustrate the variability, as shown in Figure 5. The variability increases as the precipitation increases. However, both in the present and future cases, the maximum monthly averaged precipitation occurs in August and the variability is higher in the future than in the present. The ratios of the maximum average values of the precipitation in the future to that value in the present were calculated for SFA_rcp85, c1, c2, and c3 and achieved as 1.07, 1.02, 1.07, and 1.06 for the respective scenarios. A similar pattern occurs in the 25-year averaged monthly discharge at Kratie, as shown in Figure 6. The maximum monthly average discharge for each scenario takes place in September. When the discharge increases, the variability also rises. Among all the cases, SFA_rcp85 shows the highest peak average values. The increment ratios of the monthly averaged discharges at the Kratie station for the future relative to those for the present were 1.18, 1.09, 1.14, and 1.14. SFA_rcp85 and c2 scenarios show the highest ratios in both precipitation and monthly averaged discharge at Kratie.

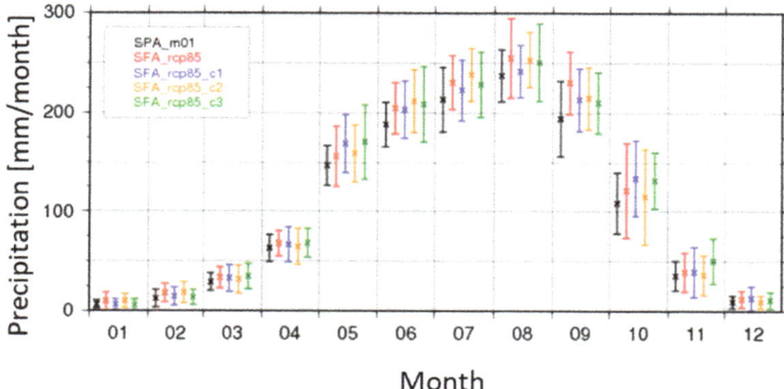

Figure 5. Monthly precipitation variability for the considered present and future climatic durations (cross: 25-year average, bar: ±1σ range).

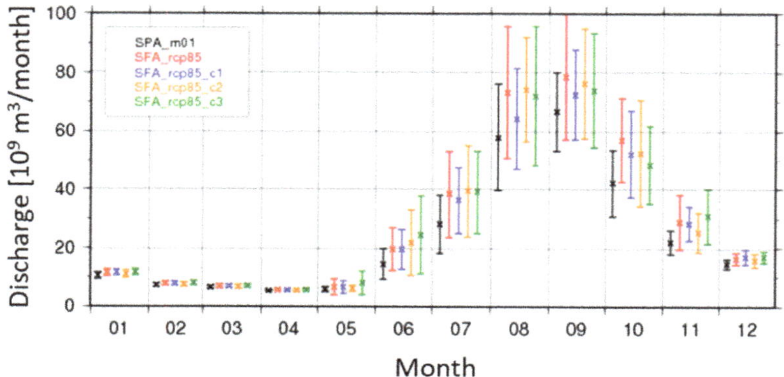

Figure 6. Monthly Kratie discharge variability for the considered present and future climatic durations (cross: 25-year average, bar: ±1σ range).

According to the RRI simulation results for the LMB, a significant increment in inundation distribution can be observed in the RCP8.5 future inundation distribution outputs compared with the AMIP-type present experiment. Each MRI-AGCM3.2S dataset comprises 25 years' worth of data. Figure 7a shows a 25-year averaged inundation distribution for the SPA, present case, while the other four figures are on the order of (b–e) illustrate the 25-year averaged inundation distributions for the scenarios of SFA_rcp85, c1, c2, and c3. The increment ratios in inundation extent in future climatic conditions were calculated by dividing the future inundation areas (SFA) by the present (SPA) inundation area. Compared to the present AMIP-type experiment, the future inundation areas are in the ratios of 1.34, 1.26, 1.35, and 1.24, respectively, for RCP8.5 experiments with four SST distributions (Table 2). The increment ratios for the inundation distributions provide an alarm for future extreme flood events which may occur due to climate change. Figure 8 illustrates the maximum inundation for each 25-year dataset. The SFA_rcp85 case shows the highest maximum inundation distribution among the simulated datasets. The increase in inundation vulnerability of Phnom Penh, the capital of Cambodia, is significant, according to the obtained inundation results for Future. The Mekong Delta (MD), which is of ~55,000 km² area and located downstream of Phnom Penh, will be at high flood risk according to the obtained results. Of the 60 million people living in the LMB, about 40% live within a 15 km range along the Mekong River with most within a 5 km range from the mainstream. The communities located within 5 km from the mainstream have high exposure to floods, and their flood risk will be higher in the future, according to the obtained results. Although minor and moderate floods are advantageous to agriculture and fisheries in the MD area, future extreme rainfalls and floods may cause severe damage to infrastructure, agriculture, transportation, and community properties in the MD area. The obtained flood increment ratios indicate possible future flood risk in the MD and provide insights for policy-makers to take actions to protect communities in the MD through the introduction of flood mitigation and risk reduction actions.

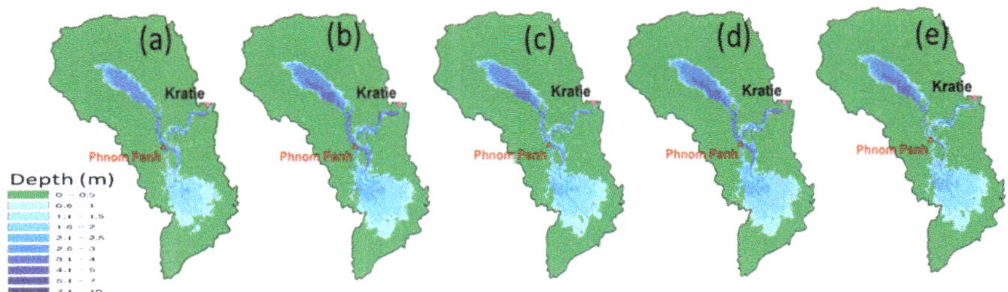

Figure 7. 25-year averaged inundation distributions: (**a**) SPA_m01; (**b**) SFA_rcp85; (**c**) SFA_rcp85-c1; (**d**) SFA_rcp85-c2 and (**e**) SFA_rcp85-c3.

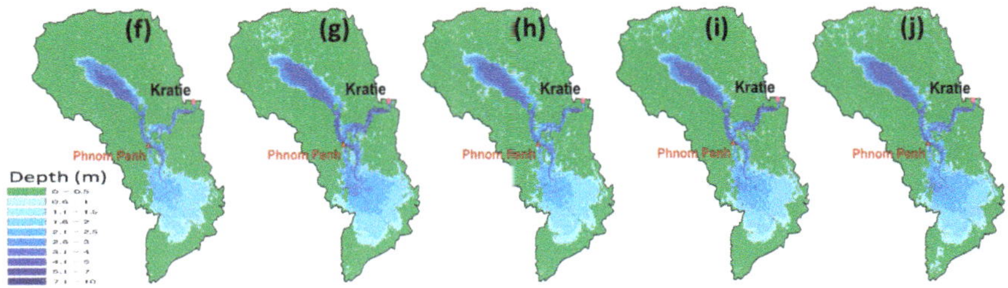

Figure 8. Maximum inundation out of 25 years simulation: (**f**) SPA_m01; (**g**) SFA_rcp85; (**h**) SFA_ rcp85-c1; (**i**) SFA_ rcp85-c2 and (**j**) SFA_ rcp85-c3.

Table 2. Comparison of present and future hydrological features of the LMB.

	SFA_rcp85 / SPA_m01	SFA_rcp85c1 / SPA_m01	SFA_rcp85c2 / SPA_m01	SFA_rcp85c3 / SPA_m01
Inundation area	1.34	1.26	1.35	1.24
Specific discharge volume at Kratie	1.25	1.16	1.21	1.21
Specific inundation volume	1.60	1.30	1.52	1.29
Cumulative rainfall	1.11	1.09	1.10	1.11

According to Figure 9, for the future experiments, the increase in cumulative rainfall is nearly the same; however, the river runoff and inundation volumes show considerable increments. The main reason for this situation is the intensity, distribution of rainfall across the basin, and the variability of rainfall, as explained in Figure 5. Even the increment ratios of cumulative rainfall spread in a narrow range due to their distributions and intensities, and the river runoff and inundation generations are different. This phenomenon should be further analyzed as the next step of the study.

The specific discharge volume for Kratie's river runoff was calculated by dividing the 25-year averaged discharge volume by the upstream basin area of the Mekong. The Kratie's river runoff volume has a significant impact on the LMB flood generation. Therefore, understanding its increment in the future is a good indicator to identify extreme climatic conditions anticipated in the future. Figure 10 illustrates the cumulative specific runoff volume at Kratie for the present and future climatic experiments. The most important indicator of the LMB flood, and the inundation volume averaged for 25 years for the present and future experiments are also summarized in Figure 10. The total volume, which can cause an inundation of over 0.5 m deep was considered in calculating the inundation volume. This was converted to a specific volume by dividing the LMB basin area. The two figures (Figures 9 and 10) show significant increments in rainfalls, river runoffs and inundation volumes relative to the Present-SPA_m01 climatic condition. Among the four future climatic experiments, SFA_rcp85 shows the highest increment compared to SPA_m01. The SFA_rcp85 cumulative basin rainfall increased by 135 mm compared to SPA_m01. The increment in discharge at Kratie under SFA_rcp85 was 115 mm, and the increment in inundation volume under SFA_rcp85 was 68 mm relative to the present. Such increased water volumes are a very clear indicator to identify future extreme flood conditions in the LMB region.

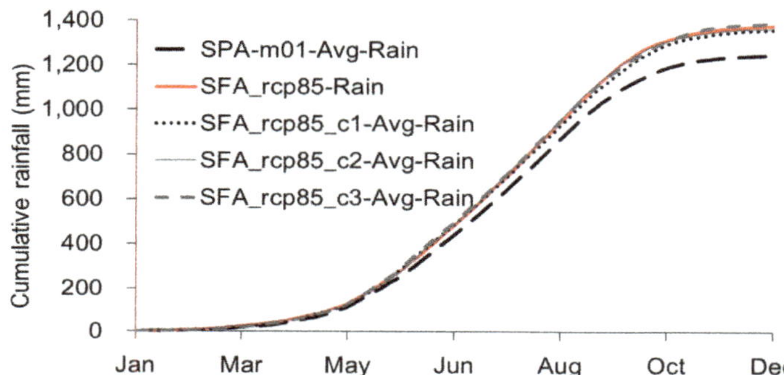

Figure 9. The 25-year averaged cumulative rainfall of the Mekong basin for the present and future climate experiments.

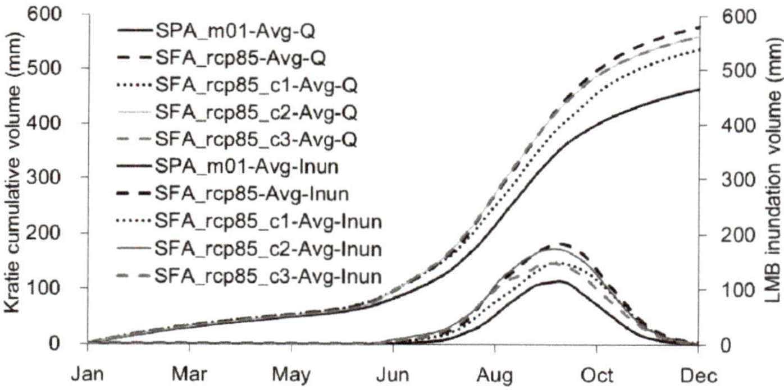

Figure 10. The 25-year averaged cumulative specific river runoff volume at the Kratie station and 25-year averaged cumulative specific inundation volume (above 0.5 m) in the LMB.

Table 2 summarizes the comparison of present and future rainfall, Kratie's river runoff, and LMB inundation volume increments as ratios. The highest ratio in each category comes from the ratio of the SFA_rcp85 scenario. The inundation increment ratio for SFA_rcp85 compared to SPA_m01 is 1.6, which shows the vulnerability of the LMB area to inundation in future climate change. The ratio of SFA_rcp85_c2 for future inundation is also critical compared to the present situation.

One of the limitations of the study is that the flood assessments were only based on precipitation outputs of MRI-AGCM3.2S for the RCP8.5 scenario, rather than several GCM or AGCM outputs. Furthermore, consideration of other RCP scenarios would be beneficial for policy-makers and river engineers to foresee future flood threats of different intensities. Different flood projections may result if several GCM outputs are considered. In the case of flood risk assessment, fine resolution inundation model would be beneficial to identify the floods more precisely. Unfortunately, this study had to limit the resolution of the inundation model to 2 km because running large-scale and long-term inundation simulations are still computationally expensive. In future research, it would be important to consider several AGCMs' outputs and other RCP scenarios, as well to view the future anticipated floods in the LMB in a broad way.

6. Conclusions

The use of climate and hydrological models to better understand extreme floods is an important method and widely used in the context of climate change impact studies. We adopted a methodology to utilize global-, regional-, and local-scale models to produce future floods under extreme climatic conditions. The impact of climate change on hydrological features in the LMB in the cross-boundary region of Cambodia and Vietnam was analyzed by feeding MRI-AGCM3.2S precipitations projected for the present and the future into two distributed hydrological models. The obtained results are alarming and provide useful information for policy-makers and river engineers to understand future extreme floods. Moreover, the results enable comparisons between present and future extreme floods considering different hydrological aspects of floods, such as inundation area, inundation volume, cumulative rainfall, and specific discharge. The results indicate a significant increase in flood severity in the LMB, predicting extreme floods possibly disastrous to the paddy cultivation in the MD area, which accounts for more than 50% of the rice production in Vietnam [68]. Although studies have been conducted in the recent past on the climate change impact on the Mekong Basin, they have limited the discussion to river runoffs. In the present study, we focused not only on river runoff, but also inundation extent. The hydrological analysis was conducted based on the 25-year bias-corrected precipitation datasets of the present and future. In the case of the future, we used four datasets of the RCP8.5 scenario considering different SST boundary conditions. The 2 km-resolution RRI model developed for the LMB area is a rather fine-resolution model applied to the Mekong Basin compared with those used in past studies. Due to the fine resolution of the inundation model, the obtained flood depths and distributions should be beneficial to understand the possible flood damage in the future. According to the obtained results, the increment ratios of averaged flood area in the future compared to that in the present are 1.34, 1.26, 1.35, and 1.24 for the SFA_rcp85 and the other three cases, respectively. The specific inundation volumes for different cases were also estimated and presented in this study. Those estimates are useful indicators to understand future extreme flood events in the LMB basin. Further research should be conducted to analyze flood risk in the LMB considering other AGCM outputs and other emission scenarios. The present study can be considered as the baseline for future studies on the LMB for extreme floods and their impacts.

Acknowledgments: This work was conducted under the framework of the "Precise Impact Assessments on Climate Change" of the Program for Risk Information on Climate Change (SOUSEI Program) supported by the Ministry of Education, Culture, Sports, Science, and Technology-Japan (MEXT). The authors sincerely thank the Mekong River Commission for providing us with river discharge data to calibrate the hydrological models. Moreover, the authors are grateful to the reviewers and journal editors for their constructive comments and helpful suggestions, which resulted in this improved manuscript.

Author Contributions: E.D.P.P., T.S. and Y.I. conceived and designed the research; E.D.P.P. and T.Y. performed the RRI simulations; J.M. conducted the BTOP simulations; A.H. carried out the bias correction and simulation data preparation for RRI and BTOP simulations; E.D.P.P. analyzed the data and results; and E.D.P.P. wrote the paper.

Conflicts of Interest: The authors declare no conflict of interest.

References

1. Solomon, S.; Qin, D.; Manning, M.; Chen, Z.; Marquis, M.; Averyt, K.B.; Tignor, M.; Miller, H.L. *The Physical Science Basis*; Contribution of Working Group I to the Fourth Assessment Report of the Intergovernmental Panel on Climate Change; Cambridge University Press: Cambridge, UK, 2007.
2. Intergovernmental Panel on Climate Change (IPCC). Summary for policymakers. In *Climate Change 2013: The Physical Science Basis*; Contribution of WGI to the Fifth Assessment Report of the Intergovernmental Panel on Climate Change; Stocker, T.F., Qin, D., Plattner, G.-K., Tignor, M., Allen, S.K., Boschung, J., Nauels, A., Xia, Y., Bex, V., Midgley, P.M., Eds.; Cambridge University Press: Cambridge, UK; New York, NY, USA, 2013.
3. Schmocker-Fackel, P.; Naef, F. More frequent flooding? Changes in flood frequency in Switzerland since 1850. *J. Hydrol.* **2010**, *381*, 1–8. [CrossRef]
4. Hirabayashi, Y.; Mahendran, R.; Koirala, S.; Konoshima, L.; Yamazaki, D.; Watanabe, S.; Kim, H.; Kanae, S. Global flood risk under climate change. *Nat. Clim. Chang.* **2013**, *3*, 816–821. [CrossRef]
5. Ramirez-Villegas, J.; Challinor, A.J.; Thornton, P.K.; Jarvis, A. Implications of regional improvement in global climate models for agricultural impact research. *Environ. Res. Lett.* **2013**, *8*, 1–12. [CrossRef]
6. Allen, M.R.; Stott, P.A.; Mitchell, J.F.B.; Schnur, R.; Delworth, T.L. Quantifying the uncertainty in forecasts of anthropogenic climate change. *Lett. Nat.* **2000**, *407*, 617–620. [CrossRef] [PubMed]
7. Webster, M.D.; Babiker, M.; Mayer, M.; Reilly, J.M.; Harnisch, J.; Hyman, R.; Sarofim, M.C.; Wang, C. Uncertainty in emissions projections for climate models. *Atmos. Environ.* **2002**, *36*, 3659–3670. [CrossRef]
8. Khoi, D.N.; Hang, P.T.T. Uncertainty Assessment of Climate Change Impacts on Hydrology: A Case Study for the Central Highlands of Vietnam. In *Managing Water Resources under Climate Uncertainty*; Shrestha, S., Anal, A., Salam, P., van der Valk, M., Eds.; Springer: Cham, Switzerland, 2015.
9. Fang, G.; Yang, J.; Chen, Y.; Li, Z.; De Maeyer, P. Impact of GCM structure uncertainty on hydrological processes in an arid area of China. *Hydrol. Res.* **2017**. [CrossRef]
10. Hurkmans, R.; Moel, H.; Aerts, J.C.J.H.; Troch, P.A. Water balance versus land surface model in the simulation of Rhine River discharges. *Water Resour. Res.* **2008**, *44*, 1–14. [CrossRef]
11. Montenegro, R.; Ragab, R. Hydrological response of a Brazilian semi-arid catchment to different land use and climate change scenarios: A modelling study. *Hydrol. Process.* **2010**, *24*, 2705–2723. [CrossRef]
12. Zhang, Y.; Wang, Y.; Niu, H. Spatio-temporal variations in the areas suitable for the cultivation of rice and maize in China under future climate scenarios. *Sci. Total Environ.* **2017**, *601*, 518–531. [CrossRef] [PubMed]
13. Lenderink, G.; Van Ulden, A.P.; Van den Hurk, B.; Van Meijgaard, E. A study on combining global and regional climate model results for generating climate scenarios of temperature and precipitation for the Netherlands. *Clim. Dyn.* **2007**, *29*, 157–176. [CrossRef]
14. Wilby, R.L.; Charles, S.P.; Zorita, E.; Timbal, B.; Whetton, P.; Mearns, L.O. *Guidelines for Use of Climate Scenarios Developed from Statistical Downscaling Methods*; IPCC Task Group on Data and Scenario Support for Impact and Climate Analysis (TGICA): Geneva, Switzerland, 2004. Available online: http://www.narccap.ucar.edu/doc/tgica-guidance-2004.pdf (accessed on 20 November 2017).
15. Mearns, L.O.; Bogardi, I.; Giorgi, F.; Matyasovszky, I.; Palecki, M. Comparison of climate change scenarios generated from regional climate model experiments and statistical downscaling. *J. Geophys. Res.* **1999**, *104*, 6603–6621. [CrossRef]
16. Yarnal, B.; Comrie, A.C.; Frakes, B.; Brown, D.P. Developments and prospects in synoptic climatology. *Int. J. Climatol.* **2001**, *21*, 1923–1950. [CrossRef]
17. Haylock, M.R.; Cawley, G.C.; Harpham, C.; Wilby, R.L.; Goodess, C.M. Downscaling heavy precipitation over the United Kingdom: A comparison of dynamical and statistical methods and their future scenarios. *Int. J. Climatol.* **2006**, *26*, 1397–1415. [CrossRef]
18. Mizuta, R.; Arakawa, O.; Ose, T.; Kusunoki, S.; Endo, H.; Kitoh, A. Classification of CMIP5 future climate responses by the tropical sea surface temperature changes. *SOLA* **2014**, *10*, 167–171. [CrossRef]
19. Kitoh, A.; Endo, H. Changes in precipitation extremes projected by a 20-km mesh global atmospheric model. *Weather Clim. Extrem.* **2016**, *11*, 41–52. [CrossRef]
20. Gosling, S.N.; Taylor, R.G.; Arnell, N.W.; Todd, M.C. A comparative analysis of projected impacts of climate change on river runoff from global and catchment-scale hydrological models. *Hydrol. Earth Syst. Sci.* **2011**, *15*, 279–294. [CrossRef]

21. Kingston, D.G.; Thompson, J.R.; Kite, G. Uncertainty in climate change projections of discharge for the Mekong River Basin. *Hydrol. Earth Syst. Sci.* **2011**, *15*, 1459–1471. [CrossRef]

22. Lauri, H.; de Moel, H.; Ward, P.J.; Räsänen, T.A.; Keskinen, M.; Kummu, M. Future changes in Mekong River hydrology: Impact of climate change and reservoir operation on discharge. *Hydrol. Earth Syst. Sci.* **2012**, *6*, 4603–4619. [CrossRef]

23. Artlert, K.; Chaleeraktrakoon, C.; Nguyen, V.T.V. Modeling and analysis of rainfall processes in the context of climate change for Mekong, Chi, and Mun River Basins (Thailand). *J. Hydro-Environ. Res.* **2013**, *7*, 2–17. [CrossRef]

24. Thompson, J.R. Assessment of uncertainty in river flow projections for the Mekong River using multiple GCMs and hydrological models. *J. Hydrol.* **2013**, *485*, 1–30. [CrossRef]

25. Takeuchi, K.; Hapuarachchi, P.; Zhou, M.; Ishidaira, H.; Magome, J. A BTOP model to extend TOPMODEL for distributed hydrological simulation of large basins. *Hydrol. Process.* **2008**, *22*, 3236–3251. [CrossRef]

26. Sayama, T.; Ozawa, G.; Kawakami, T.; Nabesaka, S.; Fukami, K. Rainfall-runoff-inundation analysis of the 2010 Pakistan flood in the Kabul River basin. *Hydrol. Sci. J.* **2012**, *57*, 298–312. [CrossRef]

27. Thilakarathne, M.; Sridhar, V. Characterization of future drought conditions in the Lower Mekong River Basin. *Weather Clim. Extrem.* **2017**. [CrossRef]

28. Mekong River Commission. *State of the Basin Report 2010*; Mekong River Commission: Vientiane, Laos, 2010.

29. Gates, W.L. AMIP—The atmospheric model inter-comparison project. *Bull. Am. Meteorol.* **1992**, *73*, 1962–1970. [CrossRef]

30. Mizuta, R.; Yoshimura, H.; Murakami, H.; Matsueda, M.; Endo, H.; Ose, T.; Kamiguchi, K.; Hosaka, M.; Sugi, M.; Yukimoto, S.; et al. Climate simulations using MRI-AGCM3.2 with 20-km grid. *J. Meteorol. Soc. Jpn.* **2012**, *90*, 233–258. [CrossRef]

31. Chen, H.; Xiang, T.; Zhou, X.; Xu, C.Y. Impacts of climate change on the Qingjiang Watershed's runoff change trend in China. *Stoch. Environ. Res. Risk Assess.* **2012**, *26*, 847–858. [CrossRef]

32. Vaze, J.; Teng, J. Future climate and runoff projections across New South Wales, Australia: Results and practical applications. *Hydrol. Process.* **2011**, *25*, 18–35. [CrossRef]

33. Yatagai, A.; Kamiguchi, K.; Arakawa, O.; Hamada, A.; Yasutomi, N.; Kitoh, A. APHRODITE: Constructing a long term daily gridded precipitation dataset for Asia based on a dense network of rain gauges. *Bull. Am. Meteorol. Soc.* **2012**, *96*, 283–296. [CrossRef]

34. Kamiguchi, K.; Arakawa, O.; Kitoh, A.; Yatagai, A.; Hamada, A.; Yasutomi, N. Development of APHRO_JP, the first Japanese high-resolution daily precipitation product for more than 100 years. *Hydrol. Res. Lett.* **2010**, *4*, 60–64. [CrossRef]

35. Yasutomi, N.; Hamada, A.; Yatagai, A. Development of a long-term daily gridded temperature dataset and its application to rain/snow discrimination of daily precipitation. *Glob. Environ. Res.* **2011**, *15*, 165–172.

36. Yasuda, T.; Nakajo, S.; Kim, S.; Mase, H.; Mori, N.; Horsburgh, K. Evaluation of future storm surge risk in East Asia based on state-of-the-art climate change projection. *Coast. Eng.* **2014**, *83*, 65–71. [CrossRef]

37. Shou, K.J.; Yang, C.M. Predictive analysis of landslide susceptibility under climate change conditions—A study on the Chingshui River Watershed of Taiwan. *Eng. Geol.* **2015**, *192*, 46–62. [CrossRef]

38. Mori, N.; Tetsuya, T. Impact assessment of coastal hazards due to future changes of tropical cyclones in the North Pacific Ocean. *Weather Clim. Extrem.* **2016**, *11*, 53–69. [CrossRef]

39. Taylor, K.E.; Stouffer, R.J.; Meehl, G.A. An overview of CMIP5 and the experiment design. *Bull. Am. Meteorol. Soc.* **2012**, *93*, 485–498. [CrossRef]

40. Inomata, H.; Takeuchi, K.; Fukami, K. Development of a statistical bias correction method for daily precipitation data of GCM20. *Ann. J. Hydraul. Eng. JSCE* **2011**, *55*, 247–252. [CrossRef]

41. Perera, E.D.P.; Hiroe, A.; Shrestha, D.; Fukami, K.; Basnyat, D.B.; Gautam, S.; Hasegawa, A.; Uenoyama, T.; Tanaka, S. Community based flood damage assessment approach for lower West Rapti River basin in Nepal under the impact of climate change. *Nat. Hazards* **2014**, *75*, 669–699. [CrossRef]

42. Magome, J.; Gusyev, M.; Hasegawa, A.; Takeuchi, K. River discharge simulation of a distributed hydrological model on global scale for the hazard quantification. In Proceedings of the 21st International Congress on Modelling and Simulation (MODSIM 2015), Broadbeach, QLD, Australia, 29 November–4 December 2015.

43. Kwak, Y.; Magome, J.; Hasegawa, A.; Iwami, Y. Global Flood Exposure Assessment under Climate and Socio-economic Scenarios for Disaster Risk Reduction. In Proceeding of the 7th International Conference on Water Resources and Environment Research—ICWRER2016, Kyoto, Japan, 5–9 June 2016.

44. Ao, T.; Ishidaira, H.; Takeuchi, K. Study of distributed runoff, simulation model based on block type TOPMODEL and Musskingum-Cunge method. *Ann. J. Hydraul. Eng. JSCE* **1999**, *43*, 7–12. [CrossRef]

45. Takeuchi, K.; Ao, T.; Ishidaira, H. Introduction of block-wise use of TOPMODEL and Muskingum-Cunge method for the hydro-environmental simulation of a large ungauged basin. *Hydrol. Sci. J.* **1999**, *44*, 633–646. [CrossRef]

46. Masutani, K.; Magome, J. An application of modified Muskingum-Cunge routing method with water conservation condition to a distributed runoff model. *J. Jpn. Soc. Hydrol. Water Resour.* **2009**, *22*, 294–300. [CrossRef]

47. Hapuarachchi, H.A.P.; Zhou, M.C.; Kiem, A.S.; Geogievsky, M.V.; Magome, J.; Ishidaira, H. Investigation of the Mekong River basin hydrology for 1980–2000 using the YHyM. *Hydrol. Process.* **2008**, *22*, 1246–1256. [CrossRef]

48. Gusyev, M.A.; Kwak, Y.; Khairul, M.I.; Arifuzzaman, M.B.; Magome, J.; Sawano, H.; Takeuchi, K. Effectiveness of water infrastructure for river flood management: Part 1—Flood Hazard Assessment using hydrological models in Bangladesh. *Proc. IAHS* **2015**, *370*, 75–81. [CrossRef]

49. Navarathinam, K.; Gusyev, M.; Hasegawa, A.; Magome, J.; Takeuchi, K. Agricultural flood and drought risk reduction by a proposed multi-purpose dam: A case study of the Malwathoya River Basin, Sri Lanka. In Proceedings of the 21st International Congress on Modelling and Simulation (MODSIM 2015), Broadbeach, QLD, Australia, 29 November–4 December 2015.

50. Nawai, J.; Gusyev, M.; Hasegawa, A.; Takeuchi, K. Flood and drought assessment with dam infrastructure: A case study of the Ba River basin, Fiji. In Proceedings of the 21st International Congress on Modelling and Simulation (MODSIM 2015), Broadbeach, QLD, Australia, 29 November–4 December 2015.

51. Odhiambo, O.; Gusyev, M.; Hasegawa, A.; Magome, J.; Takeuchi, K. Flood and drought hazard reduction by proposed dams and a retarding basin: A case study of the Upper Ewaso Ngiro North River basin, Kenya. In Proceedings of the 21st International Congress on Modelling and Simulation (MODSIM 2015), Broadbeach, QLD, Australia, 29 November–4 December 2015.

52. Kiem, A.S.; Ishidaira, H.; Hapuarachchi, H.P.; Zhou, M.C.; Hirabayashi, Y.; Takeuchi, K. Future hydro-climatology of the Mekong River basin simulated using the high-resolution Japan Meteorological Agency (JMA) AGCM. *Hydrol. Process.* **2008**, *22*, 1382–1394. [CrossRef]

53. Yoshimura, C.; Zhou, M.; Kiem, A.S.; Fukami, K.; Prasantha, H.H.; Ishidaira, H.; Takeuchi, K. 2020s scenario analysis of nutrient load in the Mekong River Basin using a distributed hydrological model. *Sci. Total Environ.* **2009**, *407*, 5356–5366. [CrossRef] [PubMed]

54. Sayama, T.; Tatebe, Y.; Iwami, Y.; Tanaka, S. Hydrologic sensitivity of flood runoff and inundation: 2011 Thailand floods in the Chao Phraya River basin. *Nat. Hazards Earth Syst. Sci.* **2015**, *15*, 1617–1630. [CrossRef]

55. Hutten, N.C.; Gifford, G.F. Using the Green and Ampt infiltration equation on native and plowed rangeland soils. *J. Range Manag.* **1988**, *41*, 159–162. [CrossRef]

56. Kidwell, M.R.; Weltz, M.A.; Guertin, P.R. Estimation of Green-Ampt effective hydraulic conductivity for rangelands. *J. Range Manag.* **1997**, *50*, 290–299. [CrossRef]

57. Hunter, N.M.; Bates, P.D.; Horritt, M.S.; Wilson, M.D. Simple spatially-distributed models for predicting flood inundation: A review. *Geomorphology* **2007**, *90*, 208–225. [CrossRef]

58. Kudo, S.; Sayama, T.; Hasegawa, A.; Iwami, A. Analysis of Flood Risk Change in Future Climate in terms of Discharge and Inundation in the Solo River Basin. In Proceedings of the 7th International Conference on Water Resources and Environment Research—ICWRER 2016, Kyoto, Japan, 5–9 June 2016.

59. Shrestha, B.B.; Okazumi, T.; Miyamoto, M.; Sawano, H. Flood damage assessment in the Pampanga river basin of the Philippines. *J. Flood Risk Manag.* **2015**, *9*, 355–369. [CrossRef]

60. Lehner, B.; Verdin, K.; Jarvis, A. New global hydrography derived from space born elevation data. *Eos Trans. AGU* **2008**, *89*, 93–94. [CrossRef]

61. Food and Agriculture Organization (FAO). *Digital Soil Map of the World*; Version 3.6; Food and Agriculture Organization (FAO): Rome, Italy, 2007.

62. Harris, I.; Jones, P.D.; Osborn, T.J.; Lister, D.H. Updated high-resolution grids of monthly climatic observations—the CRU TS3.10 Dataset. *Int. J. Climatol.* **2014**, *34*, 623–642. [CrossRef]

63. Tucker, C.J.; Pinzon, J.E.; Brown, M.E. *Global Inventory Modelling and Mapping Studies*; NA94apr15b.n11-VIg, 2.0; Global Land Cover Facility, University of Maryland: College Park, MD, USA, 2004.

64. Masutani, K.; Akai, K.; Magome, J. A new scaling algorithm of gridded river networks. *J. Jpn. Soc. Hydrol. Water Resour.* **2006**, *19*, 139–150, (In Japanese with English Abstract). [CrossRef]
65. Chow, V.T. *Open-Channel Hydraulics*; McGraw-Hill: New York, NY, USA, 1959.
66. Cochrane, T.A.; Arias, M.E.; Piman, T. Historical impact of water infrastructure on water levels of the Mekong River and the Tonle Sap system. *Hydrol. Earth Syst. Sci.* **2014**, *18*, 4529–4541. [CrossRef]
67. Gachon, P.; St-Hilaire, A.; Ouarda, T.; Nguyen, V.T.V.; Lin, C.; Milton, J.; Chaumont, D.; Goldstein, J.; Hessami, M.; Nguyen, T.D. *A First Evaluation of the Strength and Weaknesses of Statistical Downscaling Methods for Simulating Extremes over Various Regions of Eastern Canada*; Final Report, Sub-Component; Climate Change Action Fund (CCAF), Environment Canada: Montréal, QC, Canada, 2005; p. 209.
68. Ngoc Thuy, N.; Thuy, N.N.; Anh, H.H. Vulnerability of rice production in Mekong River delta under impacts from floods, salinity and climate change. *Int. J. Adv. Sci. Eng. Inf. Technol.* **2015**, *5*, 272–279. [CrossRef]

Review

Applications of Open-Access Remotely Sensed Data for Flood Modelling and Mapping in Developing Regions

Iguniwari Thomas Ekeu-wei * and George Alan Blackburn

Lancaster Environment Centre, Lancaster University, Lancaster LA1 4YQ, UK; alan.blackburn@lancaster.ac.uk
* Correspondence: i.ekeu-wei@lancaster.ac.uk; Tel.: +23-481-209-70000

Received: 28 March 2018; Accepted: 22 July 2018; Published: 31 July 2018

Abstract: Flood modelling and mapping typically entail flood frequency estimation, hydrodynamic modelling and inundation mapping, which require specific datasets that are often unavailable in developing regions due to financial, logistical, technical and organizational challenges. This review discusses fluvial (river) flood modelling and mapping processes and outlines the data requirements of these techniques. This paper explores how open-access remotely sensed and other geospatial datasets can supplement ground-based data and high-resolution commercial satellite imagery in data sparse regions of developing countries. The merits, demerits and uncertainties associated with the application of these datasets, including radar altimetry, digital elevation models, optical and radar images, are discussed. Nigeria, located within the Niger river basin of West Africa is a typical data-sparse country, and it is used as a case study in this review to evaluate the significance of open-access datasets for local and transboundary flood analysis. Hence, this review highlights the vital contribution that open access remotely sensed data can make to flood modelling and mapping and to support flood management strategies in developing regions.

Keywords: open-access remotely sensed data; flood mapping and modelling; altimetry; synthetic aperture radar; optical satellite; Digital Elevation Model (DEM); and transboundary floods

---

1. Introduction to Flood Modelling and Mapping

Managing floods effectively requires a good understanding of historical flood trends, future expectations, and identification of locations likely to be impacted by flooding. Flood mapping provides the baseline for acquiring such information, to ensure preparedness, response and recovery efficiently undertaken to mitigate the impact of flooding [1]. Flood mapping is a process that describes the expected extent of water inundation into dryland as a result of intense precipitation or river water level rise driven by natural or anthropogenic factors [2]. Flood mapping processes differ considerably from project to project, and/or country to country, depending on specific project requirements and country-specific guidelines. In addition, the scale of flood mapping is influenced by available data, resources, technical know-how and delivery timeline, and this can determine the approach deployed [3–6]. Nevertheless, the sequence of activities that lead to the final flood hazard map outcome is fundamentally the same, and involves (i) flood frequency estimation: the probability of occurrence of a flood of specific magnitude over a certain period; (ii) hydrodynamic modelling: routing of expected or known river discharge or catchment runoff over a landscape to determine water depth, velocity and inundation extent; (iii) risk mapping: determining through overlay analysis, the landscape properties (land use/cover, infrastructures, population density, socioeconomic activities, etc.) to be impacted within flooded regions [7–11].

Typical flood mapping processes are presented in Table 1, including the basic data requirements, expected outcomes and some reference case studies. These processes, if executed with reasonable accuracy, can provide the necessary information to underpin effective flood management decisions such as floodplain planning, design of flood defence structures, and implementation of disaster response and recovery measures to mitigate flood impact.

Table 1. Flood mapping process and fundamental data requirements, outcomes and case studies.

Process	Data	Outcomes	Reference Case Studies
Flood frequency estimation	Historical data: River discharge, water levels and rating curves/equations.	Flood magnitude at specific return periods (Direct and regional).	[12–15]
Hydrodynamic modelling	Flood frequency outcome River discharge Digital elevation model Land use and cover map Historical flood extent, and marks.	Inundation Extent Water depth Flood velocity and travel time	[16–19]
Flood risk and vulnerability assessment	Hydrodynamic model outcomes, demographic, socio-economic and infrastructure data.	Exposure maps Vulnerability maps Evacuation plan	[19–21]

Going forward, this review highlights that the data required for flood modelling and mapping is scarce in many developing regions (Table 1), and details how open-access remotely sensed data can compensate for ground monitoring deficiencies in local and transboundary river basins. The applications of remotely sensed data sets such as altimetry, digital elevation models, radar and optical images in each flood mapping process are discussed. To further demonstrate the usefulness of open-access remotely sensed data in developing regions, Nigeria is used as the case study for this review, which is a typical data-sparse country that has experienced severe flooding in recent years, the prospects for the use of remotely sensed data are discussed.

2. Data Limitations, Prediction of Ungauged Basins and Remote Sensing Advancements

In recent decades, floods have been perceived to be increasingly frequent, widespread and more devastating. As such, existing spatial networks of hydrological gauging stations have become inadequate for optimal data collection [22]. In some case, obsolete equipment, financial and technical challenges hamper sufficient data collection for flood modelling and mapping [23–25]. Due to increasing global data deficiency and the uncertainty associated with sparse data for hydrological and hydrodynamic modelling, the International Association of Hydrological Sciences (IAHS) launched the Prediction of Ungauged Basins (PUB) initiative to explore alternative data and techniques for improved ungauged basin modelling [26]. One of the core objectives of the PUB is to "Advance the technological capability around the world to make predictions in ungauged basins firmly based on local knowledge of the climatic and landscape that controls hydrological processes, along with access to the latest data sources, and through these means constrain the uncertainty in hydrological predictions" [27]. This objective aligns seamlessly with remote sensing (RS), considering that it provides an alternative data source to improve our understanding of local hydrology and associated uncertainties in flood mapping for data-sparse regions [28].

RS has advanced enormously in recent decades, and this has led to the availability of free datasets in many parts of the world, thereby enabling developing countries to explore its potential at little to no data acquisition cost [29]. This review focuses on the integration of open-access (freely available) satellite data into fluvial (river) flood mapping processes to compensate for data sparsity faced in developing regions, then uses a Nigerian case study to assess the possibility of leveraging on global geospatial technology for local and transboundary flood management. Inferences are drawn from previous reviews on low-cost Geographic Information System (GIS) and RS applications in hydrology, hydrodynamic modelling and flood mapping [30–32]. However, a wider range of freely available datasets and sources needed for every flood-mapping step listed in Table 1 are explored in this review.

3. Open-Access Remotely Sensed Data Sources for Flood Modelling and Management

3.1. Radar Altimetry for Water Level and Elevation Measurements

River water levels are an essential data input (initial and boundary conditions) for hydrology and hydrodynamic modelling [33], and advances in RS have improved the way changes in water surface elevation and slope can be measured since the early 90's [34]. Radar altimetry missions originally developed for ocean water level measurements now routinely measure freshwater surface elevation of large rivers [35,36]. Radar altimetry data is acquired via a process that measures the distance between the orbiting satellite and water surface in relation to a reference datum, by estimating the time it takes a sensor emitted echo pulse to be reflected by a water surface and return to satellite [37,38], using Equations (1) and (2), and the schematic of this methodology is presented in Figure 1. Altimetry water levels are usually measured at virtual stations located intermittently where altimetry satellite tracks cross path with rivers [39,40], see Figure 2. When altimetry tracks pass over dry land, the elevation of the surface intersected is measured; this is elaborated later in Section 3.1.2. A sample of altimetry time series extracted from the surface monitoring by satellite altimetry database [41] for the Niger River in Nigeria is presented in Figure 3.

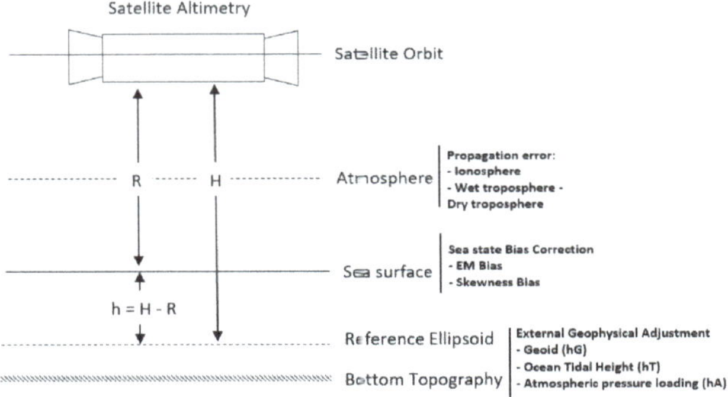

Figure 1. Graphic illustration of satellite altimetry height measurement principle (adapted from [40]).

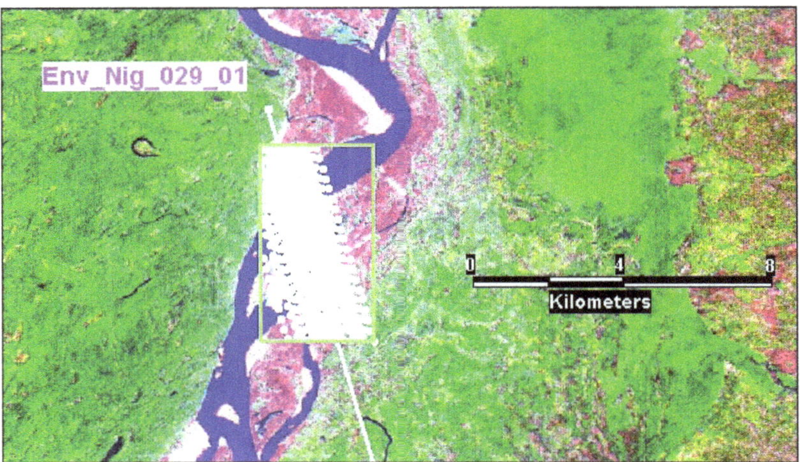

Figure 2. Illustration of a virtual station, where altimetry satellite tracks intersect the river Niger.

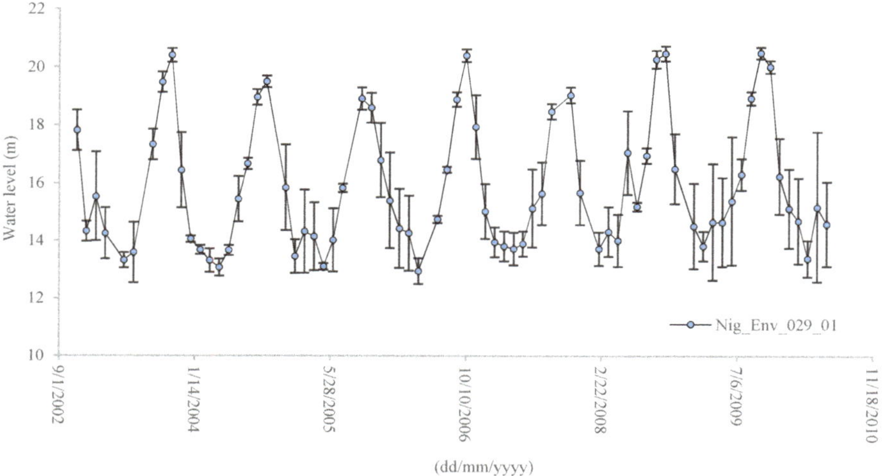

Figure 3. Typical water level time-series extracted from an altimetry virtual station along the river Niger (error bars indicate standard deviation from water level height (m), and are low during peak flooding, as altimetry measurement accuracy is improved during this season [42].

The water level at a river of interest with reference to a predefined datum (such as Earth Gravitational Model (EGM 2008)), is expressed as:

$$h = H - R_{Cor} \tag{1}$$

$$R_{Cor} = R - \left(c\frac{\Delta t}{2} \right) - \sum Cor \tag{2}$$

where, h = water surface elevation in relation to the reference ellipsoid, H = altitude of the satellite (from satellite orbit to reference ellipsoid), R = range (distance between satellite and open surface water body), R_{Cor} = corrected range, c = speed of light, $\frac{\Delta t}{2}$ = the dual direction travel time of radar signal, and $\sum Cor$ = the sum of ionospheric, tidal, wet and dry tropospheric corrections.

The vertical accuracy of altimetry water levels contributes to hydrologic and hydraulic modelling outcome uncertainties [43]. In comparison to ground (*in-situ*) measurements, altimetry water level vertical accuracy ranges from approximately 0.01 to 0.05 metres, and Root Mean Squared Error (RMSE) from 0.003 to 0.004 metres for watershed areas up to 100 km^2 [36,44–46]. In some cases, the difference between altimetry and in situ water levels can be as high as 2 metres [47]. Accuracies of altimetry water level are presented in Table 2 and these variations in accuracies are attributed to the different sensor types, the distance between in situ and virtual station, and location of altimetry track intersection with the river [29]. Other factors that affect altimetry accuracy include ionosphere, troposphere, instrument noise, geoid, tidal and water surface variations [38,48,49], as well as local topography and heterogeneity of reflecting land surfaces [50]. The river width at the location of the virtual station overpass if lower than the altimetry satellite track footprint and the presence of a tributary or distributary between in situ and virtual station have also been identified as the external factors that can contribute to altimetry water level discordancy from ground level measurements [37,51]. Despite these limitations, altimetry has been widely applied in hydrology and the four key areas of deployment, particularly in the context of hydrodynamic modelling in data-sparse regions, are discussed in the following sub-sections.

Table 2. Altimetry characteristics adapted from [43].

S/N	Mission	Ground Footprint (m)	Revisit Time (days)	Operation Timeline	Accuracy (m)	References
1	TOPEX/Poseidon	~600	9.9	1993–2003	0.35	[46]
2	ERS-1	~5000	35	1991–2000	N/A	[36]
3	ERS-2	~400	35	1995–2003	0.55	[46]
4	ENVISAT	~400	35	2002–2012	0.28	[46]
5	Jason-1	~300	10	2002–2009	1.07	[52]
6	ICE Sat/GLAS	~70	-	2003–2009	0.10	[53]
7	Cyrosat-2	~300	369	2010 *	<SRTM (30)	[54]
8	Jason-2	~300	10	2008 *	0.28	[52]
9	SARAL/Altika	~173	35	2013 *	0.11	[55]
10	Sentinel 3 SRAL	~300	27	2016 *	0.03	[36]
11	Jason-3	~300	10	2016 *	0.03	[56]
12	SWOT	~10–70	21	2020 +	0.10	[57]

S/N = Sequential Number; Current = *, Future = +, SRTM = Shuttle Radar Topography Mission.

3.1.1. Altimetry for Discharge Estimation

River discharge and water level often used as initial/boundary conditions for hydrodynamic and hydrological models are rarely available at most remote locations of many developing regions due to factors previously highlighted in Section 2 [23,39]. Radar altimetry has been explored in several studies to curb data limitation challenges and reduce the uncertainty associated with modelling ungauged rivers.

Papa et al. [58] utilised TOPEX/Poseidon, ERS-2, ENVISAT and Jason 2 altimetry water levels in combination with in situ rating curves to estimate discharge along the Ganga and Brahmaputra rivers from 1993–2011. Accuracy levels of 0.17 (mean error) and 0.28 (standard error) metres in comparison to in situ discharge at gauging stations were achieved. River discharge along the Godavari river from 2001 to 2014 was derived by combining ENVISAT (2002–2010), Jason-2 (2008–2014) and SARAL/Altika (2013–2014) radar altimeter water levels with *in-situ* rating curves at nearby gauging stations. When validated against a hydrodynamic model a correlation coefficient (R^2) of 0.9 and a standard error varying from 0.15 to 0.40 metres were achieved [59]. In an Amazon River basin study, Getirana and Peters-Lidard [60] explored the potential of estimating discharge using altimetry data from ENVISAT (2002–2005). Using the relationship between in situ water level and discharge, Getirana and Peters-Lidard, [60] successfully estimated discharge at 90 virtual stations with mean relative errors varying from 15 to 84% for small and large and river basins respectively. Discharge was estimated at transboundary rivers including the Danube (Austria, Romania, Bulgaria, Slovakia, Hungary, Ukraine, Croatia, Germany, Serbia, and Moldova), Mekong (Thailand, Cambodia, Laos, China, Myanmar (Burma and Vietnam), Amazon (Ecuador, Colombia, Peru, and Brazil), Brahmaputra (India), Amur (China and Russia), Ob (Russia), Vistula (Poland) and Niger (Nigeria, Mali, Niger, Benin, and Guinea), using a quantile function algorithm that exploits ENVISAT altimetry data [61]. This approach resulted in discharge outcomes similar to those derived from a conventional Forecast Rating Curve (FRC) approach.

The studies presented above indicate that river discharge estimation from altimetry water levels typically depends on the rating curve or river geometry data availability [62]. However, several studies have been able to demonstrate direct river discharge estimation from altimetry water levels in the absence of in situ measurements, using supplemental remotely sensed data or models. ENVISAT altimetry data from six virtual stations along the Brahmaputra river from 2008 to 2010 were assimilated into a Muskingum routing model driven by outputs of a calibrated Budyko type rainfall-runoff model derived from Tropical Rainfall Measuring Mission (TRMM) Multi-satellite Precipitation Analysis (TMPA) 3B42RT real-time products. This integrated approach improved the model's discharge predictive accuracy (Nash-Sutcliffe (NS) efficiency) from 0.78 to 0.84. Additionally, Tarpanelli et al. [63]

combined Moderate-resolution Imaging Spectroradiometer (MODIS) Terra and Aqua satellite images with ENVISAT altimetry using a pixel to water level detection approach to estimate discharge with a correlation coefficient of 0.96 and NS efficiency of 0.91 when compared to in situ discharge along the Niger and Benue rivers. Similarly, Sichangi et al. [64] integrated MODIS satellite-derived river width and altimetry water levels into Manning's equation to estimate discharge at a continental scale. The derived discharge NS efficiency varied from 0.60 to 0.97.

Although the discharge estimates derived from radar altimetry as presented above are perceived to be within acceptable levels of uncertainty, factors such as the distance between virtual and ground stations, contributing tributaries and the width of the river affect the accuracy of such estimates [51]. The studies discussed above also reveal that the availability of supplementary remotely sensed data and hydrodynamic models can enable improved discharge estimation in ungauged river basins.

3.1.2. Altimetry for Digital Elevation Model Accuracy Assessment

Once the discharge and/or flood magnitude is estimated, it is propagated longitudinally along river channels and laterally across floodplains using hydrodynamic models governed by continuity and momentum equations [65]. The accuracy of the DEM that defines the river channel and floodplain terrain upon which flow is propagated influences model outcome accuracy [66]. Therefore, in several flood modelling studies the accuracy of the primary DEM is assessed prior to usage against a higher accuracy DEM such a Light Detection and Ranging (LiDAR) or Differential Global Positioning System (GPS) elevation points [67–70]. Acquiring such detailed topography datasets for [2complexity and weather conditions that hinder logistics and field operations [71,72].

Data acquired by the National Aeronautics and Space Administration (NASA) between 12 January 2003 and 11 October 2009 using the Geoscience Laser Altimeter System (GLAS) onboard the Ice Cloud and Land Elevation Satellite (ICE Sat) provides a worthy alternative to ground elevation data due to its high accuracy in comparison to Kinematic GPS measurements [73]. The absolute accuracy of ICE Sat has been shown to range from 0.002 to 0.005 m in Bolivia [74] and French Lake [75], respectively, and depends on the slope of the terrain under scrutiny [76]. Over the years ICE Sat/GLAS has been applied in assessing various DEM accuracies including SRTM [77–79], ASTER GDEM [76,80], GPS elevation [81], Carto DEM [82], Canadian DEM [83], InSAR DEM [84], TanDEM [85] and modified/corrected DEMs [52,86,87]. The 70-m ground footprint of ICE Sat [73] coupled with its ability to penetrate gaps in vegetation canopy to capture underlying bare earth elevation [88] makes it a useful alternative to ground survey for DEM accuracy assessment.

3.1.3. Altimetry for Bathymetry Delineation

Accurate digital elevation models combined with detailed river bathymetry delineation provides the most accurate terrain data for flood routing [65,89]. Nevertheless, acquiring such data for remote locations is usually difficult as discussed earlier (Section 3.1.2). Hence, flood modellers have resorted to exploring alternative options to compensate for such deficiencies. In the Amazon and Napo Rivers in Peru, Chávarri et al. [90], examined the applicability of altimetry (ENVISAT) in constraining river cross-sections of a one-dimensional hydraulic model. The results showed reduced model uncertainty, mostly for rivers with widths less than or equal to 2.5 km. The relationship between river width and depths established using ENVISAT altimetry was combined with SRTM, Landsat, MODIS and satellite rainfall data to derive an updated river network and adjusted bed profile used in the development of Ganges, Brahmaputra, and Meghna (GBM) model suitable for large ungauged watersheds [33]. The GBM model data integration approach resulted in a reduced RMSE from 3.0 to 1.0 metres.

The proposed Surface Water and Ocean Topography (SWOT) scheduled for launch in 2020 is expected to provide some of the best altimetry data for water resource monitoring and management at a global scale [57,91]. A few studies have explored the potential of SWOT derived bathymetry for improving the accuracy of hydrodynamic modelling. For example, Durand et al. [92] experimented simulated data of the SWOT mission, applying data assimilation technique to estimate bathymetric depth and slope at five points along a 240 km reach along the Amazon river to within 0.50 m and 0.30 cm km^{-1} accuracies, respectively. These outcomes were then integrated into the LISFLOOD-FP hydrodynamic model [93] to improve estimates of inundation extent and downstream water surface elevation (WSE). SWOT WSE was also assimilated into the LISFLOOD-FP hydrodynamic model using a local ensemble batch smoother (LEnBS) method by Yoon et al. [94], to generate bathymetry, depth and discharge estimates. Bathymetry extracted from SWOT had a RMSE of 0.56 metres, improving with the inclusion of more SWOT observations in the modelling process. The proposed SWOT and recently launched Sentinel-3 provides a huge prospective dataset for future of hydrodynamic studies, and their integration into hydrodynamic models can improve flood extent, discharge and water levels predictions, particularly when multiple altimetry data are available along a modelled reach, as Yoon et al. [94] suggested.

3.1.4. Altimetry for Hydrodynamic Model Calibration and Validation

Hydrodynamic model calibration is usually undertaken by adjusting various model parameters such as floodplain roughness, channel roughness, river channel depth and river width in order to tune model outputs (water level, discharge and/or inundation extent) to observations, derived from in situ or remotely sensed measurements [38,42,95,96]. Validation, on the other hand, helps reveal how well a model represents what is found in reality [97], and is directly linked to the confidence in the flood management measures implemented as a result of the model outcome. Commercial high-resolution optical and radar satellites images, aerial images and hydrological data have been largely established as the optimal data sources for hydrodynamic model calibration and validation [98–101]. However, the high cost of acquiring such data hinders their application in developing countries [102]. Hence, radar altimetry over the past decade has been explored globally as an alternate source of data for model calibration and validation [103].

Typically, in many developing regions river measurements are manually collected using staff gauges and later converted to discharge using an established rating curve. At the peak of floods, measurement equipment may be damaged, or access roads inundated, thus impeding the observation process [32]. Therefore, radar altimetry provides an alternative river measurement option that supports hydrodynamic model calibration and validation in the absence of observed records [103].

Water level data from three ENVISAT altimetry virtual stations along a 150 km reach of the Danube river were applied in the calibration of a 2-D LISFLOOD-FP model to reconstruct the 2006 transboundary flood occurrence [104]. Yan et al. [104] achieved a Mean Average Error (MAE) of 1.53 m and 1.37 m for altimetry and in situ model calibration approaches, respectively, suggesting that both datasets can be used interchangeably to improve flood modelling in sparsely gauged river basins. Domeneghetti et al. [105] performed hydrodynamic model calibration for a 140 km reach along the Po river using ERS-2 and ENVISAT altimetry data, resulting in RMSE of 0.85 m and 0.73 m respectively, and an improved NS efficiency when altimetry is combined with in situ data for model calibration. An implementation of the Soil and Water Assessment Tool (SWAT) rainfall run-off model for the sparsely gauged Okavango transboundary river of Angola, Namibia and Botswana were calibrated using total water storage derived from Gravity Recovery and Climate Experiment (GRACE) altimetry satellite and in situ data [106]. In addition, Sun et al. [42] assessed the uncertainty associated with Hydrological Model (HYMOD) along the Mississippi River, calibrated against in situ and altimetry data. NS efficiencies of 79.05 and 64.50 were reported for in situ stream flow and radar altimetry (TOPEX/Poseidon), respectively, showing reduced uncertainty for streamflow calibration in comparison to altimetry calibration.

Notwithstanding the value of radar altimetry for hydrodynamic model calibration and validation, residual altimetry uncertainties are expected to affect flood model accuracy as Tommaso et al. [107] demonstrated. This was further emphasised by Domeneghetti et al. [105], where ENVISAT proved to provide higher accuracy than ERS-2 (See Table 2 for altimetry accuracy differences). Belaud et al. [38] applied TOPEX/Poseidon (T/P) and ENVISAT altimetry data to calibrate a propagation model and disclosed that inherent altimetry uncertainties have an effect on the model outcome. Despite these deficiencies, the importance of altimetry data in model calibration and validation in ungauged basins cannot be dismissed. However, it is advised that altimetry is applied in combination with *in-situ* data when available [105], and when there is a choice in situ data should take priority over altimetry [108].

3.2. Open-Access Digital Elevation Model Data and Applications in Flood Modelling

Topographical data is an essential requirement in hydrological and hydrodynamic modelling, especially for ungauged river basins [29,109], and accounts for a substantial portion of the uncertainty that propagates through to model outcomes [66,110]. The effect of terrain accuracy on hydrodynamic models and the need for accuracy assessment have been discussed briefly in Sections 3.1.2 and 3.1.2, revealing how improved river channel characterization using altimetry can improve flood model outcomes [90,92,94]. High-resolution topographical data such as LiDAR, TanDEM, bathymetry and differential Geographic Positioning System (dGPS) survey provides the best terrain depiction with reduced uncertainty and error [19,89,111,112]. However, the cost of acquiring such data is enormous [69] and in other cases, remote locations are inaccessible and security challenges add to the complexity of field surveys [52]. Freely available DEMs have been widely used as an alternative to commercial data in many developing regions where data is sparse, and resources limited [67,113].

The Shuttle Radar Topography Mission (SRTM) DEM is arguably one of the most widely used topographical data in developing regions, applied in improving flood modelling in data-sparse regions [52,69,103,114]. The 30 and 90 m resolution SRTM was collected during an 11-day mission in February 2000, through a collaborative effort involving NASA, the National Geospatial-Intelligence Agency (NGA) and the German Aerospace Centre (DLR), and provides near-global scale (80%) elevation data [115,116]. The 15-m Advanced Spaceborne Thermal Emission and Reflection Radiometer (ASTER) Global Digital Elevation Model (GDEM) acquired by a joint mission of the NASA and Japan's Ministry of Economy, Trade, and Industry is also widely used in flood modelling and mapping [70,117,118]. However, the ASTER GDEM is argued to be less accurate than SRTM due to extensive elevation pixel voids [68,91].

Other open-access topographic data sets such as Altimeter Corrected Elevations 2 (ACE2) GDEM, Global 30 Arc-Second Elevation (GTOPO30) and Global Multi-resolution Terrain Elevation Data 2010 (GMTED2010) are generally coarse in resolution and are therefore employed in large-scale models only [114,119]. The recently released Advanced Land Observing Satellite (ALOS) DEM [120] has been evaluated and established to provide more accurate elevation in comparison to SRTM and ASTER [121]. A recent flood extent modelling study by Courty et al [122] revealed that the ALOS DEM outperformed its SRTM counterpart. The properties of various open-access DEMs and some case studies are presented in Table 3. The discrepancies between open-access DEM and ground surveyed elevation presented in Table 3 can be attributed to inherent systemic and external factors [115]. For the SRTM, system noise, as well as beam reflection off forest canopies, water bodies and rooftops in urban areas contribute to DEM bias and elevation overestimation [66,78,123,124].

Table 3. Open source digital elevation models properties and case studies.

DEM	Spatial Resolution (m)	Vertical Error (m)	Case Study	Reference
SRTM	30, 90	±16	Damoda River, India.	[69,125]
ASTER GDEM	30	±25	Lake Tana, Ethiopia.	[126,127]
ACE 2 GDEM	1000	>10	Balkan Peninsula, Croatia	[128]
GTOPO30	1000	9–30	Balkan Peninsula, Croatia	[128]
ALOS	30	±5	Sindh and Balochistan, Pakistan	[120,129]
GMTED2010	250	26–30	Shikoku, Japan.	[130,131]

Various methods have been adopted to curb the deficiencies and reduce the uncertainty associated with open-access DEMs. For example, Baugh et al. [124] reduced STRM uncertainty by combing vegetation canopy heights [132,133] and MODIS imagery to reduce vegetation height effects. Betbeder et al. [134] reduced SRTM bias by 64 percent by adopting a systematic approach that combines vegetation height [132], Landsat land cover map and radar altimetry to produce a hydrologically corrected DEM. SRTM derived river cross-sections were adjusted using limited bathymetric surveys and applied in the one-dimensional MIKE11 model [67] and LISFLOOD-FP two-dimensional model Sanyal et al. [69] to reduce model uncertainty. Neal et al. [114] adopted an approach that reduced SRTM uncertainty by making hydrodynamic model parameters such as channel width and depth calibratable in a sub-grid LISFLOOD-FP model, thereby improving simulated water levels, wave propagation and flood extent. Biancamaria et al. [135] experimented by varying river channel depth in the SRTM DEM by 5, 10 and 15 m when modelling Obi river, and identified 10 m as the optimal average river channel depth for the best outcome. In a recent study in Australia, Jarihani et al. [52] adopted the Hydrological Correction (HC) and Vegetation Smoothening (VS) [136] approaches to reduce SRTM and ASTER DEM error and deduced that the HC DEM outperformed the VS DEM for flood modelling.

Although the DEM modification techniques described above resulted in reduced DEM and flood model uncertainty, they require specific skill sets, computational power and supplementary data that are not always readily available. Hence, there is a need to identify globally available off-the-shelf modified DEMs that can be readily applied in developing regions where such resources are seldom available. At a global scale, errors emanating from satellite system noise, and sensor beam reflection off vegetation canopy, water surfaces and urban rooftops have been treated with different techniques, resulting in the development of freely available new data sets. O'Loughlin et al. [137] reduced average vertical bias from 14.1 m to 5.9 m by systematically combining ICESat GLAS ground elevation [73], vegetation height [132], MODIS-derived forest canopy density and climate regionalization maps [138,139]. Sampson et al. [86] reduced SRTM sensor noise irregularities, urban landscape and vegetation canopy elevation overestimations using a moving window filtering technique [136]. Their approach reduced RMSE from 10.96 m to 6.05 m when compared to LiDAR, and overall flood model bias from 15.08 m to −0.1 m. The EarthEnv-DEM90 was developed by Integrating ASTER GDEM2, CGIAR-CSI SRTM V4.1 and Global Land Survey Digital Elevation Model (GLSDEM) using a combined delta surface filling [140] and adaptive DEM noise smoothing [136] methodology, resulting in minimised error compared to raw SRTM and ASTER GDEM2 [141]. A recent DEM developed by Yamazaki et al [140] was developed using a multi-error removal approach that removed error factors that include absolute bias, stripe noise, speckle noise, and tree height bias using multiple satellite datasets and filtering techniques, resulting in Multi-Error-Removed Improved-Terrain DEM (MERIT DEM). The properties of various modified SRTM DEMs and some case studies are presented in Table 4.

Table 4. Globally available Modified SRTM DEM properties and case studies.

DEM	Spatial Resolution (m)	Vertical Error (m)	Case Study	Reference
Bare-Earth SRTM (Veg/Urban)	90	6.05–12.64	Belize, Honduras	[86]
Bare-Earth SRTM (Veg)	90	4.85–8.667	Global	[87]
EarthEnv-DEM90	90	4.13–10.55	Johor River Basin, Malaysia	[141,142]
MERIT DEM	90	±2	Nile Basin, Congo and Ob rivers	[140]

Since no study currently presents a comparison of all globally available modified SRTM DEMs for a specific region, a comparative analysis of all the modified DEMs and raw SRTM is presented in Table 5, evaluated against ICE Sat/GLAS altimetry data for the Niger-South river basin of Nigeria. The result reveals that Bare-Earth SRTM corrected for vegetation provides the best elevation estimates in comparison to the ICE Sat/GLAS altimetry dataset. This is expected to support the selection of globally modified SRTM DEMs for flood modelling and mapping studies going forward.

Table 5. SRTM and Modifications comparison with ICE Sat/GLAS altimetry elevation.

Elevation	Min	Max	Mean	Std. Dev.	R^2	RMSE
Bare-Earth SRTM (Urban and Veg)	−3.89	151.00	29.65	37.66	0.99	3.21
Bare-Earth SRTM (Veg)	0.35	151.18	29.72	37.72	0.99	2.96
EarthEnv90	3.00	152.00	30.95	37.45	0.99	3.76
MERIT DEM	−1.27	148.44	28.96	37.71	0.99	3.68
Raw-SRTM	2.00	153.00	30.33	37.48	0.99	3.27
ICE Sat/GLAS	0.30	148.35	30.28	37.64	-	-

Std. Dev. = standard deviation, R^2 = Correlation coefficient, Numbers of data Points = 522.

3.3. Open-Access Optical and Radar Satellite Images and Applications in Flood Modelling and Mapping

Optical and radar images also play a crucial role in flood modelling and mapping, being used for a range of applications including (i) manning's roughness derivation [143], (ii) river width estimation [143], (iii) geomorphological properties extraction [143], (iv) inundation extent mapping [112], (v) river discharge estimation [144,145], (vi) land use/cover derivation [146], (vii) bathymetry estimation [147] and (viii) hydrodynamic model calibration and validation [148]. In this context, open-access images from Landsat, MODIS and ASTER have been widely used in developing regions [32]. Until the launch of the free high-resolution C-Band Sentinel-1 SAR mission by the European Space Agency (ESA) in 2014, the use of radar imagery in developing regions has been limited due to the cost of acquisition [149,150]. Nevertheless, other low-cost radar satellite images such as ERS-1, ERS-2, JERS1 and ALOS PALSAR have widely been applied for flood modelling [29].

Optical and radar RS each provide unique merits and demerits and are characterised based on the source of energy employed during data collection. Optical (passive) RS relies on solar energy, while radar (active) RS uses an inbuilt energy source onboard the satellite [32]. Therefore, optical remotely sensed data can only be captured in the day-time and depends on cloud-free skies [32]. However, its multispectral characteristics make it a suitable for land use/cover classification, inundation delineation, drainage mapping and flood impact assessment [40,151,152]. Flood extent is derived from the discrimination between the spectral signatures of water surface and the surrounding landscape in single or multi-temporal images, using classification or spectral indices approaches [151,153].

Radar RS has the ability to penetrate clouds and its ability to discrimination water makes it the optimal data type for flood mapping when available [149,154]. Flood maps are usually extracted by pixel discrimination, given that flooded pixels tend to have lower values of back-scatter, due to the weak return signal associated with a smooth water surface [155]. The discrimination method applied can strongly influence the accuracy of the derived flood extent [156]. Analytical techniques for flood mapping using radar data include statistical active contouring, radiometric thresholding, histogram thresholding, pixel-based segmentation, fractal dimensioning of multi-temporal images, neural networks in a grid system, image segmentation and decision tree analysis [157,158]. Despite the advantages of radar RS, sensor noise and backscatter from vegetation and buildings have been identified as factors that hamper flood discrimination potential using radar data [157,159,160]. Furthermore, the temporal resolution, spatial accuracy and flood detection precision also affect the usability of radar images, especial for near-real-time flood forecasting in data-sparse regions [161,162].

The properties of some open-access optical and radar data sources are presented in Table 6, along with some case studies in flood modelling and mapping.

Table 6. Optical and radar satellite imageries case studies.

Sat. Imagery	Res. (m)	Case Study	References
Landsat	30	Floodplain inundation delineation for 2 and 1–dimensional model calibration and validation, Inner Niger and Missouri River, Nebraska, USA	[114,163]
MODIS	200	Hydrodynamic model calibration and validation.	[113,164]
Terra ASTER	15	Urban sprawl and flood management Dhaka, Bangladesh	[113,164]
Sentinel-1	10	Sentinel-1 and Landsat-8 combination in mapping flooding at river Evros, Greece	[113,164]
Sentinel-2	10	Water bodies delineation	[113,164]

Sat. = Satellite, Res = Spatial resolution

4. Case Study: Open-Access Remotely Sensed Data Applications in Flood Monitoring and Management in Nigeria

Nigeria, used as the case study for this review is located downstream of the Niger Basin (Figure 4) which collects run-off from a 2,156,000 km^2 area and passes this through the Niger and Benue rivers [165]. Thus, Nigeria is prone to fluvial flooding, which exposes floodplain dwellers to diverse negative consequences [166–169]. Nigeria recently experienced unprecedented levels of flooding attributed to poor dam water release management and risk communication, linked to data unavailability for informed and prompt decision making [165].

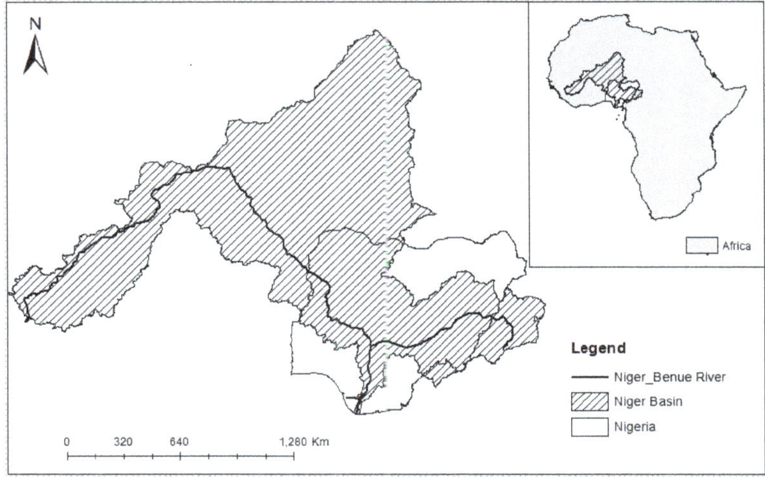

Figure 4. Map showing Nigeria, Niger Basin, Africa and the main inflow rivers (Niger and Benue).

This section focuses on identifying the causes of data deficiencies in Nigeria and presents the outcome of reviewed literature on the applications of open-access remotely sensed data in Nigeria, to identify gaps and opportunities for research based on global trends discussed in the preceding sections. This section builds upon previous reviews on GIS and RS in flood risk management in Nigeria [170–174], then expands further on data challenges, solutions and prospects for regional and national flood management using open-access remotely sensed data.

4.1. Hydro-Meteorological Data Limitations in Nigeria

Similarly to many developing countries, the lack of hydro-meteorological data in Nigeria has been widely documented, and the consequences for flood management decisions have been identified [175]. Currently, existing hydrological and meteorological gauge distributions do not meet the recommendations of the World Meteorological Organization [176] and Ngene [177], i.e., 237 hydrological stations exist out of 384 recommended and 291 meteorological stations (rain gauges) exist of 970 recommended. In addition, several of the established stations have been reported to be inactive, decommissioned or discontinued (Figure 5), contributing to the data sparsity in the country [175,176].

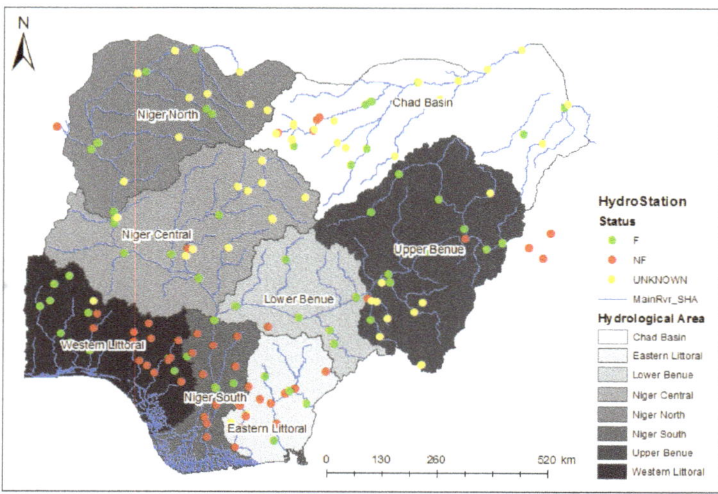

Figure 5. Status of some hydrological gauging stations in Nigeria (F = Functional, NF = Non-Functional, Unknown).

Lack of financial support, technical deficiency, poor institutional capacity and obsolete infrastructure have been identified as factors responsible for data shortages in Nigeria [178–180]. Poor hydrological data management systems and lack of standards have led to unreliable and inconsistent data (Maxwell, [24]; Ononiwu, [181]. Furthermore, Maxwell [24] and Olayinka [182] argued that even when data is available, custodians store data in paper formats, thus reducing transferability, applicability and long-term/sustainable data provision.

Hydro-meteorological data are essentially applied in estimating expected flood magnitudes based on past trends, and a restricted length of available historical data contributes to the uncertainty in the derived flood estimates [183,184] . Extended historical data result in more accurate estimates and vice versa [13]. For the purposes of the present study, in 2016 a search was conducted within the peer-reviewed literature on the Google scholar (https://scholar.google.com) database spanning the years 2000 to 2016. A combination of the search terms and keywords including "hydrology", "flood modelling", "hydrodynamic modelling", "flood frequency analysis", "vulnerability assessment", "rainfall frequency analysis", "flood mapping", and "GIS and Remote sensing of flooding", were used, with the results further refined with keywords such as "Nigeria", to represent the country of interest. A meta-analysis of these river and rainfall estimation studies (Figure 6) shows that rainfall data sets are generally longer in duration than those of streamflow data. The majority of hydrological modelling studies are based on historical data of lengths ranging from 10 to 20 years, hence there is a need for the adoption of an approach that leverages data from multiple gauging stations to reduce flood estimate uncertainty and improve flood management decision making [185].

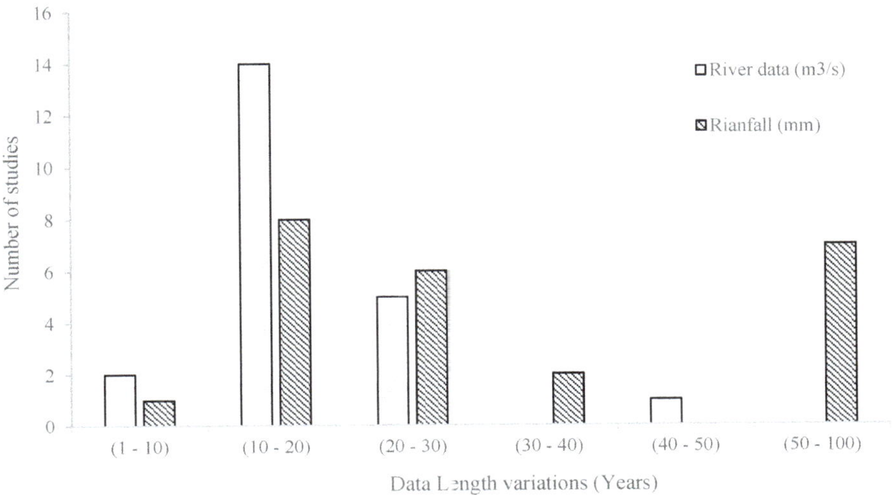

Figure 6. Rainfall and streamflow data length variation in years from previous studies in Nigeria.

4.2. Remote Sensing for Flood Management in Nigeria

RS has been applied in seven sub-categories of flood management in Nigeria: (i) vulnerability assessment: integrating socio-economic and biophysical factors to ascertain a regions' coping capacity in relation to flood exposure [186–188]; (ii) flood frequency analysis: estimating expected flood magnitudes by fitting historic flood time series to a suitable probability distribution or combining hydrological data from regions of physiographic similarity [179,189,190]; (iii) rainfall intensity-duration-Frequency: applying rainfall data to gives an idea on return period of rainfall intensity which can be expected for a defined period [191,192]; (iv) hydrodynamic modelling: once flood estimates are determined, the outcomes are routed in 1/2 dimensional models in combination with terrain data to derive flood hazard information such as inundation extent, depths and /or velocity [193]; (v) flood risk mapping: other than hydraulically modelling flood hazard, flood depths and inundation extent for a particular point in time can be directly determined using satellite images and digital elevations models [21,167]; (vi) floodplain encroachment analysis: the increasing development of industries and settlements within the floodplain increases exposure and vulnerability [188,194]; (vii) rainfall varibility assemment: understanding the degree to which the amount of rainfall across an area varies through time and space [195,196]; and RS and GIS approaches are used to monitor floodplain encroachment, to ensure adherence to, and enforcement of flood management policies [197,198]; and (viii) water resource management: adoption of GIS and RS for sustainable water resource management [199]. Figure 7 shows the flood studies application areas in Nigeria, revealing vulnerability mapping, flood frequency assessment and risk assessment are the main areas of interest.

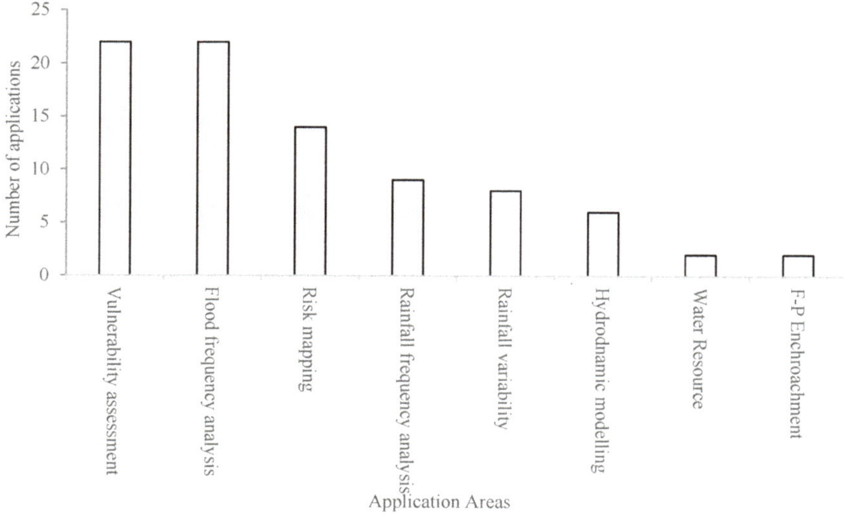

Figure 7. Flood studies in Nigeria showing specific application areas.

4.3. Applications of Open-Access Remotely Sensed Data for Flood Management in Nigeria

Meta-analysis of 100 flood research journal articles focused on Nigeria acquired from google scholar using the methodology described in Section 4.1. shows the range of data applied in flood management studies (Figure 8) and reveals high reliance on Landsat and SRTM. Various data sets provide contrasting levels of accuracy and uncertainty [110], therefore high spatial resolution data such as LiDAR and SAR are mostly recommended for flood modelling processes due to the advantages of LiDAR's ability to delineate complex terrains with high levels of details and the effective water surface discrimination capacity of SAR imagery [150,200].

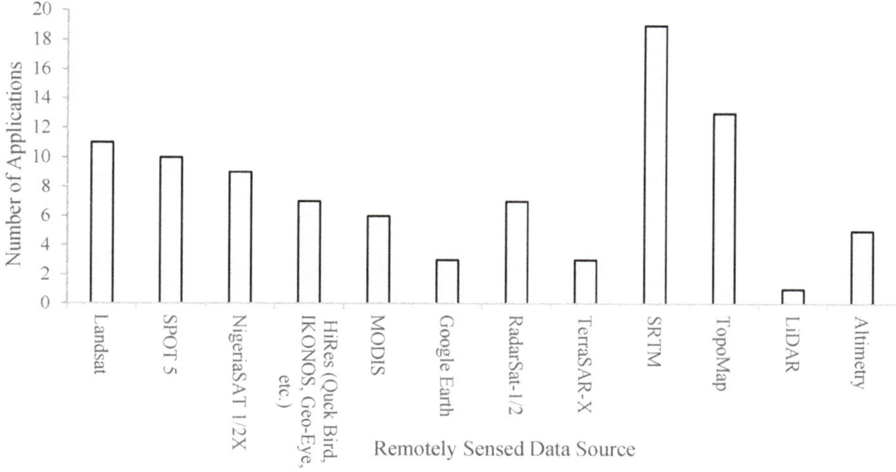

Figure 8. Remotely sensed data application in flood studies in Nigeria.

Figure 9 further demonstrates the difference between flood extent extracted from radar and optical images, revealing the optical satellite image's deficiency in delineating flood extent, especially in cloudy regions of tropical environments. The TerraSAR-X (radar) flood map was derived using a histogram thresholding approach by the Disaster Charter consortium, while the MODIS (optical) flood extent was automatically generated from the Modis Water Product (MWP) through a collaborative effort between NASA and Dartmouth Flood Observatory, University of Colorado, USA, using an algorithm that uses a ratio of MODIS 250-m Bands 1 and 2, and a threshold on Band 7 to provisionally identify pixels as water [201].

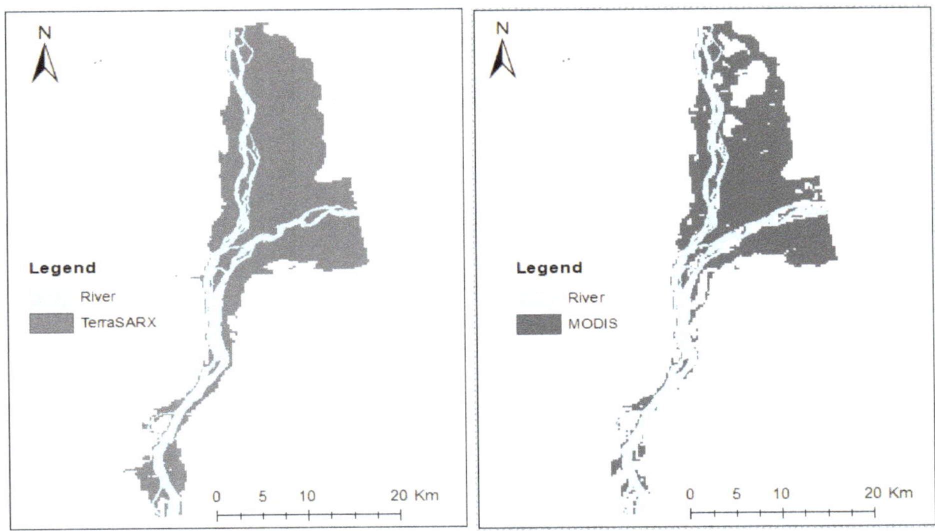

Figure 9. Radar (TerraSAR-X) and optical (MODIS) flood extent comparison at Lokoja, Nigeria.

5. Open-Access Remotely Sensed Data in Transboundary Flood Management

Managing flood occurrences in a sovereign nation is challenging enough; the complexity is increased when floods transcend borders. Floods sometimes originate from one country, and if hydraulically connected to another country within a single catchment area, this travels downstream [202] creating transboundary flooding. Poor management of excess water releases from dams triggered by variable rainfall and other anthropogenic factors have been identified as some of the leading causes of transboundary flooding [203–206]. In such situations, efforts need to be coordinated between flood origin and destination countries to ensure effective flood management. Approximately 2286 transboundary river basins exist globally (Figure 10), encircling 42% of the world's population within a 62 million Km^2 area, and they are responsible for approximately 50% of global river discharge [207,208].

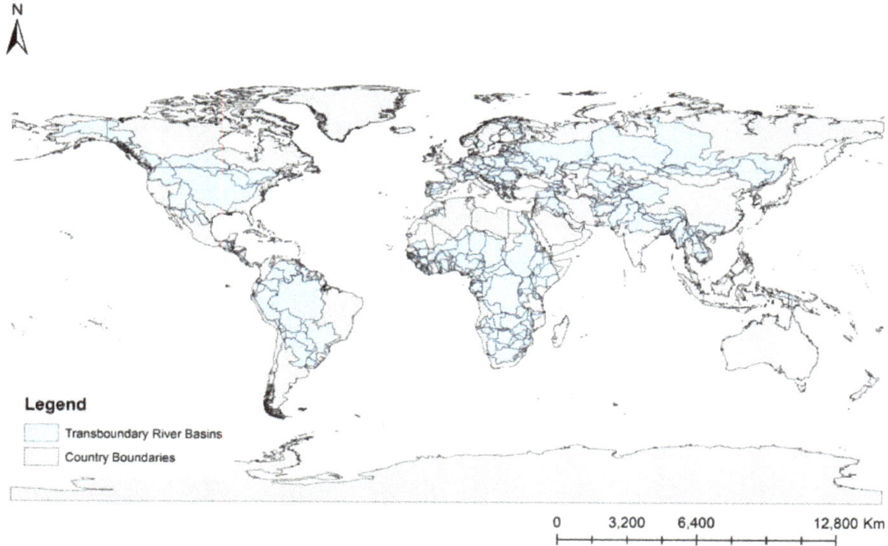

Figure 10. Global Transboundary River Basins (source: Transboundary Freshwater Dispute Database).

Coordinating the activities of individual countries within a transboundary water resource management organisation is particularly challenging due to the diverse interests, policies and activities of riparian countries [209–211], thereby prompting the need for a shift to RS approaches that enables independent data collection by riparian countries without violating administrative protocols [212]. Several RS studies have been undertaken in this regard, using radar altimetry, optical/radar imagery and hydrodynamic models to curb the data limitation challenges associated with poorly coordinated transboundary flood management efforts.

Mallinis et al. [213] delineated the transboundary Evros river (Bulgaria/Turkey) flood extent and damage caused by upstream dam water release using ENVISAT ASAR and post-flood multi-temporal LANDSAT TM images. The effect of varying flood magnitudes released from upstream Ivaylovgrad dam (Bulgaria) on the connecting Ardas River (Greece) was modelled using HEC-HMS, using in situ gauge measurements and digital terrain data [214], thereby enabling effective downstream flood planning and management. Mati et al. [215] investigated changing land use/cover impact on the Mara transboundary river (Kenya/Tanzania) hydrological regime, using remotely sensed data (Landsat MSS, TM/ETM, and SRTM), ground-collected land use/cover data, meteorological and streamflow data integrated within the Geospatial Streamflow Model (GeoSFM). Biancamaria et al. [135] established an empirical relation between downstream altimetry (TOPEX/Poseidon) water levels (India) and upstream in situ measurements (Bangladesh) for forecasting purpose along the Ganges and Brahmaputra transboundary river. Hossain et al. [216] in the same study area applied a forecasting rating curve approach combined with HEC-RAS hydraulic model to forecast downstream water levels using upstream JASON-2 altimetry, in situ water levels and rating curve. Seyler et al. [217] further demonstrated the value of altimetry and SAR satellite missions in transboundary water resource management, as remote locations along the Beni-Madeira river in the Amazon were monitored using ENVISAT altimetry and JERS-1 radar images.

The case studies discussed above illustrate the wide range of applications of open-access remotely sensed data in transboundary flood management, with radar altimetry, DEM, SAR, optical images, as well as hydrodynamic models and empirical formulas identified as alternatives for improved transboundary monitoring. These approached minimise or avoid the bureaucratic challenges of ground-based monitoring across country boundaries. In this way, RS makes it possible to forecast expected floods, estimate flood exceedance probabilities and monitor how changes in riparian land use/cover can affect downstream hydrology, across different countries.

5.1. Transboundary Flood Management Nigeria (Niger River Basin)

The unprecedented flood event of 2012 in Nigeria was attributed to (i) excess water release from dams within and outside Nigeria due to intense precipitation; (ii) inadequate risk communication; and (iii) poor stakeholder collaboration [218,219]. One instance of a lack of transboundary stakeholder collaboration is evident in Nigeria's inability to uphold part of the 1980 agreement with Cameroon, to establish Dasin Hausa dam to buffer the effect of Lagdo dam built by Cameroon along the Benue River [220,221].

The Niger transboundary river basin (Figure 11) encompasses 12 countries including Senegal, Guinea, Côte D'Ivoire Mauritania, Mali, Burkina Faso, Algeria, Niger, Benin, Nigeria, Cameroon and Chad. The basin hosts a human population of 93,617,850 within a 2,156,000 km^2 area [165,208].

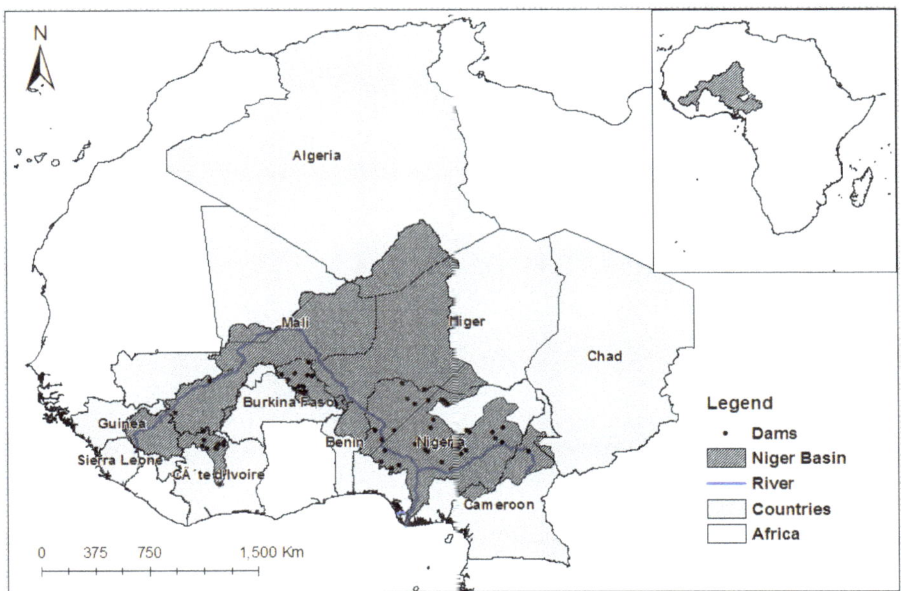

Figure 11. Map of the transboundary Niger River Basin, showing constituting countries and Dams.

Figure 11 also highlights the transboundary nature Niger River Basin, the constituent countries and characteristics. The Niger basin is largely regulated by dams, housing approximately 69 dams [222] conceived mostly as national and local projects, but these have transboundary impacts downstream [223]. To effectively manage transboundary water resource and impact on riparian countries, the Niger River Commission (NRC) was established in 1963, now the Niger Basin Authority (NBA) as reconstituted in 1980, to promote co-operation between member states and ensure sustainable Integrated Water Resource Management [224]. The Niger basin is presently controlled by several post-colonial agreements presented in Table 7.

Table 7. Niger River Basin Agreement, Nigeria. Adapted from [207,225,226].

S/N	Treaty	Function	Location	Year
1	Act regarding navigation and economic co-operation between the states of the Niger Basin.	Navigation and Joint management	Niamey, Niger	1963
2	Agreement concerning the River Niger Commission and the navigation and transport on the River Niger.	Navigation, Joint management, information exchange	Niamey, Niger	1964
3	Agreement Revising the Agreement Concerning the Niger River Commission and the Navigation and Transport on the River Niger.	Navigation, Joint management, information exchange	Niamey, Niger	1973
4	Convention Creating the Niger Basin Authority (NBA)	Water resource mgt. coordination	Faranah, Guinea	1980
5	Protocol relating to the Development Fund of the Niger Basin	Planning funds for NBA	Faranah, Guinea	1982
6	Agreement between Nigeria and Mali	Co-operation on water resource use in the Niger	-	1988
7	Agreement Nigeria and the Republic of Niger concerning the equitable sharing in the development, conservation and use of their common water resources	Environmental conservation and water resource management	Maiduguri	1990
8	Nigeria-Cameroon Protocol Agreement	Coordinate dam water release.	-	2000
9	Niger Basin Water Charter.	NBA review and update.	Niamey, Niger	2008
10	African Risk Capacity	Weather financial risk management	Pretoria, South Africa	2012

Despite these multiple cooperative frameworks, several factors have hindered effective transboundary water resource management in the Niger Basin: (i) poor and fragmented data collection, (ii) lack of coordination between riparian countries and organizations, (iii) poor communication and knowledge of legal and institutional frameworks, (iv) Funding deficiency, (v) lack of clear objectives, (vi) lingual differences, and (vii) technical limitations [226–228]. Grossmann, [229] also lamented the deplorable state of the 65 gauging stations set-up by the NBA, through the "Hydro Niger Project" initiative. Nevertheless, the emergence of the ongoing Niger-HYCOS (Hydrological Cycle Observing System) program is expected to improve river monitoring networks in the Niger river basin [178,230]. Nigeria, however, further faces specific challenges such as poor engagement, varied risk perception, lack of interest, poor communication and commitment within the Nigeria Basin Authority, which hinder effective coordination and integrated water resource implementation [178].

5.2. Application of Open-Access Remotely Sensed Data in Transboundary Flood Management, Nigeria

As transboundary floods become more prevalent and intense due to increased storms triggered by climate change and anthropogenic factors [231], sufficient hydrological data is required for planning interventions for flood impact mitigation. In addition, considering that transboundary flood management institutions are facing recurring challenges that limit their functionality and sufficient data acquisition, open-access remotely sensed data provide a low-cost and viable alternative to enable transboundary flood monitoring and management without disrupting any sovereign nation's autonomy. Open-access satellite imagery such as Landsat and MODIS have been widely applied to delineate flood extent across transboundary river basins, aiding flood impact quantification needed for prompt response, as well as risk assessment and evaluation [213,232]. Radar altimetry, on the other hand, can be applied independently or with satellite images to support planning, forecasting and flood management in riparian countries [216,217].

In Nigeria, Tarpanelli et al. [63] explored the potential of integrating MODIS imagery and ENVISAT radar altimetry to predict and forecast discharge along the Niger-Benue river. The discharge was derived from daily (MOD09GQ) and 8-day (MOD09Q1) 250 m resolution MODIS TERRA and AQUA Image pixels (BAND 2-NIR), by establishing an empirical relationship between water-free land pixels during peak flood, permanent water pixels within the river and known discharge values derived from in situ measurements. Pandey and Amarnath [51] applied a combined forecasting rating curve approach developed by Hossain et al. [216] and hydraulic (HEC-RAS) model techniques to estimate discharge from ENVISAT, Jason-2 and AltiKa altimetry virtual station water levels along the Niger and Benue rivers, resulting in NS and R^2 values of 0.7 and 0.97 respectively.

In other closely related studies in the region. Salami and Nnadi, [233] monitored Kainji Lake along the Niger river, using TOPEX/Poseidon and ENVISAT altimetry, revealing stronger correlation between altimetry and in situ measurements in the wet season (R^2 = 0.93) than the dry season (R^2 = 0.77), and RMSE varying from 0.50 m to 0.83 m for TOPEX/Poseidon and ENVISAT, respectively. Sparavigna [234] studied the water level variability of Nasser, Tana, Chad and Kainji lakes using TOPEX/POSEIDON and Jason-1 altimetry. Cretaux et al. [235] combined TOPEX/Poseidon (T/P) and ENVISAT altimetry with 8-day MODIS near-infrared band images to monitor water level variations and inundation along the Niger inner delta, Lake Tchad and Ganaga river delta.

The high correlation between altimetry and in situ water levels during the wet season along the Niger river [233] suggests that altimetry can potentially be used in flood monitoring and management in Nigeria and the Niger Basin. The varying accuracies of different altimetry missions imply that altimetry data must be applied cautiously as earlier emphasized in Section 3.1.4, due to residual uncertainty. With current radar altimetry tracks, such as Jason-2 (Figure 12), Sentinel 3A/B (Figure 13) and future SWOT (Figure 14) passing across the Niger basin, the potential for long-term acquisition of spaceborne altimetry data for flood management is considerable.

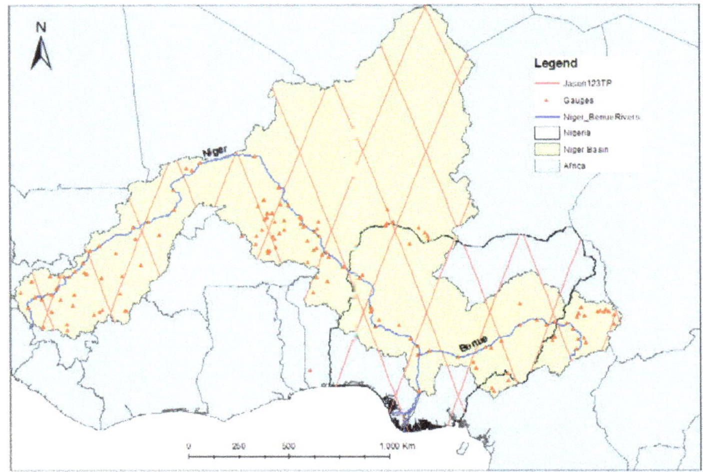

Figure 12. Jason-1/2/3/TP Altimetry Tracks within the Niger River Basin of West Africa.

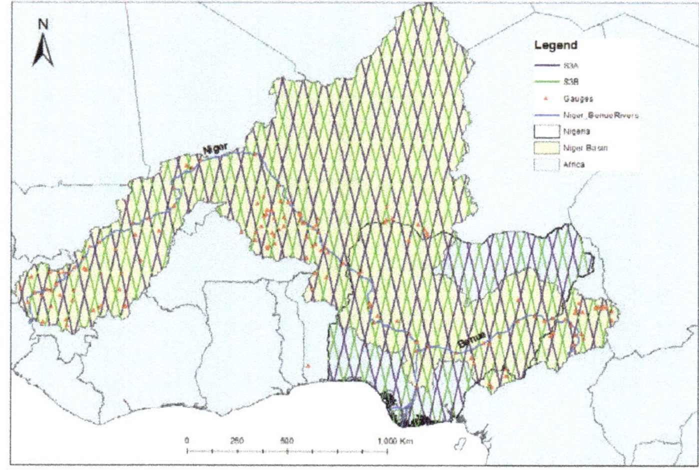

Figure 13. Sentinel 3A/B Altimetry Tracks within the Niger River Basin of West Africa.

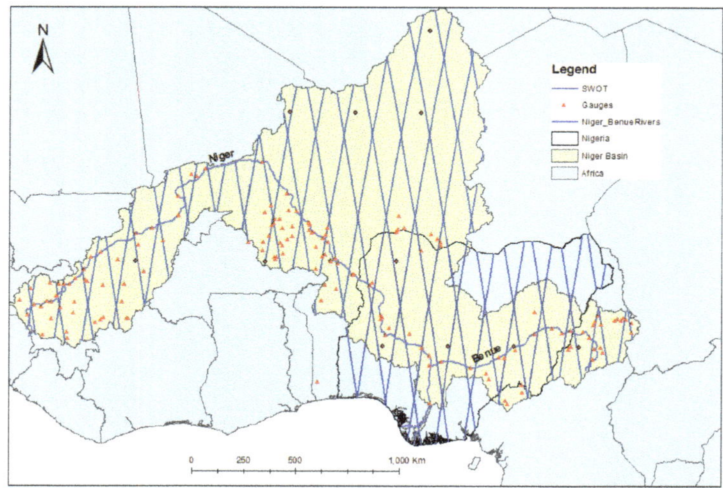

Figure 14. SWOT Altimetry Tracks within the Niger River Basin of West Africa.

6. Providers of Data for Flood Emergency Management

Other than open-access remotely sensed data, in some instances, commercial, regional and national satellite organisations collaboratively deliver high-resolution images and services to support flood response and mitigation efforts. This section discusses some of the available satellite data providers/consortia and disaster support services, as well as case study applications in Nigeria and hydraulically connected rivers in the Niger River Basin in West Africa.

6.1. International Charter "Space and Major Disasters"

The international charter "space and major disasters" (ICSMD) was established by ESA and the Centre National d'Etudes Spatiales (CNES) following the UNISPACE III conference held in Vienna in 1999 and was co-signed by the Canadian Space Agency (CSA) in 2001 [236]. The objective of the Charter is to provide data to enable critical decision making during environmental or technological disasters such as flooding, oil spills, fires, earthquake, volcanoes, hurricanes, landslides and ice hazards, thereby minimizing the impact of disasters on people and infrastructures [237]. Between 2001 and 2012, several satellite agencies joined the consortium: Japan Aerospace Exploration Agency (JAXA), Indian Space Research Organisation (ISRO), United States Geological Survey (USGS), National Oceanic and Atmospheric Administration (NOAA), Argentinean National Commission on Space Activities (CONAE), Exploration of Meteorological Satellite (EUMETSAT), German Space Agency (DLR), National Institute for Space Research (INPE) of Brazil, China National Space Administration, Disaster Monitoring Constellation International Imaging (DMCii) and Korean Aerospace Research Institute (KARI). This expansion of the consortium enhanced the Charter's ability to deliver prompt high resolution optical and SAR images when disasters strike [238]. Between 2000 and 2016 the ICSMD charter has been activated 500 times by more than 110 countries for various disasters [239]. An overview of disaster Charter activations for flood monitoring and management is presented in Figure 15, with South America, Africa and Asia showing the highest number of activations.

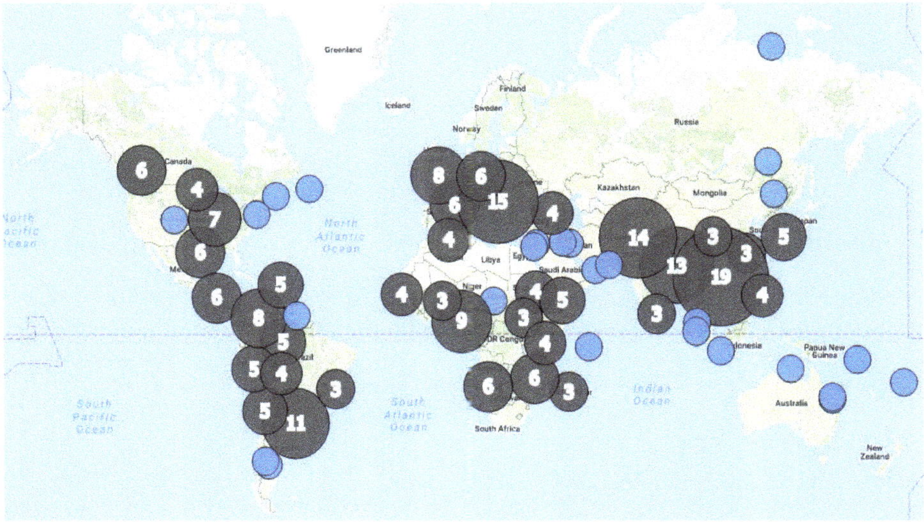

Figure 15. Map showing International Disaster Charter Flood Activations (2000–2016) (Source: Disaster Charter). Blue markers represent single activation areas, while numbers represent areas of multiple activations.

The Nigerian satellite NigeriaSat-1 joined the ICSMD in 2003, followed by NigeriaSat-2 and NigeriaSat-X in August 2011 [237] (all optical Sensors), to further enhance the Charter's capacity to deliver on its mission. Through this involvement, Nigeria provides its optical sensor images for disaster-related activities at no cost, as well as data for research purposes, and sells imagery to commercial ventures at a variable cost depending on the area of coverage. Some instances of the Nigerian Satellite activation for disaster response include Peloponnese forest fires, Greece 2007 (call 175); Beichuan Landslide and Debris flow, China 2008 (call 204); Java earthquake, Indonesia (call 269); Balkh, Kunduz, Takhar and Baghlan areas flooding, Afghanistan (call 255).

In Nigeria, the charter is usually activated by the National Emergency Management Agency's (NEMA) designated project manager. The activation follows five steps: (i) requisition by authorised person, (ii) requestor identification and request verification by a 24/7 operator, (iii) request analysis and satellite tasking for data capture, (iv) data acquisition and delivery, and (v) support in data processing throughout the emergency [240]. In Nigeria, activation of the disaster charter is relatively new, and only 6 activations have been made between 2010 and 2012 to monitor flooding events at Sokoto in 2010 (calls: 324 and 326), Ibadan in 2011 (call: 370), and in 2012 at Adamawa, Kogi and Bayelsa, (calls: 407, 415 and 416) [240]. Some of the images collected over the course of the activations in Nigeria include RADARSAT-2, SPOT-5, TerraSAR-X/TanDEM-X, Landsat ETM, KOMPSAT, ENVISAT, UK-DMC, and NIGERIASAT [219,239]. One of the lingering challenges of the Disaster Charter images is the strict license and copyright policies that prohibit re-use and distribution of the raw data [240], thus limiting the prospect of further data analysis and application in research. Nevertheless, finished products are available via the Charter Activations webpage (https://www.disasterscharter.org/web/guest/activations/charter-activations) as high-resolution maps for download, and can be digitized for use in flood mapping processes.

6.2. International Water Management Institute Emergency Response Products for Water Disasters

This is a space-based information and rapid mapping platform for emergency response aimed at providing support for disaster management in Africa and Asia. The platform was developed from a collaboration amongst the International Water Management Institute (IWMI), Asia-Pacific Regional Space Agency Forum (APRSAF), ESA, the United Nations Office for Outer Space Affairs (UNOOSA) and the United Nations Platform for Space-based Information for Disaster Management and Emergency Response (UN-SPIDER). This platform channels an impacted country's data request to the Disaster Charter, and also directly processes and analysis open-access images (i.e., Landsat, Sentinel 1, MODIS and Global Precipitation Measurement) to deliver products needed for decision making during a disaster [241]. So far, the platform has supported five countries including Sri Lanka, Myanmar, India, Bangladesh, and Nigeria [242]. In addition, a total of 37 activations to support flood information have deployed open-access satellites, as well as commercial TerraSAR-X, Radarsat-2, RISAT-1, ALOS-2 PALSAR-2, and JAXA-2 ALOS-2 satellite images [242].

Between 27th September–4th October 2015 this platform delivered 10 Sentinel-1 flood maps to support flood management efforts along the Niger and Benue rivers in Nigeria. This emanated from a collaborative effort amongst IWMI, ESA, Federal Ministry of Agriculture and Rural Development (FMARD) and Consortium of International Agricultural Research (CGIAR).

6.3. Copernicus Emergency Management Service

The European Union Copernicus Emergency Management Service (EMS) provide rapid (i.e., hours or days) free satellite-based maps to inform decision-making before, during and after natural and man-made disasters [243]. Although European nations are considered a priority for support provision, other countries can activate the Copernicus EMS. Thus far, between 1 April 2012 and 19 August 2016, the Copernicus EMS has been activated 175 times (Table 8), with flooding identified as the highest cause of activation (40%), resulting in 68% of the delineation maps generated.

Table 8. Summary of the Copernicus Emergency Management Service (EMS)-Mapping Activations.

Type of Disaster	Number of Activations	Number of Reference Maps	Number of Delineation Maps
Earthquake	9	83	31
Flood	**71**	**358**	**692**
Forest fire, wildfire	21	47	98
Industrial accident	5	12	3
Other	55	218	143
Wind storm	14	80	45
Total	175	798	1012

The Copernicus EMS has not been activated for Nigeria yet, but has been activated three times (EMSR018, EMSR019 and EMSR235) in Niger (Niamey) in 2012, Cameroon (Lake Maga, Garoua-Benue River) in 2012 and Niger (Dosso, Maradi, Niamey, and Tillaberi) in 2017. These are riparian countries within the transboundary Niger River Basin, and this could prove useful for transboundary flood monitoring in Nigeria. Authorised users France Centre Operationnel de Gestion Interministeriel de Crises (C.O.G.I.C) and EC Services I DG JRC activated the Copernicus EMS for the countries mentioned above, providing Radarsat-2, Rapid Eye, COSMO-SkyMed, and SPOT-5 satellite images flood extent maps.

6.4. Digital Globe Open Data Program

More recently, Digital Globe, a commercial satellite company launched the Open Data Program (ODP) initiative to provide high-resolution satellite imagery to support recovery from large-scale natural disasters such as flooding [244]. ODP provides pre and post-disaster images, including support via the Tomnod (http://www.tomnod.com) and Humanitarian OpenStreetMap Team (HOT, https://hotosm.org) crowdsourcing platforms for damage assessment [245]. So far, the ODP has been activated six times by Haiti, Nepal, Mexico, Ecuador, Caribbean/United States, and Madagascar, to manage disasters including earthquakes, hurricanes, and cyclones. The prospects for this initiative is substantial, as high-resolution imagery can considerably improve the detail of risk and damage assessment in remote locations. Though the ODP is yet to be deployed in Nigeria, it was deployed for post-disaster assessment of the 2017 Sierra Leone mudslide. This was the first application case on the African continent, followed with the mapping of Ebola response the Democratic Republic of Congo in 2018.

7. Synthesis

Flood disasters are becoming more frequent, intense and destructive, owing to climate change and anthropogenic factors. Managing floods requires effective decision making based on up-to-date and reliable hydrological information [8]. Typically, data needed for flood management include river discharge, water levels, terrain and land use/cover characteristics, and these are traditionally collected through the establishment of ground monitoring stations and field observations/surveys [246]. In situations where floods transcend administrative boundaries due to natural catchment delineations or river network connectivity, transboundary corporations are often set up to enable collaborative data collection, co-operation, risk communication, information sharing and planning to effectively manage flood impact in riparian countries [202,211]. However, in many developing regions both independent and transboundary data collection systems for flood management are flawed by organisational, technical, institutional, infrastructural and financial challenges that limit their effectiveness [178,202,205,247,248].

The potential of RS in supporting flood monitoring, planning and management are considerable, as it enables data collection in remote, inaccessible and data sparse locations [40]. Open-access remotely sensed data is particularly important for improving flood management in developing countries where the ground monitoring network is limited and the cost of obtaining commercial satellite data is prohibitive [29,135]. Datasets such as radar altimetry, DEMs, optical and radar imagery can be applied independently, in combination with in situ measurements or integrated into hydrodynamic models in order to reduce the uncertainty in flood estimation for ungauged river basins [39,69,89,98]. Furthermore, in the case of transboundary floods, RS allows data collection in an upstream country where the flood originates by a downstream impacted country without the need for bureaucratic authorization [59,203].

It is worth noting that the different freely available remotely sensed datasets provide varying levels of accuracy, depending on multiple factors. Altimetry mission accuracies depend on the satellite ground footprint, virtual station location, river width, tributaries discharging into the main river and satellite sensor properties [29]. The inability of C and X-band radar to penetrate vegetation canopies, and backscattering from rooftops and water surfaces, can result in over-estimation of elevation [40,66,249]. Optical imagery applications can be hampered by atmospheric conditions and spatial resolution [250], while one of the core deficiencies of radar images is the inconsistency in delineating floods in urban and forested areas [251].

Despite these deficiencies, the role of remotely sensed data in flood management is significant, especially in developing regions, as it allows for the quantification of hydrological parameters at previously undetected locations once a retrieval technique has been proven at a location where *in-situ* data is available [63]. With RS technology continuously advancing and more data becoming freely available, the reliance on ground observation data is expected to decline. Additionally, with commercial satellite companies such as Digital Globe and other satellite consortia making high-resolution images available for disaster management [237,244] will improve high-resolution flood modelling and mapping in data-sparse regions. Despite the advantages of RS, the role of ground-based data collection cannot be disregarded and must take priority or be applied in combination with remotely sensed data for enhanced flood mapping [42,105].

Planning for flood management usually requires flood magnitude estimates at varying return periods based on historical flood data. In many developing regions, such data are typically short time series if gauging stations are newly established or discontinued and contain gaps (missing data points) caused by equipment malfunction or poor data collation practices [24,182]. Altimetry can aid historical river data reconstruction where newly established and old discontinued gauging stations exist in proximity to virtual stations [252]. Nevertheless, the low revisit time of altimetry satellites [43] can result in a failure to capture peak floods needed for flood magnitude estimation [104,105] and in other instances, altimetry data is unavailable at certain locations [58]. Therefore, it is essential that the use of altimetry data is evaluated against other approaches, such as statistical techniques for infilling missing hydrological data, to ascertain the influence of both approaches on flood frequency estimates, and to understand when these individual approaches can be used.

Although this review focused on fluvial flood modelling and mapping, it is important to note that precipitation data (in situ and satellite) could also be vital in this process and has been widely applied, especially in data-sparse regions for flood modelling and hazard mapping [253–256]. However, this topic is beyond the scope of this review.

8. Conclusions

The potential of remotely sensed data such as altimetry, DEMs, optical and radar images has been highlighted in this review, with the unique merits, demerits and achievable applications being highlighted. In very remote locations of developing regions, data sparsity is so widespread that uniform data is seldom available for a whole catchment area, owing to inadequate and declining hydrological monitoring network [257]. Therefore, an integrated approach that enables the combination of all available open-access remotely sensed data is recommended in such locations, leveraging the merits of individual datasets to improve all phases of flood mapping processes, i.e., hydrological modelling, hydrodynamic modelling and inundation mapping.

Data from consortia of providers have proven to be useful in flood risk assessment when a flood occurs, where pre and post-flood images are provided for comparative analysis [219]. However, strict license and copyright policies prohibit re-use and distribution of the data [240], and this can hamper the important shift in focus from flood recovery to planning which is now imperative. Nevertheless, 3 end products (i.e., high-resolution inundation maps) are available via the Charter Activations web page and can be applied to support flood-modelling processes to inform flood planning decisions.

The deficiencies of open-source remotely sensed data for flood modelling and mapping can be quite pronounced in various landscapes, irrespective of the sensor type [157,258]. The private sector has played a vital role in advancing geo-informatics in developing regions [259], investing heavily in high-resolution satellite and airborne data needed for operational and disaster management purposes [21,260]. A significant opportunity now exists for integrating commercially sourced remotely sensed data with open-access and crowd-sourced data [261–263] to improve flood modelling and mapping in data sparse regions.

Author Contributions: I.T.E.-w. and G.A.B. conceived and designed this review study, while I.T.E.-w. undertook the review and drafted the Manuscript, and G.A.B. reviewed the Manuscript and provided suggestions as well as contributions to improve the Manuscript.

Funding: This research was funded by the Niger Delta Development Commission (NDDC), Nigeria, grant number NDDC/DEHSS/2013PGFS/BY/5.

Acknowledgments: This review is a product of part of Ekeu-wei Iguniwari Thomas's PhD research at Lancaster University, United Kingdom, funded by NDDC, Nigeria. The authors acknowledge the input of the three anonymous reviewers who provide valuable feedback that improved this review article enormously.

Conflicts of Interest: The Authors declare no conflict of interest

References

1. Plate, E. Flood risk and flood management. *J. Hydrol.* **2002**, *267*, 2–11. [CrossRef]
2. Merwade, V.; Olivera, F.; Arabi, M.; Edleman, S. Uncertainty in flood inundation mapping: Current issues and future directions. *J. Hydrol. Eng.* **2008**, *13*, 608–620. [CrossRef]
3. Moel, H.; Jongman, B.; Kreibich, H.; Merz, B.; Penning-Rowsell, E.; Ward, P. Flood risk assessments at different spatial scales. *Mitig. Adapt. Strateg. Glob. Chang.* **2015**, *20*, 865–890. [CrossRef]
4. Klijn, F.; Samuels, P.; Van Os, A. Towards flood risk management in the EU: State of affairs with examples from various European countries. *Int. J. River Basin Manag.* **2008**, *6*, 307–321. [CrossRef]
5. Büchele, B.; Kreibich, H.; Kron, A.; Thieken, A.; Ihringer, J.; Oberle, P.; Merz, B.; Nestmann, F. Flood-risk mapping: Contributions towards an enhanced assessment of extreme events and associated risks. *Nat. Hazards Earth Syst. Sci.* **2006**, *6*, 485–503. [CrossRef]
6. Ologunorisa, T.E. An assessment of flood vulnerability zones in the Niger Delta, Nigeria. *Int. J. Environ. Stud.* **2004**, *61*, 31–38. [CrossRef]
7. Valdes, H.M. *Living with Risk: A Global Review of Disaster Reduction Initiatives*; United Nations Publications: New York, NY, USA, 2004; Volume 1.
8. Els, Z. Data Availability and Requirements for Flood Hazard Mapping. Master's Thesis, Natural Sciences at Stellenbosch University, Stellenbosch, South Africa. 2013.
9. Federal Ministry of Environment. *Technical Guidelines on Soil Erosion, Flood and Coastal Zone Management*; Federal Ministry of Environment: Abuja, Nigeria, 2005.
10. Aerts, J.C.J.H.; Alphen, J.V.; Moel, H.D. Flood maps in Europe-methods, availability and use. *Nat. Hazards Earth Syst. Sci.* **2009**, *9*, 289–301.
11. Martini, F.; Loat, R. *Handbook on Good Practices for Flood Mapping in Europe*; EXCIMAP: Brussels, Belgium, 2007. [CrossRef]
12. Awokola, O.; Martins, O. Regional Flood Frequency Analysis of Osun Drainage Basin, South-Western Nigeria. *Niger. J. Sci.* **2001**, *35*, 37–44.
13. Kjeldsen, T.R.; Smithers, J.C.; Schulze, R.E. Regional flood frequency analysis in the KwaZulu-Natal province, South Africa, using the index-flood method. *J. Hydrol.* **2002**, *255*, 194–211. [CrossRef]
14. Leclerc, M.; Ouarda, T.B.M.J. Non-stationary regional flood frequency analysis at ungauged sites. *J. Hydrol.* **2007**, *343*, 254–265. [CrossRef]
15. Ahn, J.; Cho, W.; Kim, T.; Shin, H.; Heo, J.-H. Flood frequency analysis for the annual peak flows simulated by an event-based rainfall-runoff model in an urban drainage basin. *Water* **2014**, *6*, 3841–3863. [CrossRef]
16. Sarhadi, A.; Soltani, S.; Modarres, R. Probabilistic flood inundation mapping of ungauged rivers: Linking GIS techniques and frequency analysis. *J. Hydrol.* **2012**, *458–459*, 68–86. [CrossRef]
17. Di Baldassarre, G.; Schumann, G.; Bates, P.; Freer, J.; Beven, K. Flood-plain mapping: A critical discussion of deterministic and probabilistic approaches. *Hydrol. Sci. J.* **2010**, *55*, 364–376. [CrossRef]
18. Muncaster, S.; Warwick, B.; McCowab, A. Design flood estimation in small catchments using two dimensional hydraulic modelling—A case study. In *Hydrology and Water Resource Symposium*; TAS: Launceston, Australia, 2006; pp. 104–109.

19. Neal, J.; Schumann, G.; Fewtrell, T.; Budimir, M.; Bates, P.; Mason, D. Evaluating a new LISFLOOD-FP formulation with data from the summer 2007 floods in Tewkesbury, UK. *J. Flood Risk Manag.* **2011**, *4*, 88–95. [CrossRef]
20. Taubenböck, H.; Wurm, M.; Netzband, M.; Zwenzner, H.; Roth, A.; Rahman, A.; Dech, S. Flood risks in urbanized areas—Multi- sensoral approaches using remotely sensed data for risk assessment. *Nat. Hazards Earth Syst. Sci.* **2011**, *11*, 431–444. [CrossRef]
21. Eyers, R.; Obowu, C.; Lasisi, B. Niger Delta Flooding: Monitoring, Forecasting & Emergency Response Support from SPDC. In Proceedings of the FIG Working Week, Environment and Sustainability, Abuja, Nigeria, 6–10 May 2013.
22. Nigeria Hydrological Services Agency; Ankle Foot Orthosis. *Nigerian Hydrological Service Agency, 2014 Annual Flood Outlook (AFO)*; NIHSA: Sioux City, IA, USA, 2014.
23. Olayinka, D.N.; Nwilo, P.C.; Emmanuel, A. From Catchment to Reach: Predictive Modelling of Floods in Nigeria. In Proceedings of the FIG Working Week, Environment for Sustainability, Abuja, Nigeria, 6–10 May 2013.
24. Maxwell, O. Hydrological Data Banking for Sustainable Development in Nigeria: An Overview. *Aceh Int. J. Sci. Technol.* **2013**, *2*, 59–62. [CrossRef]
25. Ekeu-wei, I.T. Evaluation of Hydrological Data Collection Challenges and Flood Estimation Uncertainties in Nigeria. *Environ. Nat. Resour. Res.* **2018**, *8*, 44–54. [CrossRef]
26. Sivapalan, M. Prediction in ungauged basins: A grand challenge for theoretical hydrology. *Hydrol. Process.* **2003**, *17*, 3163–3170. [CrossRef]
27. Sivapalan, M.; Takeuchi, K.; Franks, S.W.; Gupta, V.K.; Karambiri, H.; Lakshmi, V.; Liang, X.; McDonnell, J.J.; Mendiondo, E.M.; Connell, P.E.; et al. IAHS Decade on Predictions in Ungauged Basins (PUB), 2003–2012: Shaping an exciting future for the hydrological sciences. *Hydrol. Sci. J.* **2003**, *48*, 857–880. [CrossRef]
28. Hrachowitz, M.; Savenije, H.; Blöschl, G.; McDonnell, J.; Sivapalan, M.; Pomeroy, J.; Arheimer, B.; Blume, T.; Clark, M.; Ehret, U. A decade of Predictions in Ungauged Basins (PUB)—A review. *Hydrol. Sci. J.* **2013**, *58*, 1198–1255. [CrossRef]
29. Yan, K.; Di Baldassarre, G.; Solomatine, D.P.; Schumann, G.J.P. A review of low-cost space-borne data for flood modelling: Topography, flood extent and water level. *Hydrol. Process.* **2015**, *29*, 3368–3387. [CrossRef]
30. Schumann, G.; Bates, P.D.; Horritt, M.S.; Matgen, P.; Pappenberger, F. Progress in integration of remote sensing- derived flood extent and stage data and hydraulic models. *Rev. Geophys.* **2009**, *47*. [CrossRef]
31. Mason, D.C.; Schumann, G.; Bates, P. Data utilization in flood inundation modelling. *Flood Risk Sci. Manag.* **2011**. [CrossRef]
32. Dano Umar, L.; Abdul-Nasir, M.; Ahmad Mustafa, H.; Imtiaz Ahmed, C.; Soheil, S.; Abdul-Lateef, B.; Haruna Ahmed, A. Geographic Information System and Remote Sensing Applications in Flood Hazards Management: A Review. *Res. J. Appl. Sci. Eng. Technol.* **2011**, *3*, 933–947.
33. Maswood, M.; Hossain, F. Advancing river modelling in ungauged basins using satellite remote sensing: The case of the Ganges- Brahmaputra- Meghna basin. *Int. J. River Basin Manag.* **2016**, *14*, 103–117. [CrossRef]
34. Alsdorf, D.E.; Rodríguez, E.; Lettenmaier, D.P. Measuring surface water from space. *Rev. Geophys.* **2007**, *45*, RG2002. [CrossRef]
35. Koblinsky, C.; Clarke, R.; Brenner, A.; Frey, H. Measurement of river level variations with satellite altimetry. *Water Resour. Res.* **1993**, *29*, 1839–1848. [CrossRef]
36. Da Silva, J.S.; Calmant, S.; Seyler, F.; Rotunno Filho, O.C.; Cochonneau, G.; Mansur, W.J. Water levels in the Amazon basin derived from the ERS 2 and ENVISAT radar altimetry missions. *Remote Sens. Environ.* **2010**, *114*, 2160–2181. [CrossRef]
37. Sulistioadi, Y.B.; Tseng, K.H.; Shum, C.K.; Hidayat, H.; Sumaryono, M.; Suhardiman, A.; Setiawan, F.; Sunarso, S. Satellite radar altimetry for monitoring small rivers and lakes in Indonesia. *Hydrol. Earth Syst. Sci.* **2015**, *19*, 341–359. [CrossRef]
38. Belaud, G.; Cassan, L.; Bader, J.; Bercher, N.; Feret, T. Calibration of a propagation model in large river using satellite altimetry. In Proceedings of the 6th International Symposium on Environmental Hydraulics, Athens, Greece, 23–25 June 2010; pp. 23–25.
39. Birkinshaw, S.J.; Moore, P.; Kilsby, C.G.; Donnell, G.M.; Hardy, A.J.; Berry, P.A.M. Daily discharge estimation at ungauged river sites using remote sensing. *Hydrol. Process.* **2014**, *28*, 1043–1054. [CrossRef]

40. Musa, Z.; Popescu, I.; Mynett, A. A review of applications of satellite SAR, optical, altimetry and DEM data for surface water modelling, mapping and parameter estimation. *Hydrol. Earth Syst. Sci. Discuss.* **2015**, *12*, 4857–4878. [CrossRef]
41. Crétaux, J.-F.; Jelinski, W.; Calmant, S.; Kouraev, A.; Vuglinski, V.; Bergé-Nguyen, M.; Gennero, M.-C.; Nino, F.; Del Rio, R.A.; Cazenave, A. SOLS: A lake database to monitor in the Near Real Time water level and storage variations from remote sensing data. *Adv. Space Res.* **2011**, *47*, 1497–1507. [CrossRef]
42. Sun, W.; Ishidaira, H.; Bastola, S. Calibration of hydrological models in ungauged basins based on satellite radar altimetry observations of river water level. *Hydrol. Process.* **2012**, *26*, 3524–3537. [CrossRef]
43. O'Loughlin, F.E.; Neal, J.; Yamazaki, D.; Bates, P.D. ICESat-derived inland water surface spot heights. *Water Resour. Res.* **2016**, *52*, 3276–3284. [CrossRef]
44. Birkett, C.M. The contribution of TOPEX/POSEIDON to the global monitoring of climatically sensitive lakes. *J. Geophys. Res. Oceans* **1995**, *100*, 25179–25204. [CrossRef]
45. Birkett, C.M.; Mertes, L.A.K.; Dunne, T.; Costa, M.H.; Jasinski, M.J. Surface water dynamics in the Amazon Basin: Application of satellite radar altimetry. *J. Geophys. Res. Atmos.* **2002**, *107*, LBA 26-1–LBA 26-21. [CrossRef]
46. Frappart, F.; Calmant, S.; Cauhopé, M.; Seyler, F.; Cazenave, A. Preliminary results of ENVISAT RA- 2-derived water levels validation over the Amazon basin. *Remote Sens. Environ.* **2006**, *100*, 252–264. [CrossRef]
47. Birkinshaw, S.J.; Donnell, G.M.; Moore, P.; Kilsby, C.G.; Fowler, H.J.; Berry, P.A.M. Using satellite altimetry data to augment flow estimation techniques on the Mekong River. *Hydrol. Process.* **2010**, *24*, 3811–3825. [CrossRef]
48. Ponte, R.M.; Wunsch, C.; Stammer, D. Spatial mapping of time-variable errors in Jason-1 and TOPEX/Poseidon sea surface height measurements. *J. Atmos. Ocean. Technol.* **2007**, *24*, 1078–1085. [CrossRef]
49. Chelton, D.B.; Ries, J.C.; Haines, B.J.; Fu, L.-L.; Callahan, P.S. Satellite altimetry. *Int. Geophys.* **2001**, *69*, 1–2.
50. Tourian, M.J.; Tarpanelli, A.; Elmi, O.; Qin, T.; Brocca, L.; Moramarco, T.; Sneeuw, N. Spatiotemporal densification of river water level time series by multimission satellite altimetry. *Water Resour. Res.* **2016**, *52*, 1140–1159. [CrossRef]
51. Pandey, R.; Amarnath, G. The potential of satellite radar altimetry in flood forecasting: Concept and implementation for the Niger-Benue river basin. *Proc. IAHS* **2015**, *370*, 223–227. [CrossRef]
52. Jarihani, A.A.; Callow, J.N.; McVicar, T.R.; Van Niel, T.G.; Larsen, J.R. Satellite- derived Digital Elevation Model (DEM) selection, preparation and correction for hydrodynamic modelling in large, low- gradient and data- sparse catchments. *J. Hydrol.* **2015**, *524*, 489–506. [CrossRef]
53. Urban, T.J.; Schutz, B.E.; Neuenschwander, A.L. A survey of ICESat coastal altimetry applications: Continental Coast, Open Ocean Island, and Inland River. *Terr. Atmos. Ocean. Sci.* **2008**, *19*, 1–19. [CrossRef]
54. Schneider, R.; Godiksen, P.N.; Villadsen, H.; Madsen, H.; Bauer-Gottwein, P. Application of CryoSat- 2 altimetry data for river analysis and modelling. *Hydrol. Earth Syst. Sci. Discuss.* **2016**, *19*, 1–19. [CrossRef]
55. Schwatke, C.; Dettmering, D.; Börgens, E.; Bosch, W. Potential of SARAL/AltiKa for Inland Water Applications. *Mar. Geodesy* **2015**, *38*, 626–643. [CrossRef]
56. European Space Agency. Altimetry Instrument Payload. Available online: https://sentinel.esa.int/web/sentinel/missions/sentinel-3/instrument-payload/altimetry (accessed on 26 April 2016).
57. Fu, L.-L.; Alsdorf, D.; Rodriguez, E.; Morrow, R.; Mognard, N.; Lambin, J.; Vaze, P.; Lafon, T. The SWOT (Surface Water and Ocean Topography) mission: Spaceborne radar interferometry for oceanographic and hydrological applications. *Proc. Ocean. Obs.* **2009**, *3*, 21–25.
58. Papa, F.; Durand, F.; Rossow, W.B.; Rahman, A.; Bala, S.K. Satellite altimeter- derived monthly discharge of the Ganga- Brahmaputra River and its seasonal to interannual variations from 1993 to 2008. *J. Geophys. Res. Oceans* **2010**, *115*. [CrossRef]
59. Sridevi, T.; Sharma, R.; Mehra, P.; Prasad, K.V.S.R. Estimating discharge from the Godavari River using ENVISAT, Jason- 2, and SARAL/AltiKa radar altimeters. *Remote Sens. Lett.* **2016**, *7*, 348–357. [CrossRef]
60. Getirana, A.C.V.; Peters-Lidard, C. Estimating water discharge from large radar altimetry datasets. *Hydrol. Earth Syst. Sci.* **2013**, *17*, 923–933. [CrossRef]
61. Tourian, M.; Sneeuw, N.; Bárdossy, A. A quantile function approach to discharge estimation from satellite altimetry (ENVISAT). *Water Resour. Res.* **2013**, *49*, 4174–4186. [CrossRef]
62. Michailovsky, C.I.; McEnnis, S.; Bauer-Gottwein, P.A.M.; Berry, R.; Smith, P. River monitoring from satellite radar altimetry in the Zambezi River basin. *Hydrol. Earth Syst. Sci.* **2012**, *16*, 2181–2192. [CrossRef]

63. Tarpanelli, A.; Amarnath, G.; Brocca, L.; Moramarco, T. Discharge forecasting using MODIS and radar altimetry: Potential application for transboundary flood risk management in Niger-Benue River basin. In Proceedings of the EGU General Assembly Conference Abstracts, Vienna, Austria, 17–22 April 2016.

64. Sichangi, A.W.; Wang, L.; Yang, K.; Chen, D.; Wang, Z.; Li, X.; Zhou, J.; Liu, W.; Kuria, D. Estimating continental river basin discharges using multiple remote sensing data sets. *Remote Sens. Environ.* **2016**, *179*, 36–53. [CrossRef]

65. Casas, A.; Benito, G.; Thorndycraft, V.R.; Rico, M. The topographic data source of digital terrain models as a key element in the accuracy of hydraulic flood modelling. *Earth Surf. Process. Landf.* **2006**, *31*, 444–456. [CrossRef]

66. Cook, A.; Merwade, V. Effect of topographic data, geometric configuration and modeling approach on flood inundation mapping. *J. Hydrol.* **2009**, *377*, 131–142. [CrossRef]

67. Patro, S.; Chatterjee, C.; Singh, R.; Raghuwanshi, N. Hydrodynamic modelling of a large flood-prone river system in India with limited data. *Hydrol. Process.* **2009**, *23*, 2774–2791. [CrossRef]

68. Wang, W.; Yang, X.; Yao, T. Evaluation of ASTER GDEM and SRTM and their suitability in hydraulic modelling of a glacial lake outburst flood in southeast Tibet. *Hydrol. Process.* **2012**, *26*, 213–225. [CrossRef]

69. Sanyal, J.; Carbonneau, P.; Densmore, A. Hydraulic routing of extreme floods in a large ungauged river and the estimation of associated uncertainties: A case study of the Damodar River, India. *Nat. Hazards* **2013**, *66*, 1153–1177. [CrossRef]

70. Ullah, S.; Farooq, M.; Sarwar, T.; Tareen, M.; Wahid, M. Flood modeling and simulations using hydrodynamic model and ASTER DEM—A case study of Kalpani River. *Arab. J. Geosci.* **2016**, *9*, 1–11. [CrossRef]

71. Amans, O.C.; Beiping, W.; Ziggah, Y.Y. Assessing Vertical Accuracy of SRTM Ver. 4.1 and ASTER GDEM Ver. 2 using Differential GPS Measurements–case study in Ondo State, Nigeria. *Int. J. Sci. Eng. Res.* **2013**, *4*, 523–531.

72. Isioye, O.A.; Yang, I.C. Comparison and validation of ASTER-GDEM and SRTM elevation models over parts of Kaduna State, Nigeria. *SASGI Proc.* **2013**, *1*, 1–11.

73. Zwally, H.J.; Schutz, B.; Abdalati, W.; Abshire, J.; Bentley, C.; Brenner, A.; Bufton, J.; Dezio, J.; Hancock, D.; Harding, D.; et al. ICESat's laser measurements of polar ice, atmosphere, ocean, and land. *J. Geodyn.* **2002**, *34*, 405–445. [CrossRef]

74. Fricker, H.A.; Borsa, A.; Minster, B.; Carabajal, C.; Quinn, K.; Bills, B. Assessment of ICESat performance at the salar de Uyuni, Bolivia. *Geophys. Res. Lett.* **2005**, *32*. [CrossRef]

75. Jean Stéphane, B.; Hani, A.; Nicolas, L.; Nicolas, B. The Relevance of GLAS/ICESat Elevation Data for the Monitoring of River Networks. *Remote Sens.* **2011**, *3*, 708–720.

76. Satgé, F.; Bonnet, M.P.; Timouk, F.; Calmant, S.; Pillco, R.; Molina, J.; Lavado-Casimiro, W.; Arsen, A.; Crétaux, J.F.; Garnier, J. Accuracy assessment of SRTM v4 and ASTER GDEM v2 over the Altiplano watershed using ICESat/GLAS data. *Int. J. Remote Sens.* **2015**, *36*, 465–488. [CrossRef]

77. Carabajal, C.C.; Harding, D.J. ICESat validation of SRTM C-band digital elevation models. *Geophys. Res. Lett.* **2005**, *32*, 117–137. [CrossRef]

78. Kon Joon Bhang, F.W.; Schwartz, A.; Braun, A. Verification of the Vertical Error in C- Band SRTM DEM Using ICESat and Landsat- 7, Otter Tail County, MN. *IEEE Trans. Geosci. Remote Sens.* **2007**, *45*, 36–44. [CrossRef]

79. Du, X.; Guo, H.; Fan, X.; Zhu, J.; Yan, Z.; Zhan, Q. Vertical accuracy assessment of freely available digital elevation models over low-lying coastal plains. *Int. J. Dig. Earth* **2016**, *9*, 252–271. [CrossRef]

80. Zhao, G.; Xue, H.; Ling, F. Assessment of ASTER GDEM performance by comparing with SRTM and ICESat/GLAS data in Central China. In Proceedings of the 18th International Conference on Geoinformatics, Beijing, China, 18–20 June 2010; pp. 1–5.

81. Braun, A.; Fotopoulos, G. Assessment of SRTM, ICESat, and survey control monument elevations in Canada. *Photogramm. Eng. Remote Sens.* **2007**, *73*, 1333–1342. [CrossRef]

82. Rastogi, G.; Agrawal, R.; Ajai, R. Bias corrections of CartoDEM using ICESat- GLAS data in hilly regions. *GISci. Remote Sens.* **2015**, *52*, 571–585.

83. Beaulieu, A.; Clavet, D. Accuracy assessment of Canadian digital elevation data using ICESat. *Photogramm. Eng. Remote Sens.* **2009**, *75*, 81–86. [CrossRef]

84. Yamanokuchi, T.; Doi, K.; Shibuya, K. Comparison of Antarctic Ice Sheet Elevation between ICESat GLAS and InSAR DEM. In Proceedings of the 2006 IEEE International Symposium on Geoscience and Remote Sensing, Denver, CO, USA, 31 July–4 August 2006; pp. 2712–2715.

85. Mirzaee, S.; Motagh, M.; Arefi, H. Assessment of Reference Height Models on Quality of Tandem-X dem. *Int. Arch. Photogramm. Remote Sens. Spat. Inf. Sci.* **2015**, *40*, 463–466. [CrossRef]

86. Sampson, C.C.; Smith, A.M.; Bates, P.D.; Neal, J.C.; Alfieri, L.; Freer, J.E. A high- resolution global flood hazard model. *Water Resour. Res.* **2015**, *51*, 7358–7381. [CrossRef] [PubMed]

87. O'Loughlin, F.; Paiva, R.; Durand, M.; Alsdorf, D.; Bates, P. Development of a 'bare-earth' SRTM DEM product. In Proceedings of the EGU General Assembly Conference Abstracts, Vienna, Austria, 12–17 April 2015.

88. Heyder, U. Vertical Forest Structure from ICESat/GLAS Lidar Data. Master's Thesis, Department of Geography, University College London, London, UK, 2005; pp. 12–50.

89. Trigg, M.A.; Bates, P.D.; Wilson, M.D.; Horritt, M.S.; Alsdorf, D.E.; Forsberg, B.R.; Vega, M.C. Amazon flood wave hydraulics. *J. Hydrol.* **2009**, *374*, 92–105. [CrossRef]

90. Chávarri, E.; Crave, A.; Bonnet, M.-P.; Mejía, A.; Santos Da Silva, J.; Guyot, J.L. Hydrodynamic modelling of the Amazon River: Factors of uncertainty. *J. S. Am. Earth Sci.* **2013**, *44*, 94–103. [CrossRef]

91. Bates, P.; Neal, J.; Alsdorf, D.; Schumann, G. Observing Global Surface Water Flood Dynamics. *Surv. Geophys.* **2014**, *35*, 839–852. [CrossRef]

92. Durand, M.; Andreadis, K.M.; Alsdorf, D.E.; Lettenmaier, D.P.; Moller, D.; Wilson, M. Estimation of bathymetric depth and slope from data assimilation of swath altimetry into a hydrodynamic model. *Geophys. Res. Lett.* **2008**, *35*. [CrossRef]

93. Bates, P.D.; De Roo, A.P.J. A simple raster-based model for flood inundation simulation. *J. Hydrol.* **2000**, *236*, 54–77. [CrossRef]

94. Yoon, Y.; Durand, M.; Merry, C.J.; Clark, E.A.; Andreadis, K.M.; Alsdorf, D.E. Estimating river bathymetry from data assimilation of synthetic SWOT measurements. *J. Hydrol.* **2012**, *464–465*, 363–375. [CrossRef]

95. Van Wesemael, A.; Gobeyn, S.; Neal, J.; Lievens, H.; Van Eerdenbrugh, K.; De Vleeschouwer, N.; Schumann, G.; Vernieuwe, H.; Di Baldassarre, G.; De Baets, B. Calibration of a flood inundation model using a SAR image: Influence of acquisition time. In Proceedings of the EGU General Assembly Conference Abstracts, Vienna, Austria, 17–22 April 2016.

96. Neal, J.C.; Odoni, N.A.; Trigg, M.A.; Freer, J.E.; Garcia-Pintado, J.; Mason, D.C.; Wood, M.; Bates, P.D. Efficient incorporation of channel cross-section geometry uncertainty into regional and global scale flood inundation models. *J. Hydrol.* **2015**, *529*, 169–183. [CrossRef]

97. Stephens, E.; Schumann, G.; Bates, P. Problems with binary pattern measures for flood model evaluation. *Hydrol. Process.* **2014**, *28*, 4928–4937. [CrossRef]

98. Jung, H.C.; Jasinski, M.; Kim, J.W.; Shum, C.K.; Bates, P.; Neal, J.; Lee, H.; Alsdorf, D. Calibration of two- dimensional floodplain modeling in the central Atchafalaya Basin Floodway System using SAR interferometry. *Water Resour. Res.* **2012**, *48*. [CrossRef]

99. Dung, N.V.; Merz, B.; Bárdossy, A.; Thang, T.D.; Apel, H. Multi- objective automatic calibration of hydrodynamic models utilizing inundation maps and gauge data. *Hydrol. Earth Syst. Sci.* **2011**, *15*, 1339–1354. [CrossRef]

100. Pasquale, N.; Perona, P.; Wombacher, A.; Burlando, P. Hydrodynamic model calibration from pattern recognition of non- orthorectified terrestrial photographs. *Comput. Geosci.* **2014**, *62*, 160–167. [CrossRef]

101. Wood, M.; Neal, J.; Hostache, R.; Corato, G.; Bates, P.; Giustarini, L.; Chini, M.; Matgen, P. Using time series of satellite SAR images to calibrate channel depth and friction parameters in the LISFLOOD-FP hydraulic model. In Proceedings of the EGU General Assembly Conference Abstracts, Vienna, Austria, 27 April–2 May 2014.

102. Andréfouët, S.; Ouillon, S.; Brinkman, R.; Falter, J.; Douillet, P.; Wolk, F.; Smith, R.; Garen, P.; Martinez, E.; Laurent, V.; et al. Review of solutions for 3D hydrodynamic modeling applied to aquaculture in South Pacific atoll lagoons. *Mar. Pollut. Bull.* **2006**, *52*, 1138–1155 [CrossRef] [PubMed]

103. Domeneghetti, A. On the use of SRTM and altimetry data for flood modeling in data- sparse regions. *Water Resour. Res.* **2016**, *52*, 2901–2918. [CrossRef]

104. Yan, K.; Tarpanelli, A.; Balint, G.; Moramarco, T.; Baldassarre, G.D. Exploring the Potential of SRTM Topography and Radar Altimetry to Support Flood Propagation Modeling: Danube Case Study. *J. Hydrol. Eng.* **2015**, *20*, 04014048. [CrossRef]

105. Domeneghetti, A.; Tarpanelli, A.; Brocca, L.; Barbetta, S.; Moramarco, T.; Castellarin, A.; Brath, A. The use of remote sensing- derived water surface data for hydraulic model calibration. *Remote Sens. Environ.* **2014**, *149*, 130–141. [CrossRef]

106. Milzow, C.; Bauer-Gottwein, P.E.; Krogh, P. Combining satellite radar altimetry, SAR surface soil moisture and GRACE total storage changes for hydrological model calibration in a large poorly gauged catchment. *Hydrol. Earth Syst. Sci.* **2011**, *15*, 1729–1743. [CrossRef]
107. Tommaso, M.; Angelica, T.; Luca, B.; Silvia, B. River Discharge Estimation by Using Altimetry Data and Simplified Flood Routing Modeling. *Remote Sens.* **2013**, *5*, 4145–4162.
108. Sun, W.; Song, H.; Cheng, T.; Yu, J. Calibration of hydrological models using TOPEX/Poseidon radar altimetry observations. *Proc. Int. Assoc. Hydrol. Sci.* **2015**, *368*, 3–8. [CrossRef]
109. Grimaldi, S.; Petroselli, A.; Serinaldi, F. A continuous simulation model for design-hydrograph estimation in small and ungauged watersheds. *Hydrol. Sci. J.* **2012**, *57*, 1035–1051. [CrossRef]
110. Jung, Y.; Merwade, V. Estimation of uncertainty propagation in flood inundation mapping using a 1- D hydraulic model. *Hydrol. Process.* **2015**, *29*, 624–640. [CrossRef]
111. Mason, D.C.; Trigg, M.; Garcia-Pintado, J.; Cloke, H.L.; Neal, J.C.; Bates, P.D. Improving the TanDEM- X Digital Elevation Model for flood modelling using flood extents from Synthetic Aperture Radar images. *Remote Sens. Environ.* **2016**, *173*, 15–28. [CrossRef]
112. Bates, P.D.; Wilson, M.D.; Horritt, M.S.; Mason, D.C.; Holden, N.; Currie, A. Reach scale floodplain inundation dynamics observed using airborne synthetic aperture radar imagery: Data analysis and modelling. *J. Hydrol.* **2006**, *328*, 306–318. [CrossRef]
113. Lewis, M.; Bates, P.; Horsburgh, K.; Neal, J.; Schumann, G. A storm surge inundation model of the northern Bay of Bengal using publicly available data. *Q. J. R. Meteorol. Soc.* **2013**, *139*, 358–369. [CrossRef]
114. Neal, J.; Schumann, G.; Bates, P. A subgrid channel model for simulating river hydraulics and floodplain inundation over large and data sparse areas. *Water Resour. Res.* **2012**, *48*. [CrossRef]
115. Farr, T.G.; Rosen, P.A.; Caro, E.; Crippen, R.; Duren, R.; Hensley, S.; Kobrick, M.; Paller, M.; Rodriguez, E.; Roth, L.; et al. The Shuttle Radar Topography Mission. *Rev. Geophys.* **2007**, *45*, RG2004. [CrossRef]
116. Farr, T.G.; Kobrick, M. Shuttle radar topography mission produces a wealth of data. *EOS* **2000**, *81*, 583–585. [CrossRef]
117. Gichamo, T.Z.; Popescu, I.; Jonoski, A.; Solomatine, D. River cross- section extraction from the ASTER global DEM for flood modeling. *Environ. Model. Softw.* **2012**, *31*, 37–46. [CrossRef]
118. Demirkesen, A. Flood hazard vulnerability for settlements of Turkey's province of Edirne, using ASTER DEM data and Landsat-7 ETM+ image data. *Arab. J. Geosci.* **2016**, *9*, 1–15. [CrossRef]
119. Schumann, G.P.; Neal, J.C.; Voisin, N.; Andreadis, K.M.; Pappenberger, F.; Phanthuwongpakdee, N.; Hall, A.C.; Bates, P.D. A first large-scale flood inundation forecasting model. *Water Resour. Res.* **2013**, *49*, 6248–6257. [CrossRef]
120. Tadono, T.; Ishida, H.; Oda, F.; Naito, S.; Minakawa, K.; Iwamoto, H. Precise Global DEM Generation by ALOS PRISM. *ISPRS Ann. Photogramm. Remote Sens. Spat. Inf. Sci.* **2014**, *II-4*, 71–76. [CrossRef]
121. Santillana, J.; Makinano-Santillana, M.; Ampayon, B.C.; del Norte, A. Vertical Accuracy Assessment of 30-M Resolution Alos, Aster, and Srtm Global Dems Over Northeastern Mindanao, Philippines. *ISPRS-Int. Arch. Photogramm. Remote Sens. Spat. Inf. Sci.* **2016**, *XLI-B4*, 149–156. [CrossRef]
122. Courty, L.G.; Soriano-Monzalvoa, J.C.; Pedrozo-Acuñaa, A. Evaluation of open-access global digital elevation models (AW3D30, SRTM and ASTER) for flood modelling purposes. *Zenodo* **2017**. [CrossRef]
123. Yamazaki, D.; Baugh, C.; Bates, P.D.; Kanae, S.; Alsdorf, D.; Oki, T. Adjustment of a spaceborne DEM for use in floodplain hydrodynamic modeling. *J. Hydrol.* **2012**, *436*, 81–91. [CrossRef]
124. Baugh, C.A.; Bates, P.D.; Schumann, G.; Trigg, M.A. SRTM vegetation removal and hydrodynamic modeling accuracy. *Water Resour. Res.* **2013**, *49*, 5276–5289. [CrossRef]
125. Rodriguez, E.; Morris, C.S.; Belz, J.E. A global assessment of the SRTM performance. *Photogramm. Eng. Remote Sens.* **2006**, *72*, 249–260. [CrossRef]
126. Tarekegn, T.H.; Haile, A.T.; Rientjes, T.; Reggiani, P.; Alkema, D. Assessment of an ASTER- generated DEM for 2D hydrodynamic flood modeling. *Int. J. Appl. Earth Obs. Geoinf.* **2010**, *12*, 457–465. [CrossRef]
127. Tachikawa, T.; Kaku, M.; Iwasaki, A.; Gesch, D.B.; Oimoen, M.J.; Zhang, Z.; Danielson, J.J.; Krieger, T.; Curtis, B.; Haase, J. *ASTER Global Digital Elevation Model Version 2-Summary of Validation Results*; NASA: Washington, DC, USA, 2011.
128. Varga, M.; Bašić, T. Accuracy validation and comparison of global digital elevation models over Croatia. *Int. J. Remote Sens.* **2015**, *36*, 170–189. [CrossRef]

129. Jilani, R.; Munir, S.; Siddiqui, P. Application of ALOS data in flood monitoring in Pakistan. In Proceedings of the 1st PI Symposium of ALOS Data Nodes, Kyoto, Japan, 10 July 2017.

130. Danielson, J.J.; Gesch, D.B. *Global Multi-Resolution Terrain Elevation Data 2010 (GMTED2010)*; United States Geological Survey: Reston, VA, USA, 2011.

131. Pakoksung, K.; Takagi, M. Digital elevation models on accuracy validation and bias correction in vertical. *Model. Earth Syst. Environ.* **2016**, *2*, 1–13. [CrossRef]

132. Simard, M.; Pinto, N.; Fisher, J.B.; Baccini, A. Mapping forest canopy height globally with spaceborne lidar. *J. Geophys. Res. Biogeosci.* **2011**, *116*. [CrossRef]

133. Lefsky, M.A. A global forest canopy height map from the Moderate Resolution Imaging Spectroradiometer and the Geoscience Laser Altimeter System. *Geophys. Res. Lett.* **2010**, *37*. [CrossRef]

134. Betbeder, J.; Rapinel, S.; Corgne, S.; Pottier, E.; Hubert-Moy, L. TerraSAR- X dual-pol time-series for mapping of wetland vegetation. *ISPRS J. Photogramm. Remote Sens.* **2015**, *107*, 90–98. [CrossRef]

135. Biancamaria, S.; Bates, P.D.; Boone, A.; Mognard, N.M. Large-scale coupled hydrologic and hydraulic modelling of the Ob river in Siberia. *J. Hydrol.* **2009**, *379*, 136–150. [CrossRef]

136. Gallant, J. Adaptive smoothing for noisy DEMs. *Geomorphometry* **2011**, *2011*, 7–9.

137. O'Loughlin, F.E.; Paiva, R.C.D.; Durand, M.; Alsdorf, D.E.; Bates, P.D. A multi- sensor approach towards a global vegetation corrected SRTM DEM product. *Remote Sens. Environ.* **2016**, *182*, 49–59. [CrossRef]

138. Peel, M.C.; Finlayson, B.L.; Mcmahon, T.A. Updated world map of the Köppen- Geiger climate classification. *Hydrol. Earth Syst. Sci.* **2007**, *11*, 1633–1644. [CrossRef]

139. Broxton, P.D.; Zeng, X.; Sulla-Menashe, D.; Troch, P.A. A global land cover climatology using MODIS data. *J. Appl. Meteorol. Climatol.* **2014**, *53*, 1593–1605. [CrossRef]

140. Yamazaki, D.; Ikeshima, D.; Tawatari, R.; Yamaguchi, T.; O'Loughlin, F.; Neal, J.C.; Sampson, C.C.; Kanae, S.; Bates, P.D. A high-accuracy map of global terrain elevations. *Geophys. Res. Lett.* **2017**, *44*, 5844–5853. [CrossRef]

141. Robinson, N.; Regetz, J.; Guralnick, R.P. EarthEnv-DEM90: A nearly-global, void-free, multi-scale smoothed, 90 m digital elevation model from fused ASTER and SRTM data. *ISPRS J. Photogramm. Remote Sens.* **2014**, *87*, 57–67. [CrossRef]

142. Tan, M.L.; Ficklin, D.L.; Dixon, B.; Yusop, Z.; Chaplot, V. Impacts of DEM resolution, source, and resampling technique on SWAT-simulated streamflow. *Appl. Geogr.* **2015**, *63*, 357–368. [CrossRef]

143. Medeiros, S.C.; Hagen, S.C.; Weishampel, J.F. Comparison of floodplain surface roughness parameters derived from land cover data and field measurements. *J. Hydrol.* **2012**, *452–453*, 139–149. [CrossRef]

144. Tarpanelli, A.; Brocca, L.; Lacava, T.; Melone, F.; Moramarco, T.; Faruolo, M.; Pergola, N.; Tramutoli, V. Toward the estimation of river discharge variations using MODIS data in ungauged basins. *Remote Sens. Environ.* **2013**, *136*, 47–55. [CrossRef]

145. Gleason, C.J.; Smith, L.C. Toward global mapping of river discharge using satellite images and at-many-stations hydraulic geometry. *Proc. Natl. Acad. Sci. USA* **2014**, *111*, 4788–4791. [CrossRef] [PubMed]

146. Sanyal, J.; Densmore, A.L.; Carbonneau, P. Analysing the effect of land-use/cover changes at sub-catchment levels on downstream flood peaks: A semi-distributed modelling approach with sparse data. *Catena* **2014**, *118*, 28–40. [CrossRef]

147. Karimi, N.; Bagheri, M.H.; Hooshyaripor, F.; Farokhnia, A.; Sheshangosht, S. Deriving and Evaluating Bathymetry Maps and Stage Curves for Shallow Lakes Using Remote Sensing Data. *Water Resour. Manag.* **2016**, *30*, 5003–5020. [CrossRef]

148. Wood, M.; Hostache, R.; Neal, J.; Wagener, T.; Giustarini, L.; Chini, M.; Corato, G.; Matgen, P.; Bates, P. Calibration of channel depth and friction parameters in the LISFLOOD- FP hydraulic model using medium resolution SAR data. *Hydrol. Earth Syst. Sci. Discuss.* **2016**, *20*, 4983–4997. [CrossRef]

149. Townsend, P.A.; Walsh, S.J. Modeling floodplain inundation using an integrated GIS with radar and optical remote sensing. *Geomorphology* **1998**, *21*, 295–312. [CrossRef]

150. Qasim, A.-A.M.S.M. *Assessment of High Resolution SAR Imagery for Mapping Floodplain Water Bodies: A Comparison between Radarsat-2 and TerraSAR-X*; Durham University: Durham, UK, 2011.

151. Stephen, M.C.; Ryan, S.A.; Paul, H.E.; Melinda, J.L.; David, M.M. Multi-Temporal Independent Component Analysis and Landsat 8 for Delineating Maximum Extent of the 2013 Colorado Front Range Flood. *Remote Sens.* **2015**, *7*, 9822–9843.

152. Alexakis, D.D.; Gryllakis, M.G.; Koutroulis, A.G.; Agapiou, A.; Themistocleous, K.; Tsanis, I.K.; Michaelides, S.; Pashiardis, S.; Demetriou, C.; Aristeidou, K.; et al. GIS and remote sensing techniques for the assessment of land use changes impact on flood hydrology: The case study of Yialias Basin in Cyprus. *Nat. Hazards Earth Syst. Sci. Discuss.* **2013**, *1*, 4833–4869. [CrossRef]

153. Zhang, F.; Zhu, X.; Liu, D. Blending MODIS and Landsat images for urban flood mapping. *Int. J. Remote Sens.* **2014**, *35*, 3237–3253. [CrossRef]

154. Schnebele, E.; Cervone, G. Improving remote sensing flood assessment using volunteered geographical data. *Nat. Hazards Earth Syst. Sci.* **2013**, *13*, 669–677. [CrossRef]

155. Henderson, F.M.; Lewis, A.J. *Principles and Applications of Imaging Radar. Manual of Remote Sensing*; John Wiley and Sons: Hoboken, NJ, USA, 1998; Volume 2.

156. Veljanovski, T.; Lamovec, P.; Ostir, K.; Pehani, P. Comparison of three techniques for detection of flooded areas on Envisat and Radarsat-2 satellite images. In Proceedings of the GEOSS Era: Towards Operational Environmental Monitoring, Sydney, Australia, 10–15 April 2011.

157. Long, S.; Fatoyinbo, T.E.; Policelli, F. Flood extent mapping for namibia using change detection and thresholding with SAR. *Environ. Res. Lett.* **2014**, *9*, 035002. [CrossRef]

158. Im, J.; Jensen, J.R.; Tullis, J.A. Object- based change detection using correlation image analysis and image segmentation. *Int. J. Remote Sens.* **2008**, *29*, 399–423. [CrossRef]

159. Lamovec, P.; Veljanovski, T.; Mikoš, M.; Oštir, K. Detecting flooded areas with machine learning techniques: Case study of the Selška Sora river flash flood in September 2007. *J. Appl. Remote Sens.* **2013**, *7*, 073564. [CrossRef]

160. Giustarini, L.; Hostache, R.; Matgen, P.; Schumann, G.J.P.; Bates, P.D.; Mason, D.C. A Change Detection Approach to Flood Mapping in Urban Areas Using TerraSAR- X. *IEEE Trans. Geosci. Remote Sens.* **2013**, *51*, 2417–2430. [CrossRef]

161. Grimaldi, S.; Li, Y.; Pauwels, V.; Walker, J.P. Remote Sensing-Derived Water Extent and Level to Constrain Hydraulic Flood Forecasting Models: Opportunities and Challenges. *Surv. Geophys.* **2016**, *37*, 977–1034. [CrossRef]

162. Di Baldassarre, G.; Schumann, G.; Brandimarte, L.; Bates, P. Timely low resolution SAR imagery to support floodplain modelling: A case study review. *Surv. Geophys.* **2011**, *32*, 255–269. [CrossRef]

163. Seung Oh, L.; Yongchul, S.; Kyudong, Y.; Younghun, J.; Venkatesh, M. An Approach Using a 1D Hydraulic Model, Landsat Imaging and Generalized Likelihood Uncertainty Estimation for an Approximation of Flood Discharge. *Water* **2013**, *5*, 1598–1621.

164. Sanyal, J. *Flood Prediction and Mitigation in Data-Sparse Environments*; Durham University: Durham, UK, 2013.

165. Aich, V.; Koné, B.; Hattermann, F.F.; Müller, E.N. Floods in the Niger basin – analysis and attribution. *Nat. Hazards Earth Syst. Sci. Discuss.* **2014**, *2*, 5171–5212. [CrossRef]

166. Nkeki, F.; Henah, P.; Ojeh, V. Geospatial Techniques for the Assessment and Analysis of Flood Risk along the Niger- Benue Basin in Nigeria. *J. Geogr. Inf. Syst.* **2013**, *5*, 123–135. [CrossRef]

167. Akinbobola, A.; Okogbue, E.C.; Olajiire, O. A GIS based flood risk mapping along the Niger-Benue river basin in Nigeria using watershed approach. *Ethiop. J. Environ. Stud. Manag.* **2015**, *8*, 616–627. [CrossRef]

168. Agada, S.; Nirupama, N. A serious flooding event in Nigeria in 2012 with specific focus on Benue State: A brief review. *Nat. Hazards* **2015**, *77*, 1405–1414. [CrossRef]

169. Tami, A.G.; Moses, O. Flood Vulnerability Assessment of Niger Delta States Relative to 2012 Flood Disaster in Nigeria. *Am. J. Environ. Protect.* **2015**, *3*, 76–83.

170. Komolafe, A.A.; Adegboyega, S.A.; Akinluyi, F.O. A Review of Flood Risk Analysis in Nigeria. *Am. J. Environ. Sci.* **2015**, *11*, 157–166. [CrossRef]

171. Ugonna, C. A Review of Flooding and Flood Risk Reduction in Nigeria. *Glob. J. Hum. Soc. Sci. Res.* **2016**, *16*, 1–21.

172. Opolot, E. Application of remote sensing and geographical information systems in flood management: A review. *Res. J. Appl. Sci. Eng. Technol.* **2013**, *6*, 1884–1894. [CrossRef]

173. Adeaga, O.; Oyebande, L.; Balogun, I. PUB and Water Resources Management Practises in Nigeria. *Water Energy Abstr.* **2008**, *18*, 58.

174. Ologunorisa, T.; Abawua, M. Flood risk assessment: A review. *J. Appl. Sci. Environ. Manag.* **2005**, *9*, 57–63.

175. Ngene, B.U.; Agunwamba, J.C.; Nwachukwu, B.A.; Okoro, B.C. The Challenges to Nigerian Raingauge Network Improvement. *RJEES* **2015**, *7*, 68–74. [CrossRef]

176. Federal Ministry of Water Resources. *The Project for Review and Update of Nigeria National Water Resources Master Plan*; Federal Ministry of Water Resources: Abuja, Nigeria, 2013.

177. Ngene, B.U. *Optimization of Rain Gauge Stations in Nigeria*; Federal University of Technology: Owerri, Nigeria, 2009.

178. Olomoda, I. Challenges of Continued River Niger Low Flow into Nigeria. *Spec. Publ. Niger. Assoc. Hydrol. Sci.* **2012**, *2012*, 145–155.

179. Izinyon, O.; Ehiorobo, J. L-moments approach for flood frequency analysis of river Okhuwan in Benin-Owena River basin in Nigeria. *Niger. J. Technol.* **2014**, *33*, 10–18. [CrossRef]

180. Ertuna, C. Water Resources Development and Management in Asia and the Pacific. *Environ. Soil Water Manag.* **1996**, *10*, 32–53.

181. Ononiwu, N. Appraisal of the role of satellite systems in acquisition of data for monitoring and evaluating global climatic changes with respect to reservoir energy generation. *Glob. Clim. Chang. Impact Energy Dev.* **1994**, *1994*, 1.

182. Olayinka, D.N. *Modelling Flooding in the Niger Delta*; Lancaster University: Lancaster, UK, 2012.

183. Merz, R.; Blöschl, G. Flood frequency regionalisation—Spatial proximity vs. catchment attributes. *J. Hydrol.* **2005**, *302*, 283–306. [CrossRef]

184. Reed, D. *Procedures for Flood Freequency Estimation, Volume 3: Statistical Procedures for Flood Freequency Estimation*; Institute of Hydrology: Parker, CO, USA, 1999.

185. Federal Ministry of Environment. *Action Plan for Erosion and Flood Control*; Federal Ministry of Environment: Abuja, Nigeria, 2005.

186. Musa, Z.N.; Popescu, I.; Mynett, A. The Niger Delta's vulnerability to river floods due to sea level rise. *Nat. Hazards Earth Syst. Sci.* **2014**, *14*, 3317–3329. [CrossRef]

187. Adelekan, I. Vulnerability assessment of an urban flood in Nigeria: Abeokuta flood 2007. *Nat. Hazards* **2011**, *56*, 215–231. [CrossRef]

188. Tamuno, P.; Ince, M.; Howard, G. Understanding vulnerability in the Niger floodplain. In Proceedings of the 29th (Water, Engineering and Development Centre) Conference WEDC towards the Millennium Development Goals-Actions for Water and Environmental Sanitation, Abuja, Nigeria, 23–26 September 2003; pp. 358–361.

189. Izinyon, O.; Ajumka, H. Regional Flood Frequency Analysis of Catchments in upper Benueriver Basin Using Index Flood Procedure. *Niger. J. Technol.* **2013**, *32*, 159–169.

190. Fasinmirin, J.T.; Olufayo, A.A. Comparison of Flood Prediction Models for River Lokoja, Nigeria. *Geophys. Res. Abstr.* **2006**, *8*.

191. Isikwue, M.O.; Onoja, S.B.; Laudan, K.J.; Bauchi, F. Establishment of an empirical model that correlates rainfall-intensity-duration-frequency for Makurdi Area, Nigeria. *Int. J. Adv. Eng. Technol.* **2012**, *5*, 40–46.

192. Ologunorisa, T.E.; Tersoo, T. The changing rainfall pattern and its implication for flood frequency in Makurdi, Northern Nigeria. *J. Appl. Sci. Environ. Manag.* **2005**, *10*, 97–102. [CrossRef]

193. Adewale, P.O.; Sangodoyin, A.Y.; Adewale, J.; Adamowski, J. Flood routing in the Ogunpa River in nigeria using HEC- RAS. *J. Environ. Hydrol.* **2010**, *18*, 1.

194. Padi, P.T.; Baldassarre, G.D.; Castellarin, A. Floodplain management in Africa: Large scale analysis of flood data. *Phys. Chem. Earth* **2011**, *36*, 292–298. [CrossRef]

195. Balogun, I.I.; Sojobi, A.O.; Oyedepo, B.O. Assessment of rainfall variability, rainwater harvesting potential and storage requirements in Odeda local government area of Ogun State in Southwestern Nigeria. *Cogent Environ. Sci.* **2016**, *2*, 1138597. [CrossRef]

196. Ogungbenro, S.B.; Morakinyo, T.E. Rainfall distribution and change detection across climatic zones in Nigeria. *Weather Clim. Extremes* **2014**, *5–6*, 1–6. [CrossRef]

197. Oyinloye, M.A.; Olamiju, O.I.; Oyetayo, B.S. Combating flood crisis using GIS: Empirical evidences from ala river floodplain, Isikan Area, Akure, Ondo State, Nigeria. *Commun. Inf. Sci. Manag. Eng.* **2013**, *3*, 439–447.

198. Ndabula, C.; Jidauna, G.; Oyatayo, K.; Averik, P.; Iguisi, E. Analysis of urban floodplain encroachment: Strategic approach to flood and floodplain management in Kaduna metropolis, Nigeria. *J. Geogr. Geol.* **2012**, *4*, 170. [CrossRef]

199. Okeke, I.C. Geographic Information Systems and Sustainable Water Resources Management in Nigeria. In *Coastal and Marine Geospatial Technologies*; Springer: Dordrecht, The Netherlands, 2010; pp. 219–226.

200. Hunter, N.M.; Bates, P.D.; Neelz, S.; Pender, G.; Villanueva, I.; Wright, N.G.; Liang, D.; Falconer, R.A.; Lin, B.; Waller, S.; et al. Benchmarking 2D hydraulic models for urban flooding. *Proc. ICE Water Manag.* **2008**, *161*, 13–30. [CrossRef]

201. Nigro, J.; Slayback, D.; Policelli, F.; Brakenridge, G. NASA/DFO MODIS Near Real-Time (NRT) Global Flood Mapping Product Evaluation of Flood and Permanent Water Detection. Available online: https://floodmap.modaps.eosdis.nasa.gov/documents/NASAGlobalNRTEvaluationSummary_v4.pdf (accessed on 30 July 2018).

202. Bakker, M.H.N. Transboundary River Floods and Institutional Capacity. *J. Am. Water Resour. Assoc.* **2009**, *45*, 553–566. [CrossRef]

203. Angelidis, P.; Kotsikas, M.; Kotsovinos, N. Management of Upstream Dams and Flood Protection of the Transboundary River Evros/Maritza. *Water Resour. Manag.* **2010**, *24*, 2467–2484. [CrossRef]

204. Clement, A.R. Causes of seasonal flooding in flood plains: A case of Makurdi, Northern Nigeria. *Int. J. Environ. Stud.* **2012**, *69*, 904–912. [CrossRef]

205. Zeitoun, M.; Goulden, M.; Tickner, D. Current and future challenges facing transboundary river basin management. *WIREs Clim. Chang.* **2013**, *4*, 331–349. [CrossRef]

206. Cooley, H.; Gleick, P. Climate- proofing transboundary water agreements. *Hydrol. Sci. J.* **2011**, *56*, 711–718. [CrossRef]

207. Wolf, A.T. *Atlas of International Freshwater Agreements*; UNEP: Nairobi, Kenya; Earthprint: Stevenage, UK, 2002; Volume 4.

208. Transboundary Water Assessment Programme. The Global Transboundary River Basins. Available online: http://twap-rivers.org/#global-basins (accessed on 10 August 2016).

209. ECOWAS-SWAC/OECD. *Transboundary River Basins*; ECOWAS-SWAC/OECD: Paris, France, 2008.

210. Hooper, B.P.; Lloyd, G.J. *Report on Iwrm in Transboundary Basins*; UNEP-DHI Centre for Water Environment: Hørsholm, Denmark, 2011.

211. Chikozho, C. Pathways for building capacity and ensuring effective transboundary water resources management in Africa: Revisiting the key issues, opportunities and challenges. *Phys. Chem. Earth* **2014**, *76–78*, 72–82. [CrossRef]

212. Klemas, V. Remote Sensing of Floods and Flood-Prone Areas: An Overview. *J. Coast. Res.* **2015**, *31*, 1005–1013. [CrossRef]

213. Mallinis, G.; Gitas, I.Z.; Giannakopoulos, V.; Maris, F.; Tsakiri-Strati, M. An object-based approach for flood area delineation in a transboundary area using ENVISAT ASAR and LANDSAT TM data. *Int. J. Dig. Earth* **2013**, *6*, 124–136. [CrossRef]

214. Serbis, D.; Papathanasiou, C.; Mamassis, N. Flood mitigation at the downstream areas of a transboundary river. In Proceedings of the 8th International Conference of EWRA "Water Resources Management in an Interdisciplinary and Changing Context", Porto, Portugal, 26–29 June 2013.

215. Mati, B.M.; Mutie, S.; Gadain, H.; Home, P.; Mtalo, F. Impacts of land-use/cover changes on the hydrology of the transboundary Mara River, Kenya/Tanzania. *Lakes Reserv. Res. Manag.* **2008**, *13*, 169–177. [CrossRef]

216. Hossain, F.; Siddique-E-Akbor, A.H.; Mazumder, L.C.; Shahnewaz, S.M.; Biancamaria, S.; Lee, H.; Shum, C.K. Proof of Concept of an Altimeter- Based River Forecasting System for Transboundary Flow Inside Bangladesh. *IEEE J. Sel. Top. Appl. Earth Obs. Remote Sens.* **2014**, *7*, 587–601. [CrossRef]

217. Seyler, F.; Calmant, S.; da Silva, J.; Filizola, N.; Roux, E.; Cochonneau, G.; Vauchel, P.; Bonnet, M.-P. Monitoring water level in large trans-boundary ungauged basins with altimetry: The example of ENVISAT over the Amazon basin. In Proceedings of the Asia-Pacific Remote Sensing, Noumea, New Caledonia, 19 December 2008; p. 715017.

218. Ojigi, M.; Abdulkadir, F.; Aderoju, M. Geospatial mapping and analysis of the 2012 flood disaster in central parts of Nigeria. In Proceedings of the 8th National GIS Symposium, Dammam, Saudi Arabia, 15–17 April 2013; pp. 1–14.

219. Olojo, O.O.; Asma, T.I.; Isah, A.A.; Oyewumi, A.S.; Adepero, O. The Role of Earth Observation Satellite during the International Collaboration on the 2012 Nigeria Flood Disaster. In Proceedings of the 64th International Astronautical Congress, Beijing, China, 22–27 September 2013.

220. Erekpokeme, L.N. Flood Disasters in Nigeria: Farmers and Governments' Mitigation Efforts. *J. Biol. Agric. Healthc.* **2015**, *5*, 150–154.

221. Daura, M.; Mayomi, I. Geo-Spatial Assessments of Flood Disaster Vulnerability of Benue and Taraba States. *Acad. Res. Int.* **2015**, *1*, 166–183.

222. Lehner, B.; Liermann, C.R.; Revenga, C.; Vörösmarty, C.; Fekete, B.; Crouzet, P.; Döll, P.; Endejan, M.; Frenken, K.; Magome, J. *Global Reservoir and Dam (GRanD) Database*; Technical Documentation, Version 1; NASA: Washington, DC, USA, 2011.

223. The Great Rivers Partnership. Niger River Basin. Available online: http://www.greatriverspartnership.org/en-us/africa/niger/pages/default.aspx (accessed on 11 August 2016).

224. Global Water Partnership. *West Africa—Iwrm in the Niger River Basin Case #46*; Global Water Partnership: Stockholm, Sweden, 2016.

225. Bossard, L. *West African Studies Regional Atlas on West Africa*; OECD Publishing: Paris, France, 2009.

226. International Waters Governance. Niger Basin. Available online: http://www.internationalwatersgovernance.com/niger-basin.html (accessed on 11 August 2016).

227. Morand, P.; Mikolasek, O. Review of the present state of knowledge of environment, fish stocks and fisheries of the River Niger (West Africa). In Proceedings of the Second International Symposium on the Management of Large Rivers for Fisheries: Sustaining Livelihoods and Biodiversity in the New Millenium, Phnom Penh, Cambodge, 11–14 February 2003.

228. Olomoda, I. Integrated Water Resources Management: Niger Authority's Experience. In Proceedings of the From Conflict to Co-Operation in International Water Resources Management: Challenges and Opportunities, Delft, The Netherlands, 20–22 November 2002.

229. Grossmann, M. Cooperation on Africa's internat onal waterbodies: Information needs and the role of information-sharing. *Editors* **2006**, *1*, 173.

230. Pilon, P.J.; Asefa, M.K. *Comprehensive Review of the World Hydrological Cycle Observing System*; World Meteorological Organization: Geneva, Switzerland, 2011.

231. Earle, A.; Cascão, A.E.; Hansson, S.; Jägerskog, A.; Swain, A.; Öjendal, J. *Transboundary Water Management and the Climate Change Debate*; Routledge: Abingdon, UK, 2015.

232. Skakun, S.; Kussul, N.; Shelestov, A.; Kussul, O. Flood Hazard and Flood Risk Assessment Using a Time Series of Satellite Images: A Case Study in Namibia. *Risk Anal.* **2014**, *34*, 1521–1537. [CrossRef] [PubMed]

233. Salami, Y.D.; Nnadi, F.N. Seasonal and interannual validation of satellite-measured reservoir levels at the Kainji dam. *Int. J. Water Resour. Environ. Eng.* **2012**, *4*, 105–113.

234. Sparavigna, A.C. Recurrence plots from altimetry data of some lakes in Africa. *Int. J. Sci.* **2014**, *3*, 19–27. [CrossRef]

235. Cretaux, J.-F.; Berge-Nguyen, M.; Leblanc, M.; Abarca Del Rio, R.; Delclaux, F.; Mognard, N.; Lion, C.; Pandey, R.K.; Tweed, S.; Calmant, S. Flood mapping inferred from remote sensing data. *Int. Water Technol. J.* **2011**, *1*, 48–62.

236. Bessis, J.L.; Béquignon, J.; Mahmood, A. The International Charter "Space and Major Disasters" initiative. *Acta Astronaut.* **2004**, *54*, 183–190. [CrossRef]

237. ICSMD. *The International Charter: Space and Major Disasters*; ICSMD: Konya, Turkey, 2015.

238. UNOOSA. *International Charter 'Space and Major Disasters', Towards Universl Access*; UNOOSA: Vienna, Austria, 2013.

239. National Centre for Space Studies. International Charter Space and Major Disasters. In *Charter Geographic Tool*; CNES: Paris, France, 2016.

240. James, G.; Shaba, H.; Zubair, O.; Teslim, A.G. Space–Based Disaster Management in Nigeria: The Role of the International Charter "Space and Major Disasters" In Proceedings of the FIG Working Week, Environment for Sustainability, Abuja, Nigeria, 6–10 May 2013.

241. Backhaus, R.; Czaran, L.; Epler, N.; Leitgab, M.; Lyu, Y.S.; Ravan, S.; Stevens, D.; Stumpf, P.; Szarzynski, J.; de Leon, J.-C.V. Support from space: The United Nations platform for space-based information for disaster management and emergency response (UN-SPIDER). In *Geoinformation for Disaster and Risk Management: Examples and Best Practices. Copenhagen, Denmark Joint Board of Geospatial Information Societies*; UNITED NATIONS: New York, NY, USA, 2010.

242. International Water Mangement Institute. *Emergency Response Products for Water Disasters*; International Water Mangement Institute: Colombo, Sri Lanka, 2016; Available online: http://www.iwmi.cgiar.org/resources/emergency-response-products-for-water-disasters/ (accessed on 17 August 2016).

243. Copernicus. *The Emergency Management Service—Mapping*; Copernicus: Keilor Downs, Australia, 2016.

244. Price, R. Digital Globe Open Data Program. Available online: http://blog.digitalglobe.com/news/launching-our-open-data-program-for-disaster-response/ (accessed on 20 January 2017).

245. Baruch, A.; May, A.; Yu, D. The motivations, enablers and barriers for voluntary participation in an online crowdsourcing platform. *Comput. Hum. Behav.* **2016**, *64*, 923–931. [CrossRef]
246. Kite, G.; Pietroniro, A. Remote sensing applications in hydrological modelling. *Hydrol. Sci. J.* **1996**, *41*, 563–591. [CrossRef]
247. Chikozho, C. Towards best-practice in transboundary water governance in Africa: Exploring the policy and institutional dimensions of conflict and cooperation over water. In *Rethinking Development Challenges for Public Policy*; Palgrave Macmillan: London, UK, 2012; pp. 155–200.
248. Tilleard, S.; Ford, J. Adaptation readiness and adaptive capacity of transboundary river basins. *Clim. Chang.* **2016**, *137*, 575–591. [CrossRef]
249. Sandro, M.; Christoph, R. Backscatter Analysis Using Multi-Temporal and Multi-Frequency SAR Data in the Context of Flood Mapping at River Saale, Germany. *Remote Sens.* **2015**, *7*, 7732–7752.
250. Asner, G.P. Cloud cover in Landsat observations of the Brazilian Amazon. *Int. J. Remote Sens.* **2001**, *22*, 3855–3862. [CrossRef]
251. Veljanovski, T.; Kanjir, U.; Oštir, K. Object-based image analysis of remote sensing data. *Geod. Vestnik* **2011**, *55*, 678–688. [CrossRef]
252. Escloupier, E.; Becker, M.; Marie-Joseph, I.; Linguet, L.; Timmermann, P.; Calmant, S.; Seyler, F. Reconstruction of Hydrological Archives in French Guiana by Radar Altimetry, Hydrodynamic Modeling and Nonlinear Analysis of Time Series. In Proceedings of the 20 Years of Progress in Radar Altimetry Symposium, Venice, Italy, 24–29 September 2012.
253. Yoshimoto, S.; Amarnath, G. Applications of Satellite-Based Rainfall Estimates in Flood Inundation Modeling-A Case Study in Mundeni Aru River Basin, Sri Lanka. *Remote Sens.* **2017**, *9*, 998. [CrossRef]
254. Komi, K.; Neal, J.; Trigg, M.A.; Diekkrüger, B. Modelling of flood hazard extent in data sparse areas: a case study of the Oti River basin, West Africa. *J. Hydrol.* **2017**, *10*, 122–132. [CrossRef]
255. Yu, D.; Yin, J.; Liu, M. Validating city-scale surface water flood modelling using crowd-sourced data. *Environ. Res. Lett.* **2016**, *11*, 124011. [CrossRef]
256. Revilla-Romero, B.; Beck, H.E.; Burek, P.; Salamon, P.; de Roo, A.; Thielen, J. Filling the gaps: Calibrating a rainfall-runoff model using satellite-derived surface water extent. *Remote Sens. Environ.* **2015**, *171*, 118–131. [CrossRef]
257. Di Baldassarre, G.; Uhlenbrook, S. Is the current flood of data enough? A treatise on research needs for the improvement of flood modelling. *Hydrol. Process.* **2012**, *26*, 153–158. [CrossRef]
258. Corcoran, J.; Knight, J.; Brisco, B.; Kaya, S.; Cull, A.; Murnaghan, K. The integration of optical, topographic, and radar data for wetland mapping in northern Minnesota. *Can. J. Remote Sens.* **2012**, *37*, 564–582. [CrossRef]
259. African Association of Remote Sensing of the Environment; European Association of Remote Sensing Companies. *A Survey into the Africanprivate Sector in Earthobservation Andgeospatial Fields*; African Association of Remote Sensing of the Environment: Kampala, Uganda; European Association of Remote Sensing Companies: Brussels, Belgium, 2016.
260. Nwilo, P.; Osanwuta, D. National Spatial Data Infrastructure for Nigeria-Issues to Be Considered. In Proceedings of the FIG Working Week, Athens, Greece, 22–27 May 2004.
261. Degrossi, L.C.; de Albuquerque, J.P.; Fava, M.C.; Mendiondo, E.M. Flood Citizen Observatory: A crowdsourcing-based approach for flood risk management in Brazil. In Proceedings of the 26th International Conference on Software Engineering and Knowledge Engineering (SEKE 2014), Vancouver, BC, Canada, 1–3 July 2014; pp. 570–575.
262. Ekeu-wei, I.T.; Blackburn, G.A. Evaluation of Crowd-Sourcing (Volunteered GIS) and NRT-MODIS Flood Map in Monitoring Flood in Nigeria. In Proceedings of the 7th International Conference of the Nigerian association of Hydrological Sciences (NAHS), Abuja, Nigeria, 18–21 October 2016.
263. Ekeu-wei, I.T. Application of Open-Access and 3rd Party Geospatial Technology for Integrated Flood Risk Management in Data Sparse Regions of Developing Countries. In *Lancaster Environmental Centre*; Lancaster University: Lancaster, UK, 2018.

Technical Note

Merging Real-Time Channel Sensor Networks with Continental-Scale Hydrologic Models: A Data Assimilation Approach for Improving Accuracy in Flood Depth Predictions

Amir Javaheri [1], Mohammad Nabatian [2], Ehsan Omranian [3],*, Meghna Babbar-Sebens [1] and Seong Jin Noh [2]

1 Department of Civil and Construction Engineering, Oregon State University, Corvallis, OR 97331, USA; javaheam@oregonstate.edu (A.J.); meghna@oregonstate.edu (M.B.-S.)
2 Department of Civil Engineering, University of Texas at Arlington, Arlington, TX 76019, USA; mohammad.nabatian@mavs.uta.edu (M.N.); seongjin.noh@uta.edu (S.J.N.)
3 Department of Civil & Environmental Engineering, University of Texas at San Antonio, San Antonio, TX 78256, USA
* Correspondence: seyedehsan.omranian@utsa.edu; Tel.: +1-210-803-3847

Received: 27 October 2017; Accepted: 18 January 2018; Published: 21 January 2018

Abstract: This study proposes a framework that (i) uses data assimilation as a post processing technique to increase the accuracy of water depth prediction, (ii) updates streamflow generated by the National Water Model (NWM), and (iii) proposes a scope for updating the initial condition of continental-scale hydrologic models. Predicted flows by the NWM for each stream were converted to the water depth using the Height Above Nearest Drainage (HAND) method. The water level measurements from the Iowa Flood Inundation System (a test bed sensor network in this study) were converted to water depths and then assimilated into the HAND model using the ensemble Kalman filter (EnKF). The results showed that after assimilating the water depth using the EnKF, for a flood event during 2015, the normalized root mean square error was reduced by 0.50 m (51%) for training tributaries. Comparison of the updated modeled water stage values with observations at testing locations showed that the proposed methodology was also effective on the tributaries with no observations. The overall error reduced from 0.89 m to 0.44 m for testing tributaries. The updated depths were then converted to streamflow using rating curves generated by the HAND model. The error between updated flows and observations at United States Geological Survey (USGS) station at Squaw Creek decreased by 35%. For future work, updated streamflows could also be used to dynamically update initial conditions in the continental-scale National Water Model.

Keywords: data assimilation; ensemble Kalman filter; flood inundation maps; National Water Model (NWM)

---

1. Introduction

Flooding is among the most destructive natural disasters globally. In the United States, based on a U.S. National Weather Service (NWS) report, the average annual property/human losses are estimated to be more than \$8 billion [1–3] (Federal Emergency Management Agency (FEMA), 2013). Flood inundation maps help to detect flood-prone regions and can prevent major catastrophe by providing reliable information to the public about the flood-risk [4,5]. However, these maps are based on predictions of water depth values in the stream and on the landscape for extreme rainfall events [6]. Hydrology and hydraulic models can be useful for predicting these depth values, although, like many other models, these models are only as reliable as the underlying assumptions in the model's structure and parameters [7,8]. The assimilation of water depth measurements, as a post-processing technique, has the potential to reduce the error between model predictions and observations in order to generate more reliable flood inundation maps [9].

Sequential data assimilation techniques are often used for updating a model's state variables when new observations become available [10]. A data assimilation process is also able to reduce the uncertainty in prediction by integrating real-time observations from a variety of monitoring technologies [11,12]. The Kalman filter [13] is a commonly used data assimilation technique that was initially developed to update the state variables of linear systems [14]. However, this method has been used for nonlinear problems as well [15]. The ensemble Kalman filter (EnKF) is another data assimilation technique that was introduced by Evensen [16]. In the case of non-linear models for which the assumption of linearity is not satisfied, EnKF can be used as an effective technique. Multiple studies have also used this method in the past to update hydraulic, hydrologic, and hydrodynamic models [17–19]

The main objectives of this study include:

(1) Assimilating water depth measurements to dynamically update water level predictions.
(2) Improving streamflow predicted by the National Water Model (NWM) using updated water levels.
(3) Proposing a scope to update continental-scale hydrologic models (e.g., NWM).

2. Study Area and Methodology

2.1. Study Area Characteristics and Data Collection

The study domain is the Squaw Creek watershed (Hydrologic Unit Code (HUC) ID = 0708010503), located in Jasper County, Iowa (Figure 1), which drains 40.3 km^2 and discharges into the Skunk River. It is located on Southern Iowa Drift Plain and characterized by steeply rolling hills and well developed drainage [20]. This watershed has experienced several flooding events in the past [21], and it contains a dense network of bridge sensors that measure the water level along the channel. Flow measurements are collected at a USGS station (# 5470500) in this watershed. Water surface elevation were gathered from the Iowa Flood Information System (IFIS). The Iowa Flood Center (IFC) developed and maintains nearly 250 stream stage sensors across the state. Sensors are attached to the sides of bridges and are designed to measure the distance from the sensor to the water surface. Data is transmitted automatically and frequently to the Iowa Flood Information System. Water surface elevation at each time step can be calculated by subtracting the sensor measurements from the elevation of each sensor.

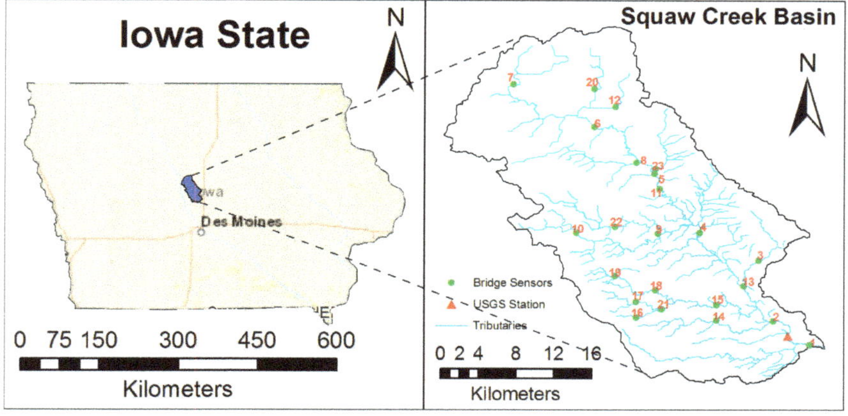

Figure 1. Squaw Creek watershed (Iowa State).

2.2. Proposed Approach

Figure 2 illustrates the overall proposed methodology:

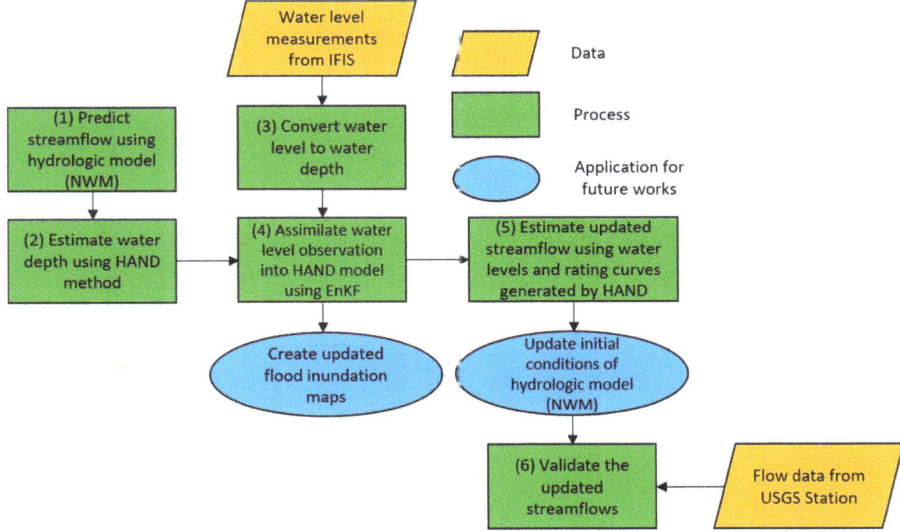

Figure 2. Schematic methodology.

Hourly predicted flows for each tributary were first obtained from the NWM (Step 1). The Height Above Nearest Drainage (HAND) [22] method was then used to predict the water depth in river channels (Step 2). Incoming hourly water level observations from the Iowa Flood Information System (IFIS) were converted to water depth (Step 3) to become consistent with HAND water depth predictions, and then assimilated into the HAND model via an ensemble Kalman filter technique (Step 4). Finally, using back calculation, updated water levels were converted to flow using rating curves generated by the HAND (Step 5). For future work, results from Steps 4 and 5 can be used to create flood inundation maps and update the initials conditions of the hydrologic model (NWM), respectively. Finally, updated streamflows were validated using flow measurements at USGS stations (Step 6).

There are 85 tributaries and 20 bridge sensors that measure the water levels. Since the cross-sections were not available for all of these sensor locations, measurements of only 10 sensors were used for data assimilation. Six sensors were randomly selected for training and the other four sensors were selected for testing to make sure that this approach is also suitable for tributaries with no observation. The NWM predicts the flow at the outlet of each sub-basin, however, bridge sensors are located upstream of outlets. Assuming equal soil type and land use for each sub-basin, the flow at the outlet was linearly distributed along the river.

$$Q_{outlet} = \frac{L_{Total}}{L_{Total} - L_{Partial}} \times Q_B \qquad (1)$$

where Q_{outlet} is the flow at the outlet of each sub-basin, L_{Total} is the length of river, $L_{Partial}$ is the river length from bridge sensor to the outlet of each sub-basin, and Q_B is the flow measured at each bridge sensor.

2.3. National Water Model

The National Water Model (MWM) is a continental-scale hydrologic model that generates forecasts for multiple variables [23]. The NWM was released by the National Weather Service (NWS) Office of Water Prediction (OWP) in collaboration with the National Center for Atmospheric Research (NCAR) and the National Center for Environmental Prediction (NCEP). The model generates streamflows for 2.7 million reaches of the National Hydrography Dataset (NHD) with four forecast products (i.e., analysis and assimilation, short range forecast, medium range forecast, and long range forecast) [24]. Table 1 shows the forecast step, frequency, and forecast duration of each product.

Table 1. National Water Model details (http://water.noaa.gov/about/nwm).

Forecast Product	Forecast Latency	Frequency	Forecast Duration
Analysis and assimilation	1 h	Hourly	0–3 h
Short range forecast	1 h	Hourly	0–15 h
Medium range forecast	3 h	Daily	0–10 days
Long range forecast	6 h	4× Daily	0–30 days

2.4. The HAND Method

The Height Above Nearest Drainage (HAND) [22] is a model generated by the National Science Foundation (NSF) and the National Water Center for developing continental-scale inundation mapping [25]. It is based on hydrological terrain analysis to produce flood inundation maps [26,27]. The Digital Elevation Model (DEM) of watershed is used to drive the flow direction and flow accumulation area. A stream grid is then generated to be used as an input to Terrain Analysis Using Digital Elevation Models [28], along with the flow-direction grid, to create the height of each grid cell above the nearest drainage. The NWM predicts flow at the outlet, which can be converted to water depth with a rating curve, and 10 sensors also provide information of water depth. The HAND is a raster-based method and specifies the inundation zone according to the corresponding river segment. It produces an inundation map according to the water depth of tributaries [25].

2.5. Ensemble Kalman Filter

The ensemble Kalman filter algorithm was used in the proposed methodology (Step 4) to assimilate water depth observations into the model. In the ensemble Kalman filter, the prediction model is represented by:

$$h_k = F(Q) + w_k \tag{2}$$

where h denotes the vector of state variables (water depth), Q is the flows from the National Water Model, w_k is stationary zero-mean white noises, F represents the prediction model (HAND model), and the subscript "k" denotes the time step. If an ensemble of n predicted state variables is available, h_k^f can be written as:

$$h_k^f = (h_k^{f_1}, \ldots, h_k^{f_n}) \tag{3}$$

where the superscript "f_i" represents the ith forecast ensemble member. Generally, initial condition and inputs are perturbed and ensemble of predicted state variables is generated by running the model for each initial condition. This technique needs to run the model several times for each realization. However, due to computational cost, it is not possible to run the NWM several times to create the ensemble. Instead of perturbing the initial conditions (IC) and inputs, ensemble was created by taking several predicted flows by the NWM up to 4 h (outflow changes by 10% during this period) before the selected events. Outflow was then converted to water depth using rating curves generated by the HAND. For this study, 10 realizations were created to estimate the error matrix in the model. The average of the ensemble is defined by:

$$\overline{h}_k^f = \frac{1}{n}\sum_{i=1}^{n} h_k^{f_i} \tag{4}$$

Since true states are not known, we estimate them using the average of realizations in the ensemble. Then the error matrix can be estimated by:

$$P_k^f = \frac{1}{n-1}\langle (h_k^f - \overline{h}_k)(h_k^f - \overline{h}_k)^T \rangle \tag{5}$$

The error matrix is then used to calculate the Kalman gain matrix by:

$$K_k = P_k^f H_k^T \left(H_k P_k^f H_k^T + R_k \right)^{-1} \tag{6}$$

where H is the linear transformation which relates the state variables to observations. The updated state vector (h^a) is taken to be a linear combination of the forecast and the observations. The observations should be treated as random variables to obtain consistent error propagation in the ensemble Kalman filter [29]. Hence, the actual measurements were used as reference and random noise with zero mean and covariance R was added to measurements. The updating equation is given by:

$$h_k^a = h_k^f + K_k \left(h^*_k - H_k h_k^f \right) \tag{7}$$

where h* is the water level observations after adding random noise.

2.5.1. Undersampling

The accuracy of the ensemble Kalman filter is highly dependent on the size of the ensemble. However, due to the complexity of models and computational cost, it is not always possible to generate a large ensemble. If the number of the ensemble is relatively small, it will not be able to accurately estimate the model covariance matrix and system may be undersampled. Inbreeding, filter divergence, and spurious correlation are the main issues caused by undersampling [30]. Covariance inflation and covariance localization are the most common methods used to solve the undersampling problem [31].

2.5.2. Covariance Inflation

Equation (7) is used to inflate the underestimated covariance [32]:

$$h_k^f \leftarrow r \left(h_k^f - \overline{h}_k^f \right) + \overline{h}_k^f \tag{8}$$

where $\leftarrow$ denotes the replacement of a previous value and r is the inflation factor. The optimal inflation factor is related to the ensemble size and may vary between 1.01 and 1.07 [31]. Several methods have been proposed to estimate the inflated forecast and observational error covariance matrices [33]. The adjusted forms of forecast and observational error covariance matrices are $\lambda_k P_k$ and $\mu_k R_k$, respectively. Wu et al. (2013) proposed Equations (9) and (10) to estimate λ_k and μ_k [34].

$$\lambda_k = \frac{\mathrm{Tr}\left(d_k^T H_k P_k H_k^T d_k\right)\mathrm{Tr}(R_k^2) - \mathrm{Tr}\left(d_k^T R_k d_k\right)\mathrm{Tr}\left(H_k P_k H_k^T R_k\right)}{\mathrm{Tr}\left(H_k P_k H_k^T H_k P_k H_k^T\right)\mathrm{Tr}(R_k^2) - \mathrm{Tr}\left(H_k P_k H_k^T R_k\right)2} \tag{9}$$

$$\mu_k = \frac{\mathrm{Tr}\left(H_k P_k H_k^T H_k P_k H_k^T\right)\left(d_k^T R_k d_k\right) - \mathrm{Tr}\left(d_k^T H_k P_k H_k^T d_k\right)\mathrm{Tr}\left(H_k P_k H_k^T R_k\right)}{\mathrm{Tr}\left(H_k P_k H_k^T H_k P_k H_k^T\right)\mathrm{Tr}(R_k^2) - \mathrm{Tr}\left(H_k P_k H_k^T R_k\right)2} \tag{10}$$

where $d_k \equiv h^*{}_k - \overline{h}^f_k$.

2.5.3. Covariance Localization

Covariance localization indicated the cutting of the covariance matrix at a specific length [35]. Correlation function suggested by Gaspari and Cohn (1999) was applied to the covariance matrix by using the Schur product to eliminate the spurious correlation [36,37]. The application of covariance inflation and localization methods transform the updating Equation (7) to the one below (Equation (11).

$$h^a_k = h^f_k + [\rho o(\lambda_k P^f_k H^T_k)][\rho o(H_k \lambda_k P^f_k H^T_k) + \mu_k R_k]^{-1}(h^*{}_k - H_k h^f_k) \tag{11}$$

3. Results

Streamflow obtained from the NWM for all 85 tributaries were converted to water depth using rating curves generated by the HAND model. Figure 3 shows examples of the rating curves derived from the HAND for two tributaries in the watershed.

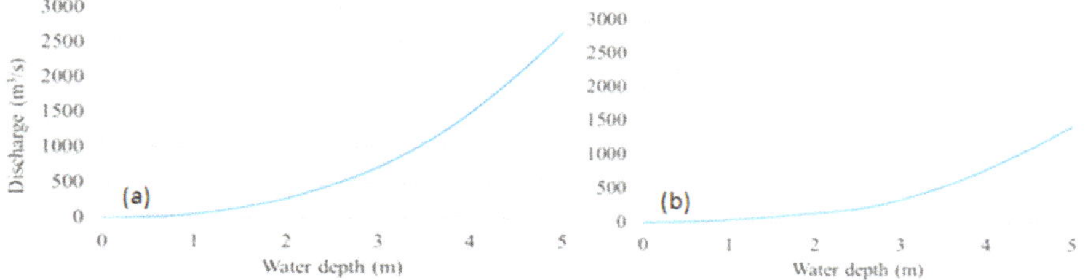

Figure 3. Rating curves created by the Height Above Nearest Drainage (HAND) for (**a**) Squaw Creek River and (**b**) Prairie Creek River.

Water level measurements were assimilated into the model for seven bridge sensors selected as training points. Figure 4 shows the water depth at the training locations (a) before data assimilation and (b) after data assimilation, compared with observations from the IFIS. Green triangles are model predictions from the HAND before data assimilation, red squares are updated water depths after data assimilation, and blue dots are water depth observations from the IFIS.

We found that after the assimilation of water depth, the overall error for training locations reduced from 98 cm to 48 cm. We also calculated the error for the testing locations to make sure that the Kalman filter is able to improve the model accuracy for the sites with no observations. It was found that the overall error decreased for testing locations as well (from 89 to 44 cm). Table 2 compares the water depths before and after data assimilation with observations from IFIS for training and testing tributaries. Results show that overall error for both training and testing locations decreased by 48 cm (51%). However, the error has not been improved or error improvement is not significant in some of the sensor locations. This could be due to a small ensemble size and error in the observations (especially error in cross-sections of bridge sensor). The general Root Mean Square Error (RMSE) formula is indicated in Equation (13).

$$\text{Root Mean Square Error (RMSE)} = \sqrt{\frac{1}{N}\sum_{n=1}^{N}(Predicted_n - Observation_n)^2} \tag{13}$$

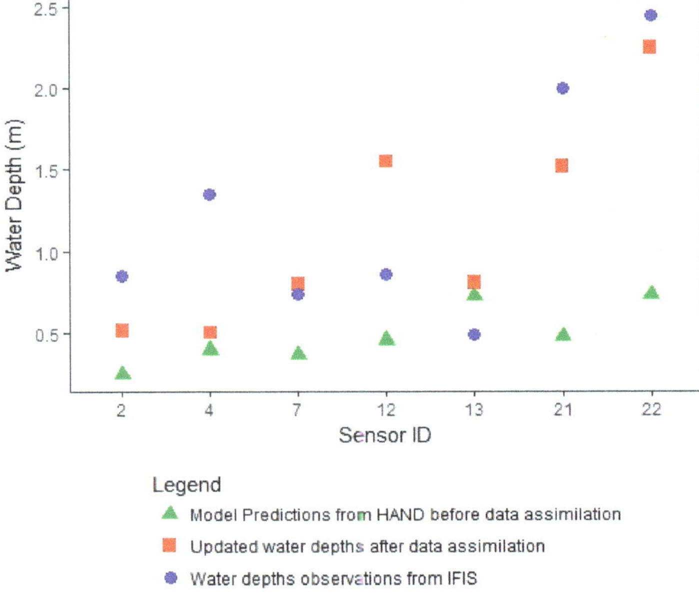

Figure 4. Water depth at each sensor location.

Table 2. Water depth before and after data assimilation compared with water depth measurement from the Iowa Flood Information System (IFIS) for training and testing tributaries.

	Water Depth (m)		
	Observation	Before Data Assimilation	After Data Assimilation
	0.85	0.24	0.55
	1.34	0.37	0.52
	0.72	0.36	0.81
Training Tributaries	0.85	0.44	1.27
	0.48	0.69	0.84
	2.00	0.49	1.25
	2.39	0.71	2.26
	1.30	2.01	1.67
Testing Tributaries	0.32	1.01	0.71
	1.50	0.30	0.81
	1.86	1.01	1.99
Overall RMSE (m)		0.95	0.47

Figure 5 illustrates the time series of flow predicted by the NWM versus flow observation at the USGS station at Squaw Creek at Ames. Water depth observations were assimilated at 6:00, 11:00, and 17:00 on 24 June 2015. After updating the water depth at selected times, the rating curve from the HAND was used to estimate the corresponding flow (green dots in Figure 5). It was found that the overall error between USGS measurements and updated flows from the NWM was reduced by 35% for selected times. Since continental-scale hydrologic models such as the National Water Model are computationally expensive and it is not possible to run them several times to create a larger ensemble, the proposed methodology can be used as an effective method for future studies to update the initial condition of such models.

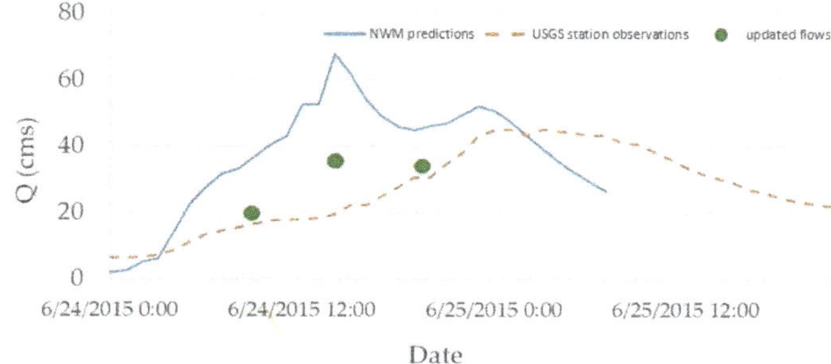

Figure 5. River flow at a USGS station at Squaw Creek at Ames. Solid blue line shows the NWM predictions, dashed-red line shows the observations at the USGS station, and green dots show the updated flows.

4. Conclusions

The National Water Model provides flow rates for 85 tributaries in the study area. A data assimilation framework was proposed to (i) reduce the error of water depth prediction in tributaries, (ii) update the streamflow prediction, and (iii) introduce a scope for updating the initial conditions of continental-scale hydrologic models. Streamflows from the NWM were first converted to water depth using the HAND model and then assimilated to the model. After data assimilation, the root mean square errors of the estimated depth were reduced by 50 cm (51%). However, the error was not reduced at all sensor locations. The main reason for this is the error in bridge cross-sections. For future works, it is highly recommended to collect accurate cross-sections of rivers at each bridge equipped with a sensor. In this study, cross-sections were estimated based on measurements taken downstream and upstream of each bridge. Another reason could be the small ensemble size. It was also found that the proposed methodology is effective for the tributaries without observation. The model predictions improved by 45 cm for tributaries for which observations were not assimilated to the model. Finally, updated water depth at the outlet of the watershed was converted to streamflow using rating curves generated by the HAND. The proposed methodology could also be used to dynamically update initial conditions in the continental-scale National Water Model.

Acknowledgments: The authors would like to thank the Consortium of Universities for the Advancement of Hydrologic Science, Inc. (CUAHSI) and the Office of Water Prediction (OWP) for coordination of Summer Institute. The authors would also like to thank Ibrahim Demir for providing water level observation for state of Iowa.

Author Contributions: Amir Javaheri and Mohammad Nabatian conceived and designed the methodology; Amir Javaheri and Mohammad Nabatian analyzed the data; Amir Javaheri and Ehsan Omranian prepared the graphs; Amir Javaheri and Ehsan Omranian wrote the paper with the multiple inputs and reviews by Meghna Babbar-Sebens and Seong Jin Noh.

Conflicts of Interest: The authors declare no conflict of interest.

References

1. Saksena, S.; Merwade, V. Incorporating the effect of DEM resolution and accuracy for improved flood inundation mapping. *J. Hydrol.* **2015**, *530*, 180–194. [CrossRef]
2. Zarekarizi, M.; Rana, A.; Moradkhani, H. Precipitation extremes and their relation to climatic indices in the Pacific Northwest USA. *Clim. Dyn.* **2017**. [CrossRef]
3. Omranian, E.; Sharif, H.O. Evaluation of the Global Precipitation Measurement (GPM) Satellite Rainfall Products Over the Lower Colorado River Basin, Texas. *J. Am. Water Resour. Assoc.* **2018**. [CrossRef]
4. Follum, M.L.; Tavakoly, A.A.; Niemann, J.D.; Snow, A.D. AutoRAPID: A Model for Prompt Streamflow Estimation and Flood Inundation Mapping over Regional to Continental Extents. *JAWRA J. Am. Water Resour. Assoc.* **2017**, *53*, 280–299. [CrossRef]

5. Afshari, S.; Omranian, E.; Feng, D. *Relative Sensitivity of Flood Inundation Extent by Different Physical and Semi-Empirical Models*; CUAHSI Technical Report No. 13; Consortium of Universities for the Advancement of Hydrologic Science, Inc.: Cambridge, MA, USA, 2016; pp. 19–24.

6. Jafarzadegan, K.; Merwade, V. A DEM-based approach for large-scale floodplain mapping in ungauged watersheds. *J. Hydrol.* **2017**, *550*, 650–662. [CrossRef]

7. McHugh, P.A.; Saunders, W.C.; Bouwes, N.; Wall, C.E.; Bangen, S.; Wheaton, J.M.; Nahorniak, M.; Ruzycki, J.R.; Tattam, I.A.; Jordan, C.E. Linking models across scales to assess the viability and restoration potential of a threatened population of steelhead (*Oncorhynchus mykiss*) in the Middle Fork John Day River, Oregon, USA. *Ecol. Model.* **2017**, *355*, 24–38. [CrossRef]

8. Hamidi, S.A.; Hosseiny, H.; Ekhtari, N.; Khazaei, B. Using MODIS remote sensing data for mapping the spatio-temporal variability of water quality and river turbid plume. *J. Coast. Conserv.* **2017**, *21*, 939–950. [CrossRef]

9. Durand, M.; Andreadis, K.M.; Alsdorf, D.E.; Lettenmaier, D.P.; Moller, D.; Wilson, M. Estimation of bathymetric depth and slope from data assimilation of swath altimetry into a hydrodynamic model. *Geophys. Res. Lett.* **2008**, *35*. [CrossRef]

10. Evensen, G. *Data Assimilation: The Ensemble Kalman Filter*; Springer: Berlin/Heidelberg, Germany, 2009.

11. Moradkhani, H. Hydrologic Remote Sensing and Land Surface Data Assimilation. *Sensors* **2008**, *8*. [CrossRef] [PubMed]

12. Yan, H.; Moradkhani, H.; Zarekarizi, M. A probabilistic drought forecasting framework: A combined dynamical and statistical approach. *J. Hydrol.* **2017**, *548*, 291–304. [CrossRef]

13. Kalman, R.E. A New Approach to Linear Filtering and Prediction Problems. *J. Basic Eng.* **1960**, *82*, 35–45. [CrossRef]

14. Maybeck, P.S. *Stochastic Models, Estimation and Control*; Academic Press: Cambridge, MA, USA, 1979.

15. Krener, A.; Duarte, A. A Hybrid Computational Approach to Nonlinear Estimation. In Proceedings of the Decision and Control, Kobe, Japan, 13 December 1996.

16. Evensen, G. Sequential data assimilation with a nonlinear quasi-geostrophic model using Monte Carlo methods to forecast error statistics. *J. Geophys. Res.* **1994**, *99*, 10143–10162. [CrossRef]

17. Crow, W.; Wood, E. The assimilation of remotely sensed soil brightness temperature imagery into a land surface model using Ensemble Kalman filtering: A case study based on ESTAR measurements during SGP97. *Adv. Water Resour.* **2003**, *26*, 137–149. [CrossRef]

18. Noh Seong, J.; Tachikawa, Y.; Shiiba, M.; Kim, S. Ensemble Kalman Filtering and Particle Filtering in a Lag-Time Window for Short-Term Streamflow Forecasting with a Distributed Hydrologic Model. *J. Hydrol. Eng.* **2013**, *18*, 1684–1696. [CrossRef]

19. Miller, R.N.; Cane, M.A. A Kalman Filter Analysis of Sea Level Height in the Tropical Pacific. *J. Phys. Oceanogr.* **1989**, *19*, 773–790. [CrossRef]

20. Schilling, K.E.; Thompson, C.A. Walnut creek watershed monitoring project, iowa monitoring water quality in response to Prairie restoration. *JAWRA J. Am. Water Resour. Assoc.* **2000**, *36*, 1101–1114. [CrossRef]

21. Zogg, J. *The Top Five Iowa Floods*; National Weather Service WFO: Des Monies, IA, USA, 2014.

22. Liu, Y.; Maidment, D.; Tarboton, D.; Zheng, X.; Yıldırım, A.; Sazib, N.; Wang, S. A CyberGIS Approach to Generating High-Resolution Height Above Nearest Drainage (HAND) Raster for National Flood Mapping. In Proceedings of the Third International Conference on CyberGIS and Geospatial Data Science, Urbana, IL, USA, 26–28 July 2016.

23. National Oceanic and Atmospheric Administration (NOAA). *NOAA Launches America's First National Water Forecast Model*; National Oceanic and Atmospheric Administration: Silver Spring, MD, USA, 2016.

24. Souffront Alcantara, M.A.; Crawley, S.; Stealey, M.J.; Nelson, E.J.; Ames, D.P.; Jones, N.L. Open Water Data Solutions for Accessing the National Water Model. *Open Water J.* **2017**, *4*, 1–14.

25. Afshari, S.; Tavakoly, A.A.; Rajib, M.A.; Zheng, X.; Follum, M.L.; Omranian, E.; Fekete, B.M. Comparison of new generation low-complexity flood inundation mapping tools with a hydrodynamic model. *J. Hydrol.* **2018**, *556*, 539–556. [CrossRef]

26. Nobre, A.D.; Cuartas, L.A.; Hodnett, M.; Rennó, C.D.; Rodrigues, G.; Silveira, A.; Waterloo, M.; Saleska, S. Height Above the Nearest Drainage—A hydrologically relevant new terrain model. *J. Hydrol.* **2011**, *404*, 13–29. [CrossRef]

27. Rennó, C.D.; Nobre, A.D.; Cuartas, L.A.; Soares, J.V.; Hodnett, M.G.; Tomasella, J.; Waterloo, M.J. HAND, a new terrain descriptor using SRTM-DEM: Mapping terra-firme rainforest environments in Amazonia. *Remote Sens. Environ.* **2008**, *112*, 3469–3481. [CrossRef]

28. Fan, Y.; Liu, Y.; Wang, S.; Tarboton, D.; Yildirim, A.; Wilkins-Diehr, N. Accelerating TauDEM as a Scalable Hydrological Terrain Analysis Service on XSEDE. In Proceedings of the 2014 Annual Conference on Extreme Science and Engineering Discovery Environment, Atlanta, GA, USA, 13–18 July 2014; pp. 1–2.

29. Burgers, G.; Jan van Leeuwen, P.; Evensen, G. Analysis Scheme in the Ensemble Kalman Filter. *Mon. Weather Rev.* **1998**, *126*, 1719–1724. [CrossRef]

30. Petrie, R.E. *Localization in the Ensemble Kalman Filter*; University of Reading: Reading, UK, 2008.

31. Hamill, T.M.; Whitaker, J.S.; Snyder, C. Distance-Dependent Filtering of Background Error Covariance Estimates in an Ensemble Kalman Filter. *Mon. Weather Rev.* **2001**, *129*, 2776–2790. [CrossRef]

32. Anderson, J.L.; Anderson, S.L. A Monte Carlo Implementation of the Nonlinear Filtering Problem to Produce Ensemble Assimilations and Forecasts. *Mon. Weather Rev.* **1999**, *127*, 2741–2758. [CrossRef]

33. Wang, X.; Bishop, C.H. A Comparison of Breeding and Ensemble Transform Kalman Filter Ensemble Forecast Schemes. *J. Atmos. Sci.* **2003**, *60*, 1140–1158. [CrossRef]

34. Wu, G.; Zheng, X.; Wang, L.; Zhang, S.; Liang, X.; Li, Y. A new structure for error covariance matrices and their adaptive estimation in EnKF assimilation. *Q. J. R. Meteorol. Soc.* **2013**, *139*, 795–804. [CrossRef]

35. Houtekamer, P.L.; Mitchell, H.L. Data Assimilation Using an Ensemble Kalman Filter Technique. *Mon. Weather Rev.* **1998**, *126*, 796–811. [CrossRef]

36. Gaspari, G.; Cohn, S.E. Construction of correlation functions in two and three dimensions. *Q. J. R. Meteorol. Soc.* **1999**, *125*, 723–757. [CrossRef]

37. Javaheri, A.; Babbar-Sebens, M.; Miller, R.N. From skin to bulk: An adjustment technique for assimilation of satellite-derived temperature observations in numerical models of small inland water bodies. *Adv. Water Resour.* **2016**, *92*, 284–298. [CrossRef]

MDPI

St. Alban-Anlage 66

4052 Basel

Switzerland

Tel. +41 61 683 77 34

Fax +41 61 302 89 18

www.mdpi.com

Hydrology Editorial Office

E-mail: hydrology@mdpi.com

www.mdpi.com/journal/hydrology

www.ingramcontent.com/pod-product-compliance
Lightning Source LLC
LaVergne TN
LVHW070954170726
843515LV00005B/985